Neural Representation of Temporal Patterns

Neural Representation of Temporal Patterns

Edited by

Ellen Covey

Duke University Medical Center
Durham, North Carolina

Harold L. Hawkins

Office of Naval Research
Arlington, Virginia

and

Robert F. Port

Indiana University
Bloomington, Indiana

Springer Science+Business Media, LLC

Library of Congress Cataloging-in-Publication Data

Neural representation of temporal patterns / edited by Ellen Covey,
  Harold L. Hawkins, and Robert F. Port.
       p.   cm.
    Proceedings of a symposium held April 29-May 2, 1993, in Durham,
  N.C.
    Includes bibliographical references and index.
    ISBN 978-1-4613-5785-8   ISBN 978-1-4615-1919-5 (eBook)
    DOI 10.1007/978-1-4615-1919-5
    1. Time perception--Congresses. 2. Neural networks
  (Neurophysiology)--Congresses.   I. Covey, Ellen. II. Hawkins,
  Harold L., 1938-   . III. Port, Robert F.
    [DNLM: 1. Neurophysiology--congresses. 2. Perception--physiology-
  -congresses. 3. Memory--physiology--congresses. 4. Learning-
  -physiology--congresses.   WL 705 N494 1995]
  QP445.N48   1995
  152--dc20
  DNLM/DLC
  for Library of Congress   95-48964
                                                                     CIP

Proceedings of a symposium on Neural Representation of Temporal Information,
held April 29–May 2, 1993, in Durham, North Carolina

© 1995 Springer Science+Business Media New York
Originally published by Plenum Press, New York in 1995
Softcover reprint of the hardcover 1st edition 1995

10 9 8 7 6 5 4 3 2 1

CONTRIBUTORS

Geoffrey P. Bingham
Department of Psychology
Indiana University
Bloomington, IN 47405

John H. Casseday
Department of Neurobiology
Duke University Medical Center
Durham, NC 27710

Joydeep Ghosh
Department of Electrical and Computer
Engineering
The University of Texas
Austin, TX 78712-1084

Richard Granger
Center for the Neurobiology of Learning and
Memory
University of California, Irvine
Irvine, CA 92717

Stephen Grossberg
Center for Adaptive Systems and
Department of Cognitive and Neural Systems
Boston University
Boston, MA 02215

Scott L. Hooper
Department of Biological Science
Ohio University
Athens, OH 45701

Steven Keele
Department of Psychology
University of Oregon
Eugene, OR 97403

Robert F. Port
Departments of Linguistics and Computer
Science
Indiana University
Bloomington, IN 47405

Gary J. Rose
Department of Biology
University of Utah
Salt Lake City, UT 84112

De Liang Wang
Laboratory for AI Research
Department of Computer and Information
Science and Center for Cognitive Science
Ohio State University
Columbus, OH 43210-1277

PREFACE

Both the analysis and generation of temporal patterns are fundamental tasks of biological systems. Throughout the animal kingdom, every sensory modality is designed to analyze patterns of information distributed over time. Human speech and music, a visual scene in which objects move or a stationary scene that we scan with our eyes, a pattern of pressure that changes as we move our fingertips over an object, the electrical field detected by a fish as it swims past objects in a stream, or the pattern of ultrasonic echoes detected by a bat as it flies through a cave, all depend upon specific distributions of information over time. It is perhaps even more obvious that all forms of action must include a time dimension. Walking, running, talking, reaching for an object, writing, or pressing keys in a particular order all require the generation of specific patterns of muscle contractions distributed over time. Finally, most forms of behavior require a transformation from a temporal pattern of sensory input to a temporal pattern of motor output, as well as interactive modulation of sensory input systems by motor output systems and vice versa.

Despite the fact that the processes of animal life are inseparable from the time dimension, most experimental and theoretical research on neural circuitry has emphasized the encoding of static or spatially distributed information. Within recent years, a large body of data has become available regarding the time course of fundamental neural processes, so that we finally have at our disposal some of the necessary tools and information to discover mechanisms used by neural circuitry to deal with time. The study of how the nervous system represents information distributed over time is currently an exciting new frontier in neurobiology, and one in which rapid progress is likely to be made over the next decade.

In order to provide an opportunity for a multidisciplinary group of scientists to present their approaches and findings on questions related to temporal pattern representation and to discuss possible future strategies for studying temporal pattern analysis, a workshop on "Neural Representation of Temporal Patterns," sponsored by the Office of Naval Research, was held at Duke University in April 1993. The resulting volume contains a sampling of the information that was presented at this conference, and it is designed to provide an idea of the many approaches that can be taken to answering the basic question of how time-varying information is represented in the nervous system. The chapters in this book review the problem of temporal processing from many different points of view including vertebrate and invertebrate neurobiology, experimental psychology of perception and memory, and computer science techniques for processing information distributed over time.

Because of the ubiquitous nature of the topic, it is impossible to provide a comprehensive treatment of every aspect of temporal pattern analysis in a volume this size. Instead, the intention is to offer a multidisciplinary selection of topics, presented in a form that is readily accessible to professionals and students alike, regardless of their field of study. It is hoped that this book will help foster an understanding and appreciation for the work being done in widely diverse fields by engineers, mathematicians, computer scientists, experimental neurobiologists, and cognitive psychologists, and will inspire more scientists to think about the time dimension in the systems that they study.

We gratefully acknowledge the financial support provided by the Office of Naval Research for the workshop on which this book is based. In particular, we would like to express our thanks to Drs. Teresa McMullen and Tom McKenna for their help in organizing the workshop. In addition, we wish to thank all of the contributors to this volume for their hard work and cooperation. Finally, we wish to thank Tomas Ayala, Boma Fubara, and Sherrie James for their assistance in preparing the manuscript for publication.

Ellen Covey (Duke University)
Harold Hawkins (Office of Naval Research)
Robert Port (Indiana University)

CONTENTS

INTRODUCTION

As pointed out in the preface to this book, the generation and analysis of temporal patterns are crucial aspects of everyday life for any member of the animal kingdom. Elucidating the mechanisms of production, processing, recognition, and storage of these temporal patterns forms one of the major challenges in both experimental and computational neurobiology and behavioral sciences today. In the chapters that follow, the problems addressed range from patterns involving only a few elements to rhythmic sequences that can contain an unlimited number of elements. The authors take many different approaches in their investigation of these mechanisms. The experimental work ranges from the study of a very simple neural circuit in vitro, through neurophysiological experiments on whole animals, to behavioral work on humans. The computational studies use models that represent the nervous system at many different levels of abstraction. The models range from those that simulate specific identified neurons and their known interconnections to those that are very abstract and do not map directly onto any known neural structures.

In spite of the differences in approach, a number of common themes can be found in this diverse set of chapters on temporal patterns. Many of the authors use the terms temporal pattern and temporal sequence interchangeably. By sequence, we mean that the order of the elements in the pattern are fixed, one element following directly after another. To be more precise, the order of the element onsets are fixed so as to allow for elements that overlap in time. In some cases, the authors discuss representations in which time plays no role and only the order of the elements is preserved.

The use of terms like element, sequence and rhythm by the authors also implies a common conceptual framework for temporal patterns. This framework is hierarchical. Primitive or simple elements are assembled into basic sequences and then into higher-level sequences made up of the basic sequences. In some cases, the higher-level sequences are periodic or rhythmic, as in the case of central pattern generators (Chapter 3). In other cases, the higher-level sequences are more complex, such as the motor movements associated with handwriting (Chapter 5). For a sensory system to analyze a hierarchically organized pattern, it must be able to segment the incoming sensory stream into elements and then appropriately group elements from the same source into a pattern or sequence (Chapters 6, 10).

This hierarchical theme is also present in many of the models of pattern generation, storage, and analysis (Chapters 7, 9, 10). It is generally accepted in neurobiology today that sensory and motor systems are made up of subsystems or modules, and that these subsystems are in turn made up of specialized neural circuits. The hierarchical control of systems consisting of many muscles by central pattern generators that produce rhythmic patterns at different rates and with different low-level sequences relieves higher-level motor control centers from the burden of controlling each muscle individually and yet allows higher-level control centers to specify the details of rate and pattern. In sensory systems, as in human-engineered systems, features must be extracted and then events or elements detected and sequences of elements classified and/or stored. In these sensory systems the features are extracted by parallel pathways, which themselves are made up of a series of modules.

A module that is often used with models of neural pattern generators and analyzers is the oscillator. In the case of central pattern generators it often takes the form of a pacemaker neuron (Chapter 3). In models of rhythmic pattern perception nonlinear oscillators are often used which synchronize to the stimulus (Chapters 6, 8).

Another common theme in this volume is the assumption that generation and analysis of patterns coevolved, since sensory systems most often analyze signals that result from muscle activity produced by pattern generators. These signals may be self-generated as in the case of echolocation systems (Chapter 2), electrosensory systems (Chapter 1), and in task-dynamic models of motor control (Chapter 4), or may be generated by other organisms, as is the case in anuran communication sounds (Chapter 1). In yet other cases, such as predator-prey interactions, auditory or visual patterns produced by central pattern generators associated with locomotion provide sensory information with high survival value to predator and prey alike.

Although it is possible to identify some common themes that cut across the many efforts aimed at revealing the neural representations of temporal patterns, the unsolved problems in the field are numerous. Perhaps one of the most fundamental questions is, how does the nervous system generate and analyze temporal patterns in a manner that is insensitive to time scale? Even very simple neural circuits are able to generate rhythmic patterns over a range of frequencies and yet maintain the phase relationships between the activity in the individual neurons (Chapter 3). Human observers find it easy to recognize specific rhythms over a range of time scales and musical melodies and human speech are recognized at many different production rates. The generation and recognition of temporal patterns across multiple time scales also require   short- and long-term memory systems that can store patterns in a time-independent manner.

Not only do we not understand the origin of the time scale independence of motor and sensory systems, but there is also much to be learned about how sensory and motor systems arc intcgratcd. How is thc tcmporal information coming into thc scnsory systcms uscd to control and/or modify the motor patterns produced during a task like directed arm movement? Much of the work on temporal patterns in sensory systems assumes that there is a transformation from the temporal domain to a place code. Does this make sense in the case of sensorimotor integration, since the desired output is also a temporal pattern? Could it be, as Casseday and Covey ask at the end of Chapter 2, that even for recognition tasks the neural representation of the input pattern is itself another temporal pattern?

The biggest challenge to those of us working on the representation of temporal patterns is, however, to obtain relevant physiological data for model development and verification. Most of the existing models are implemented at a relatively high level of abstraction that makes it difficult to generate predictions which are testable with the tools currently available to the neurobiologist. Neurobiologists, psychologists, and modelers need to work towards a common ground where some of the most robust behavioral experiments are repeated by neurobiologists and then modeled by the mathematicians and engineers at appropriate levels of abstraction.

OVERVIEW

This book is organized with the experimental work presented first followed by the modeling studies. The first two chapters cover temporal pattern representations in sensory systems followed by two chapters that cover temporal pattern representations in motor systems. The experimental work in general is coupled to models, some of which are largely conceptual but in other cases are computational. The last six chapters cover a range of topics on modeling. Most focus on issues related to sensory systems, such as segmentation and classification. Learning is also a common theme in the modeling section. For those readers who wish to focus on a particular theme, such as pattern representation in sensory systems or the storage of temporal patterns, an overview or guide to specific chapters is provided below.

Temporal Pattern Representations in Sensory Systems

    1. Experimental results
        Chapter 1: AM coding in frogs and electric fish
        Chapter 2: Duration and FM coding in bats

Temporal Pattern Representations in Motor Systems

Storage of Temporal Patterns

David C. Mountain

Department of Biomedical Engineering
Boston University
Boston, MA 02215

REPRESENTATION OF TEMPORAL PATTERNS OF SIGNAL AMPLITUDE IN THE ANURAN AUDITORY SYSTEM AND ELECTROSENSORY SYSTEM

Gary J. Rose

Department of Biology
University of Utah
Salt Lake City, UT 84112
USA

INTRODUCTION

The temporal structure of sensory signals plays an important role in the biology of many animals including man (Emlen, 1972; Gerhardt, 1982; Heiligenberg, 1991; Kay, 1982; Rose, 1986). This chapter focuses on how temporal variations in the amplitude of acoustic or electric signals are represented in auditory systems of frogs and toads (anurans) and in the electrosensory system of a weakly electric fish.

To avoid heterospecific matings, many anurans must discriminate their conspecific vocalizations from those of closely related species. In some cases, such as the frogs *Hyla versicolor* and *Hyla chrysoscelis*, these calls are virtually identical in spectral structure, but differ markedly in their temporal structure (Gerhardt, 1982). Female *H. versicolor* are able to discriminate between signals that differ in their pulse rate and/or pulse shape (Gerhardt and Doherty, 1988). Given the choice, females prefer calls with species-typical pulse shape and rate.

Electric fish of the genus *Eigenmannia* sense modulations of the amplitude and phase of their electric organ discharges (EODs) during electrolocation (Heiligenberg, 1977) and while experiencing the EODs of neighboring fish (Heiligenberg, 1989; 1991). In the latter case, the signals summate to produce a waveform that "beats" at a rate equal to their frequency difference (Df); the amplitude and phase of the resultant signal are modulated at the rate Df.

Amplitude and phase modulations resulting from the summation of a fish's EODs with those of its neighbor can impair or "jam" a fish's electrolocation abilities (Matsubara and Heiligenberg 1978). Modulation (beat) rates of approximately 4 Hz are most detrimental. Fish minimize jamming by increasing the frequency difference of their EODs. The lower-frequency fish decreases its EOD frequency and the higher-frequency fish increases its EOD frequency. These jamming avoidance responses are greatest for beat rates of approximately 3-8 Hz, and minimal for rates of 20 Hz or more (Bullock et al., 1972; Heiligenberg et al., 1978; Partridge et al., 1981; Bastian and Yuthus, 1984).

Thus, temporal patterns in acoustic and electric signals are of clear importance in the biology of anurans and electric fish. These systems, therefore, are well suited for investigating 1) How temporal variations in signal amplitude are represented in nervous systems, and 2) The neural basis of behavioral selectivity for particular temporal patterns of amplitude modulation.

Why a Comparative Approach?

Invariably, investigations of brain function and the evolution of behavior are plagued by the multiplicity of theoretically plausible solutions to particular problems. The purpose of neuroethology is to conduct behavioral analyses of specialized systems to elucidate general computational algorithms and hypothetical neural mechanisms for these behaviors. This process is an important first step in exploring the neural basis of behavior (Heiligenberg, 1991). While systematic behavioral studies can narrow the possible solutions to a problem, the specific cellular or molecular mechanisms that might underlie these algorithms and hypothetical mechanisms are still numerous. Identification of the neural correlates of a particular behavior does not prove a causative link, nor does it rule out involvement of other plausible mechanisms.

How do comparative studies address this problem? By investigating instances where a similar problem has been solved independently in different sensory systems (i.e. cases of convergent evolution), the "uniqueness" of particular neural implementations of algorithms can be tested. For example, both anurans and electric fish have independently evolved central processing mechanisms for filtering temporal patterns of amplitude modulation in signals, "emphasizing" relevant patterns and rejecting others. While these organisms selectively attend to different parts of the temporal spectrum, both employ a similar transformation in the central representation of this temporal information. Finding that different sensory systems have converged on similar neural mechanisms for processing temporal information in signals increases our confidence in the importance, and, possibly, uniqueness, of these solutions.

The comparative analysis of cases of *divergent* evolution also provides a powerful test of the relations between neural specializations and function. For example, the vocalizations of different species of frogs vary considerably in their temporal patterns of amplitude modulation; neural mechanisms important for the processing of these signals should show corresponding variation.

Theoretical Mechanisms of Temporal Selectivity

How might temporal variations in the amplitude of signals (Fig. 1) be represented in the nervous system?

Secondly, how might selective behavioral responses to particular temporal patterns be achieved? Because the firing rate of a primary sensory neuron is generally a function of the amplitude of its effective stimulus, one could expect modulations of signal amplitude over time to be coded in the periodic fluctuations of their discharge rates i.e., a "periodicity code", in which the time intervals between successive amplitude maxima are coded in the time intervals between neural response maxima. In the central nervous system, individual neurons might preferentially code particular amplitude modulation (AM) rates by exhibiting maximal periodicity coding, i.e., the strongest synchronization of their discharges to a particular phase of the amplitude envelope, at some AM rates and poor synchronization at others. Selective behavioral responses to particular AM rates might be realized if neurons that preferentially synchronized their discharges to those rates provided input to "motivational" and/or sensorimotor areas through "accommodating" connections with follower cells. For example, the poorly synchronized neural discharges at nonoptimal AM rates might depress transmitter release and thereby be less effective in eliciting postsynaptic responses. Alternatively, temporal patterns of AM might be represented in the spatial pattern of activity of central neurons that are tuned, i.e., respond maximally, to particular rates of AM. Selective behavioral responses could be generated if particular "AM-tuned" neurons projected to appropriate motivational and/or sensorimotor areas. The most dramatic transformation in the representation of temporal patterns of AM in the central nervous system might involve a conversion to a temporal filter code accompanied by a loss of periodicity coding.

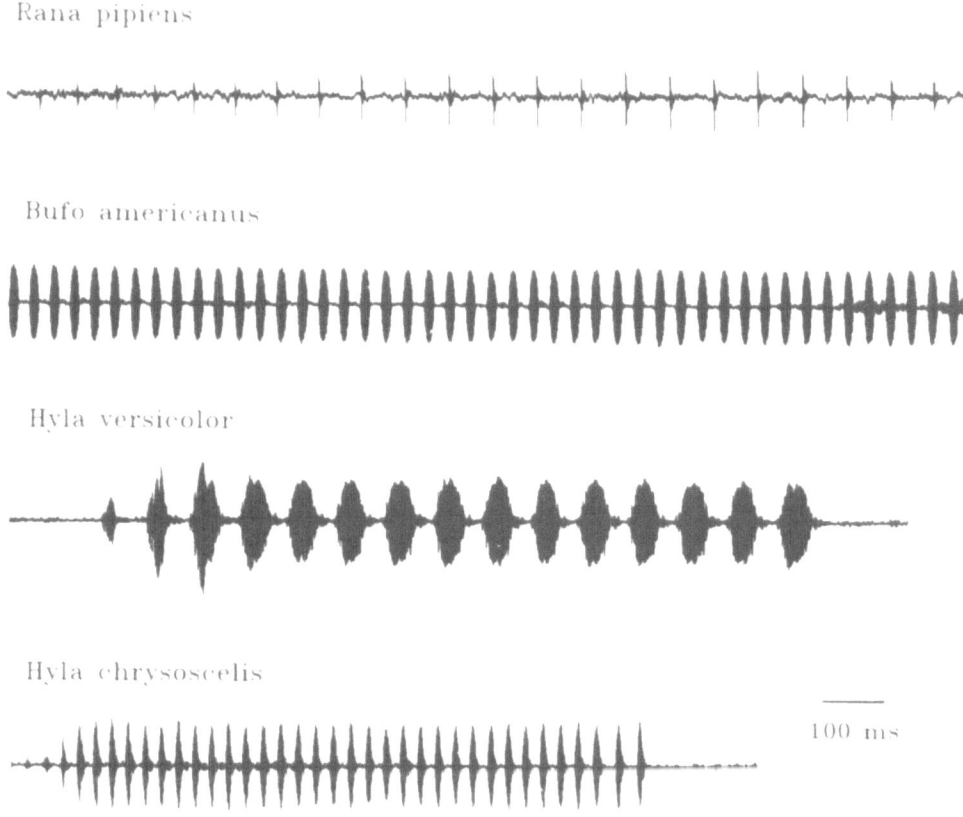

Figure 1. Oscillograms of advertisement calls of several species of frogs and toads (anurans) discussed in this chapter. Time is represented on the horizontal axis, amplitude on the vertical axis.

Such a complete transformation in the representation of AM rates occurs in anuran auditory systems and electrosensory systems. While not yet fully understood, recent investigations have provided some insight into the mechanisms underlying these transformations.

The representation of the amplitude envelope of the stimulus by the temporal fluctuations of group discharges of primary afferents also raises the question of whether the central nervous system uses an analog representation of the amplitude envelope of the stimulus and, if so, for what purpose?

Behavior

Many frogs and toads are nocturnal breeders and rely on acoustic signals to communicate with conspecifics (Wells, 1977). Males use calls to advertise their presence to females, and females can choose among males based on the properties of their calls (Gerhardt, 1982;1988; Ryan, 1985). An important role of the anuran communication system is to minimize the chances of heterospecific mating i.e., mating with a male of the wrong species. In addition, males can communicate aggressive arousal and defend calling territories through acoustic signals. In many cases, the various calls of a particular species differ little in spectral structure but differ markedly in temporal structure i.e., the changes in signal amplitude and/or

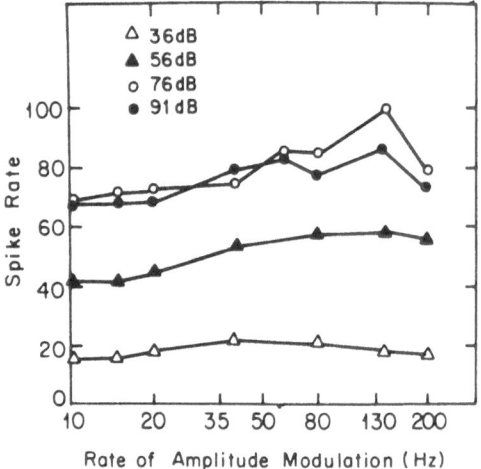

Figure 2. Relation between average spike rate (spikes/s) and the rate of amplitude modulation (AM) of a white noise stimulus for a single auditory nerve fiber in *Rana pipiens.* Average sound levels ranging from 36 to 91 dB SPL were tested. The best excitatory frequency (tones) was 650 Hz. Stimulus was white noise amplitude modulated at a depth of 100%.

frequency over time. In cases where closely related species breed at similar times and places, their "advertisement" calls differ primarily in temporal structure (Gerhardt, 1982). Both amplitude and frequency may vary over time; however, most calls are differentiated primarily in only one of these temporal dimensions. Amplitude modulated calls of some North American anurans are shown in Figure 1. Behavioral studies have shown that anurans are able to discriminate among synthetic calls differing only in their pattern of frequency or amplitude modulations (Rose et al., 1988; Gerhardt and Dougherty, 1988). With regard to amplitude modulated calls, this discrimination appears to be intensity dependent. Females may actually prefer calls with heterospecific patterns of amplitude modulation over conspecific calls of much lower amplitude (Gerhardt ,1982;1988). These studies, therefore, support the existence of temporal filters for amplitude modulation and suggest that these filters are rather broadly tuned.

Periphery

In anuran auditory systems, temporal patterns of AM in acoustic signals are represented in the periodic fluctuations of group discharges of primary afferents (Rose and Capranica, 1985). A unit's average firing rate is largely independent of the rate of AM (Fig. 2), provided that broad-band noise is the "carrier", i.e., the signal that is being amplitude modulated. If a pure tone centered near the unit's best excitatory frequency (BEF) is amplitude modulated, the mean spike rate can change as the rate of AM is varied (Fig. 3).

Two factors contribute to this difference: Amplitude modulating a pure tone at a rate, fm, generates a signal that has energy at the original carrier frequency, fc, and in the sidebands

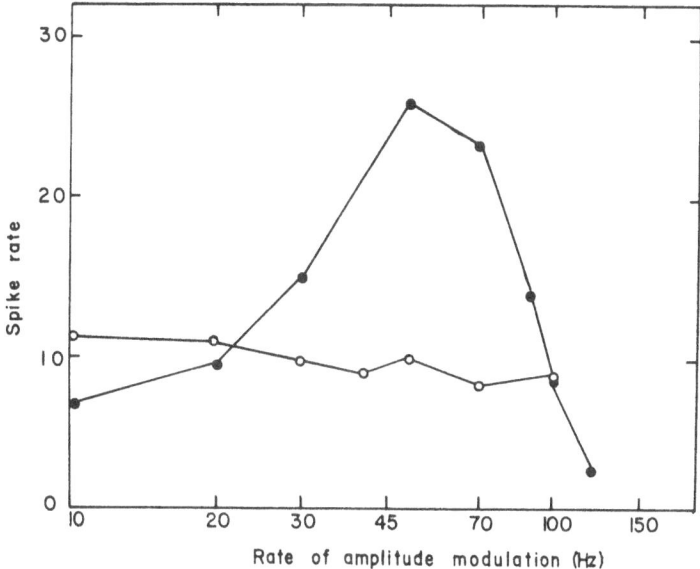

Figure 3. Spike rate of an auditory nerve (AN) fiber in *Rana pipiens* vs rate of AM of a tone at 400 Hz (the unit's best excitatory frequency, BEF) and 90 dB SPL (closed symbols), or AM white noise at 105 dB SPL (open symbols).

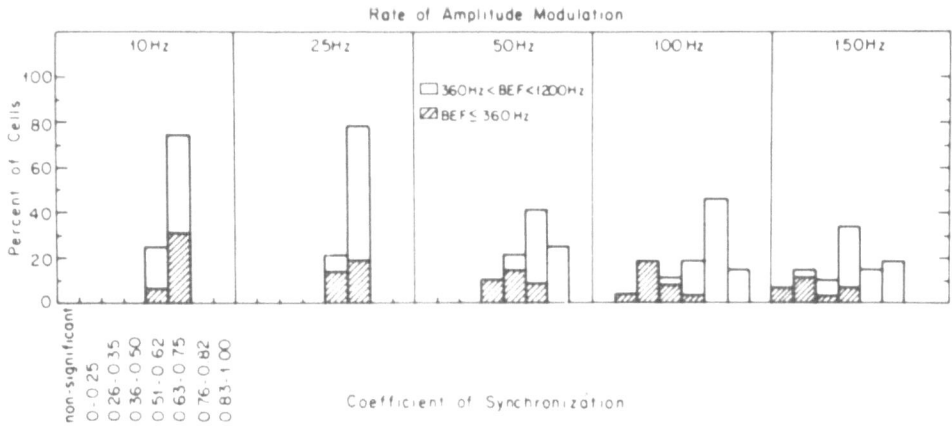

Figure 4. Response synchronization characteristics of 10 low-BEF AN units and 18 mid-BEF units from *Rana pipiens*. Five modulation rates of white noise were used (denoted at top of each frame). Hatched area of bar reflects proportion of 28 units with low BEF (<360 Hz). Open area of bar represents proportion of units with mid BEF (Between 360 and 1200 Hz).

fc-fm and fc+fm. As the modulation rate is increased to high rates, progressively more of the energy in the signal will be displaced outside the neuron's frequency tuning curve. The long-term power spectrum of AM noise, in contrast, is independent of the AM rate, permitting the temporal structure of the stimulus to be varied while holding spectral properties constant. Secondly, unlike AM noise, the AM pure tone generates stronger adaptation of the unit's responses because energy is maintained at a constant level at or near the unit's BEF (Hall and Feng, 1991).

The representation of periodic fluctuations of stimulus amplitude in the periodicity of a neuron's discharges can be quantified by computing a "synchronization coefficient". This statistic reflects the concentration (locking) of a unit's discharges to a particular phase of the modulation cycle (coefficients near 1.0 indicate strong synchrony, coefficients near 0.0 reflect poor synchrony). Each auditory nerve fiber uniformly represents AM rate in its periodicity of discharges for rates of at least 80 Hz. At higher rates of AM, units with low BEFs show weak periodicity coding while those with higher BEFs exhibit stronger periodicity coding (Fig. 4).

Midbrain

The torus semicircularis is an important auditory region in the midbrain of anurans (Fig. 5). In the torus, there is evidence of a transformation in the representation of temporal patterns of AM (Rose and Capranica, 1983; 1985). In contrast to auditory nerve fibers, approximately two thirds of the single units recorded in the torus (possibly including toral afferents) show a change in their mean discharge level as the rate of amplitude modulated noise is varied (Fig. 6).

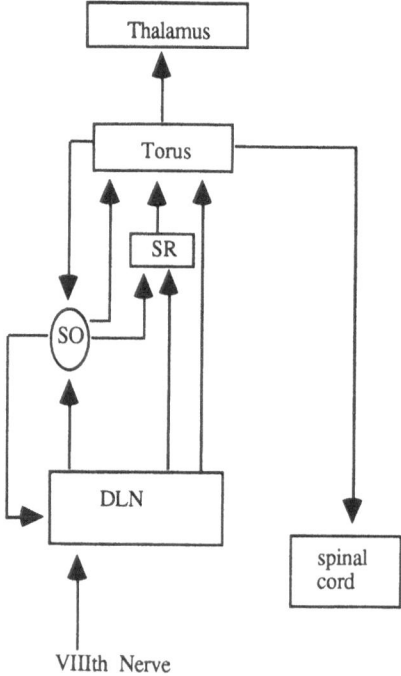

Figure 5. Schematic of the anuran auditory pathway. This figure summarizes the main connections between the various auditory areas, but does not identify them with regard to laterality. DLN, dorsal lateral nucleus; SO, superior olivary nucleus; SR, superficial isthmal reticular nucleus.

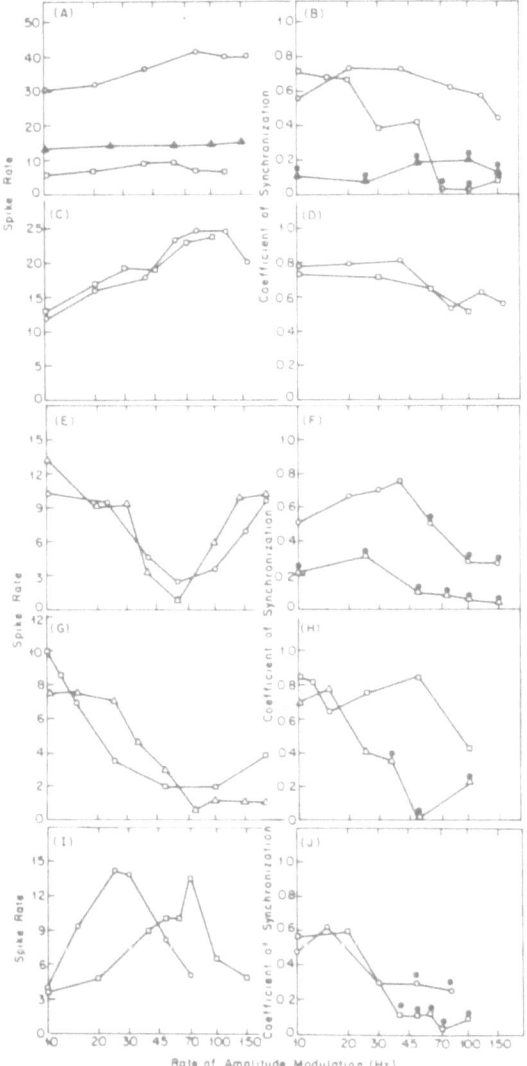

Figure 6. Spike-rate (spikes/s) vs AM rate (**A,C,E,G,I**), and coefficient of synchronization vs. AM rate (**B,D,F,J**). Data are from 11 representative toral neurons in *Rana pipiens*. Each curve represents data from a different unit. Five types of response to AM noise are illustrated: AM non-selective (**A,B**); AM high-pass (**C,D**); AM band-suppression (**E,F**); AM low-pass (**G,H**); and AM tuned (**I,J**). All stimuli were 100% AM white noise, at 10-20 dB above each unit's threshold. Coefficient of synchronization reflects the concentration of each unit's spikes at a particular phase of the modulation cycle (time resolution = 100 μs). All values were computed on the basis of at least 10 stimulus presentations. Nonsignificant synchronization values are indicated by filled circles.

"Band-pass" neurons (Fig. 6 I, J) are particularly interesting in that they respond best over a narrow range of modulation rates, i.e., they are "AM-tuned". The AM tuning of these neurons appears to be unrelated to their frequency tuning (Rose and Capranica, 1984;1985). As will be shown below, these neurons generally respond best to AM rates at or near those of conspecific calls. The response properties of low-pass, high-pass and band-suppression units are not obviously correlated with the band-pass properties of the behavior, which indicates a preference for temporal patterns of amplitude modulation characteristic of the advertisement

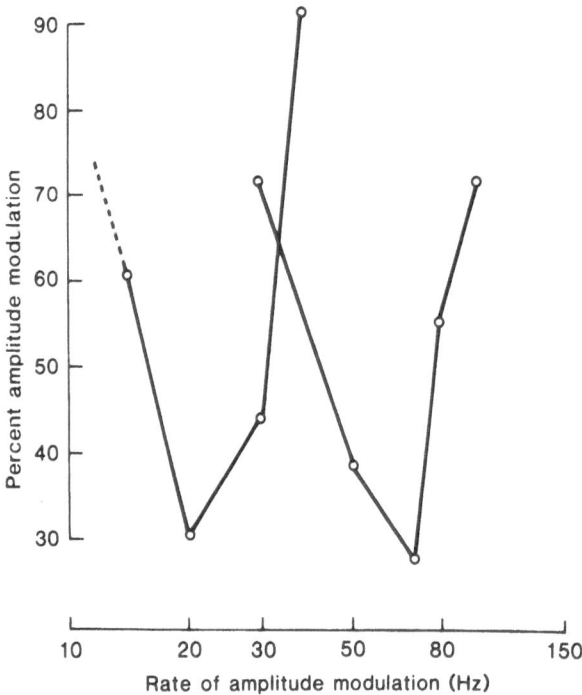

Figure 7. Temporal tuning curves of two auditory units in the torus semicircularis of *Rana pipiens*. One unit is tuned to 20 Hz AM, and the other to 70 Hz AM. In all of these measurements the stimulus intensity was held constant while the depth of modulation was varied to maintain a threshold criterion. The dashed line below 14 Hz indicates that threshold criterion was not reached. At 100% AM, the signal envelope varies sinusoidally from full amplitude to zero amplitude. At lower percentages of amplitude modulation, the amplitude of the modulation envelope decreases. Zero percent AM corresponds to unmodulated white noise.

call. These neurons may represent components required for constructing band-pass neurons and/or for processing other vocalizations and sounds. The temporal tuning of these neurons can be displayed, in analogy with frequency tuning curves, by constructing isoresponse functions ("AM tuning curves", Fig. 7).

The range of AM rates that best excite band-pass neurons is largely independent of the overall (root-mean-square) amplitude (Fig. 8) or depth of modulation (Fig. 9) of the stimulus, and is not evident when their response synchronization is plotted as a function of AM rate (Figs. 6, 9).

Further, units that show the sharpest AM tuning, as seen in their firing levels, also exhibit little, if any, significant representation of the AM rate in their periodicity of discharges (Figs. 10, 11).

Thus, temporal patterns of AM are represented in the midbrain by an ensemble of temporal filter neurons, such that this information is coded in the relative activity of neurons within this area. If the behavioral selectivity for AM rate results primarily from the activity of AM band-pass cells in the midbrain, then the different preferences of different species for AM rate should be paralleled by differences in the temporal selectivities of these cells.

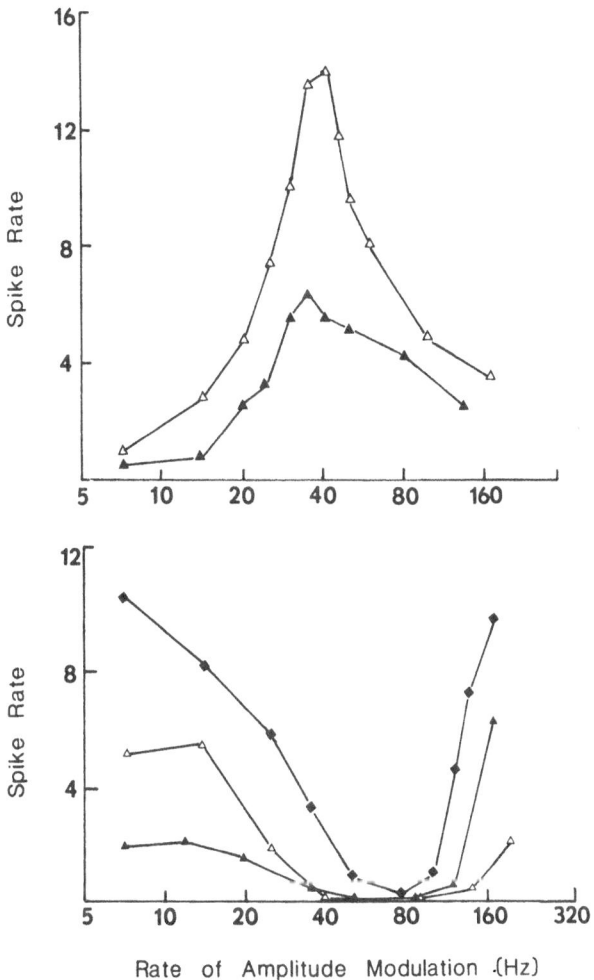

Figure 8. Intensity independence of temporal selectivity. A band-pass unit, recorded from *Hyla chrysoscelis* (upper) was tested at 54 dB SPL (filled triangles) and 64 dB SPL. A band suppression neuron, recorded from *Hyla versicolor*, was tested at 72 dB SPL (filled triangles), 82 dB SPL (open triangles) and 92 dB SPL.

Such species-specificity of behavioral preference and neural representation of temporal patterns of AM is evident for the cryptic species pair, *H. versicolor* and *H. chrysoscelis* . Male *H. versicolor* amplitude modulate their advertisement calls at lower rates (20 Hz at 20° C) than do *H. chrysoscelis* (35 Hz for the eastern type; Fig. 1). Female *H. versicolor* show a clear preference for the temporal pattern of AM characteristic of their conspecifics (Gerhardt, 1978; Gerhardt and Dougherty, 1988). Correspondingly, the distributions of toral AM band-pass neurons differ significantly, and in the expected direction, for the two species (Fig. 12) (Rose et al., 1985).

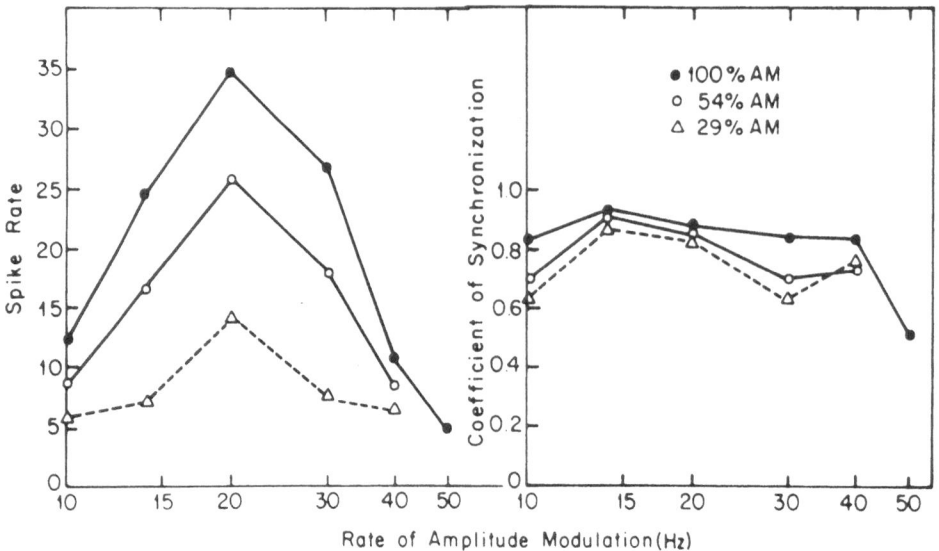

Figure 9. Spike rate vs AM rate (left) and coefficient of synchronization vs AM rate (right) for a representative toral neuron with band-pass AM rate selectivity. This unit, recorded from *R. pipiens*, was tested with 100%, 54%, and 29% AM white noise.

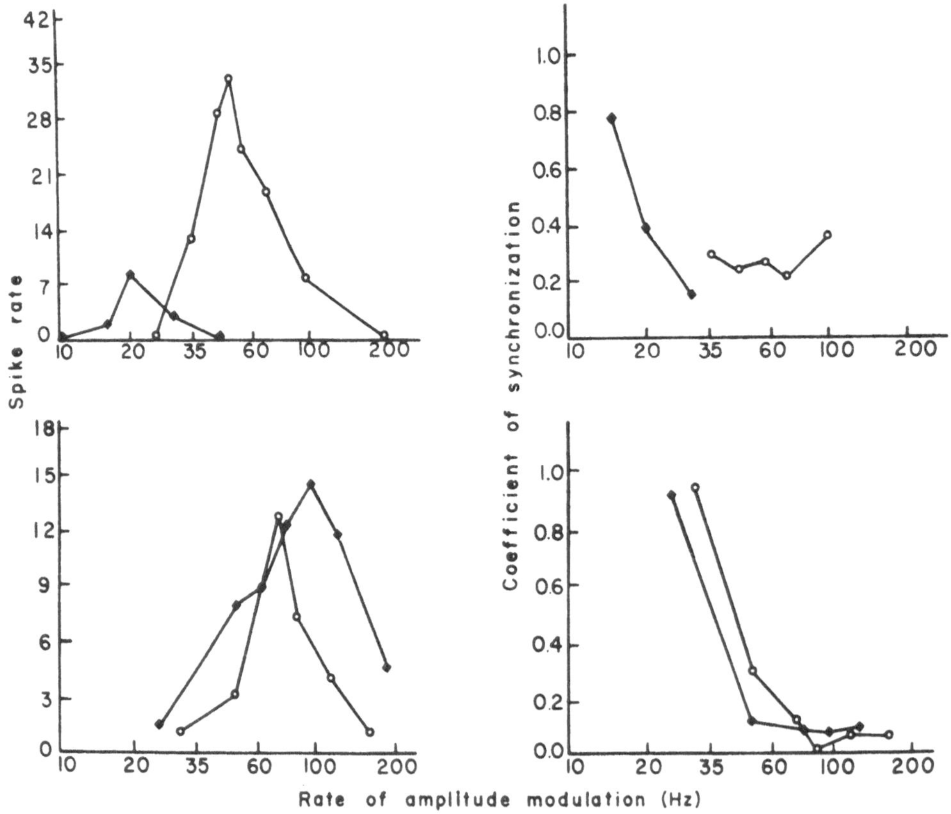

Figure 10. Spike rate vs. rate of AM (left) and coefficient of synchronization vs rate of AM (right) for two neurons recorded in the torus semicircularis of the American toad (upper) and two neurons recorded in the torus semicircularis of Fowler's toad (lower). Latencies of response following onset of white noise, amplitude modulated at the best rate for each unit were greater than 40 ms.

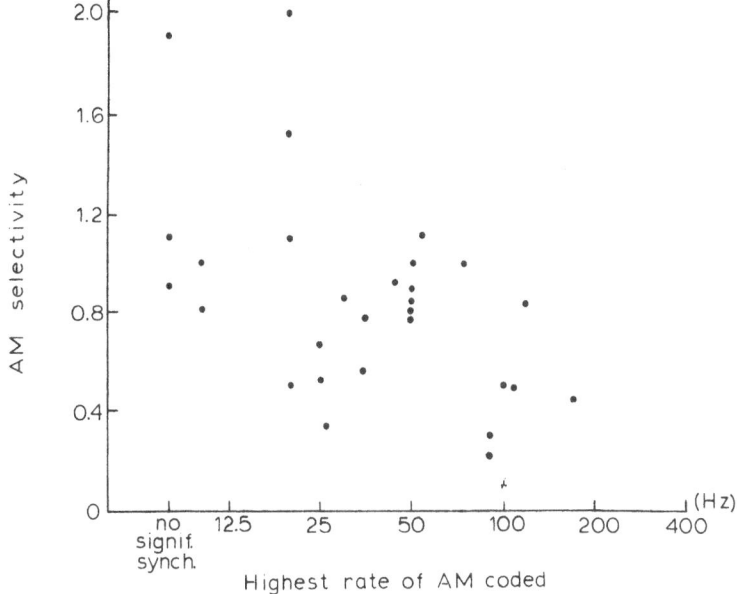

Figure 11. Relation between sharpness of AM tuning and limits of periodicity coding in 29 neurons recorded in the torus semicircularis of *Rana pipiens, Bufo americanus* and *Bufo fowleri*. Sharpness of tuning was quantified by dividing the rate of AM noise to which the unit was tuned by the bandwidth of its isoamplitude function (e.g., Fig. 6I) at half-maximal response.

Although there is insufficient behavioral data for making a quantitative comparison between the temporal selectivity seen at the behavioral and neural levels, the relative attractiveness of particular AM rates appears to be related to the proportion of neurons that are tuned to those rates. Also, the temporal tuning of band-pass neurons shifts to higher rates as the frog's body temperature is increased (Brenowitz et al. 1985). This temperature dependency closely parallels that of the female preference for modulation rate (Gerhardt, 1978).

The relation between the temporal selectivities of toral neurons and the temporal preferences of the animal were addressed further in the American toad *B. americanus* (Rose and Capranica, 1984). In playback studies, the relative effectiveness of particular rates of AM in evoking advertisement calls from males is correlated with the relative number of toral units tuned to those rates (Fig. 13).

Ultimately, it may be possible to relate the attractiveness of particular rates of AM to the aggregate level of activity within the ensemble of band-pass toral neurons.

ELECTROSENSORY SYSTEM

Behavior

The electric fish, *Eigenmannia*, detects and utilizes temporal variation in the amplitude of its electric organ discharges (EOD) during electrolocation and social behavior. In contrast to the anurans described above, these fish almost exclusively extract comparatively slow variations in signal amplitude, e.g., below 10 Hz. This filtering is evident in the context of the jamming avoidance response. As described earlier, when jammed by the EODs of a

11

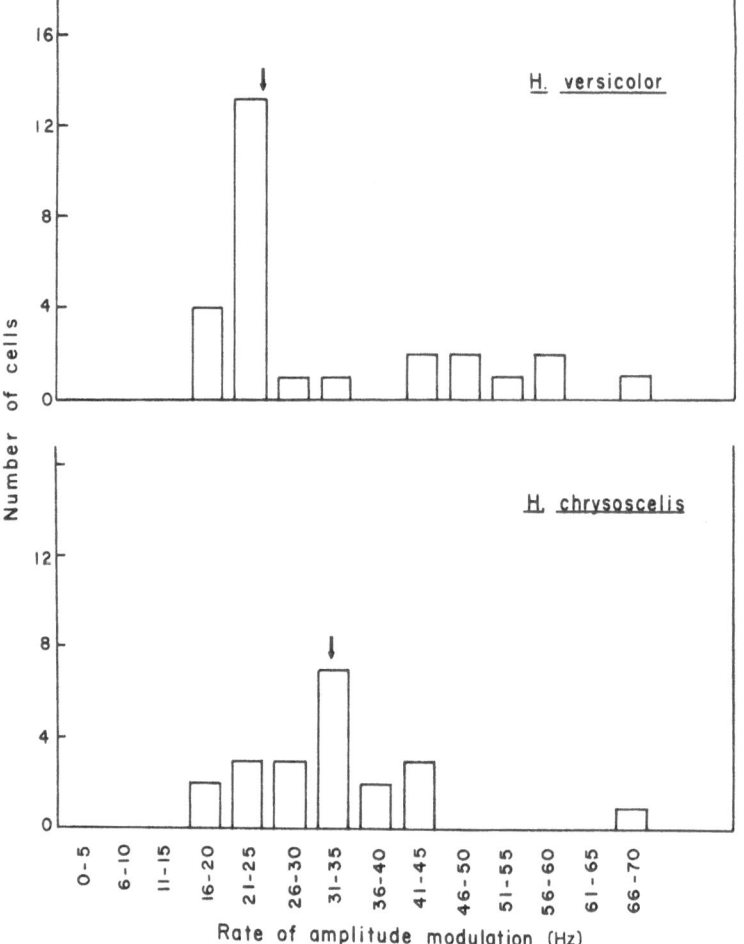

Figure 12. Distribution of preferred AM rates for AM-tuned neurons recorded from *H. versicolor* and *H. chrysoscelis*. Arrows indicate median values for each distribution. The frogs' body temperature was held at 21° C during data collection. At this temperature and within the populations studied, male *H. versicolor* amplitude modulate their calls at approximately 20 pulses/s and male *H. chrysoscelis* modulate their calls at about 35 pulses/s.

neighboring fish, *Eigenmannia* changes its EOD frequency in the direction appropriate to increase the "beat rate" (see Heiligenberg, 1989; 1991 for review). Behavioral studies have shown that *Eigenmannia* is able to filter out, or reject, fast AM information and selectively attend to slower modulations. This filtering process is evident in the jamming avoidance response itself, where beat rates of 3-8 Hz elicit the largest responses and rates above 20 Hz have virtually no effect (Fig. 14). As in anurans, this temporal selectivity is temperature dependent.

Neural Correlates

As in the anuran auditory system, temporal patterns of amplitude modulation in sensory signals are represented by a periodicity code in primary afferents and by a temporal filter ensemble in the torus semicircularis, or midbrain (Partridge et al., 1981).

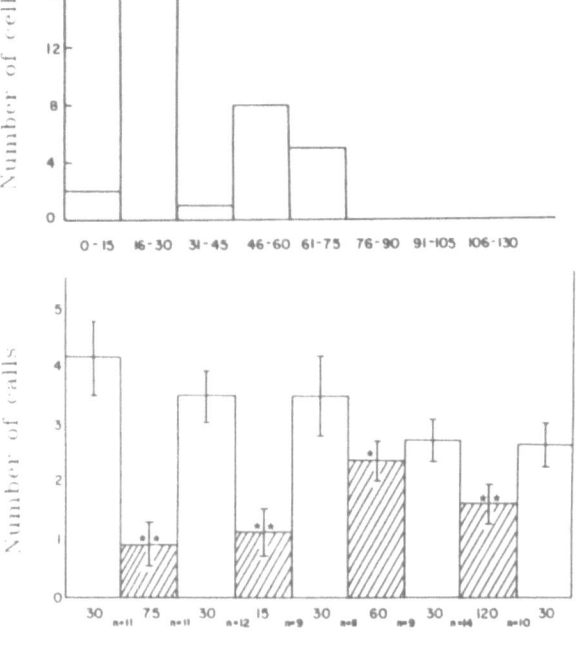

Rate of amplitude modulation (Hz)

Figure 13. Top: Distribution of best rates of AM for AM-tuned toral neurons recorded from the American toad. Bottom: Effectiveness of synthetic calls, differing in temporal structure, in evoking advertisement calls from male American toads. Ten isolated males were tested in the field. Synthetic calls consisted of band-passed noise, centered at 1410 Hz, and amplitude modulated at 7.5, 15, 30, 60 and 120 Hz. The reference modulation rate of 30 Hz (open bars) was presented before and after each of the other experimental calls (shaded bars). Each bar represents mean number of calls evoked by each stimulus type. A Wilcoxon matched-pairs signed ranks test was performed on each set of reference-experimental pairs; n = sample size after ties. One-tailed tests were used. ** indicates that the experimental trial was significantly less effective (p<0.05) than either reference trial; * denotes a significant difference between the experimental trial and preceding reference trial only.

Amplitude modulations of the fish's EODs are coded by "P-type" electrosensory afferents. For P-type units, the probability of spike generation increases with stimulus amplitude. Amplitude modulations of at least 64 Hz are represented in the periodicity of afferent group discharges (Bastian 1981a), and the mean firing rate of these units is largely independent of beat rate.

Electrosensory afferents terminate in the electrosensory lateral line lobe (ELL) (Fig. 15). In this first-order electrosensory area of the hindbrain, the mean firing rates of neurons are more dependent on stimulus AM rate than are those of primary afferents, however, low-pass or band-pass properties comparable to those of toral neurons are not observed in the ELL (Bastian, 1981b; Partridge et al., 1981). Further, as a population, the activity level of ELL neurons is not maximal for beat rates that best drive the jamming avoidance response (3-8 Hz).

The temporal selectivities (low-pass, band-pass and high-pass) of ELL neurons may result from descending feedback involving the nucleus praeeminentialis and cerebellum (Bastian, 1986a,b). The negative feedback system governs the adaptation properties of ELL neurons and has been implicated in gain control.

Most neurons in the dorsal torus semicircularis of the midbrain do respond best to beat rates of 3-8 Hz (Fig. 16) (Partridge et al., 1981; Rose and Heiligenberg, 1986), and exhibit

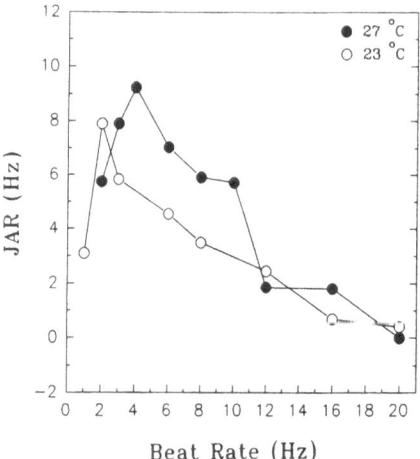

Figure 14. *Eigenmannia* jamming avoidance response (JAR) at two temperatures, as a function of the magnitude of frequency difference between a signal mimicking the fish's own electric organ discharge and a jamming signal.

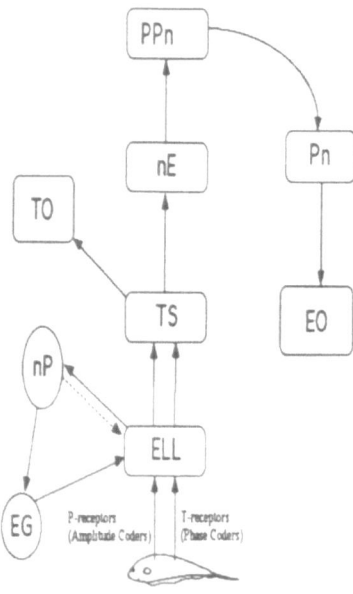

Figure 15. Schematic of the electrosensory pathway in *Eigenmannia*. ELL, electrosensory lateral line lobe; nP, nucleus praeeminentialis; EG, emminentialis granularis of cerebellum; TS, torus semicircularis; TO, optic tectum; nE, nucleus electrosensorius; PPn, prepacemaker nucleus; Pn, pacemaker nucleus; EO, electric organ.

pronounced low-pass or band-pass temporal selectivities. Over this range, the discharges of toral neurons are strongly locked to a particular phase of the beat cycle. The beat rates that are behaviorally preferred, therefore, are also most effective in exciting toral cells.

Beyond the torus, neurons in two other electrosensory regions are excited reliably by beat stimuli, the N. electrosensorius (NE) and prepacemaker nucleus (PPN) (Fig. 15). Neurons in the NE show band-pass temporal selectivity (Keller, 1988), but poor locking of their discharges to a particular phase of the beat cycle. Thus, the periodic fluctuations of stimulus amplitude are poorly represented in the periodicity of firing of NE neurons (Fig. 17). In the PPN, the transformation in representation is completed (Rose et al. 1988). PPN neurons respond maximally to beat rates most effective in eliciting JARs, yet fail to code these slow modulations in their periodicity of discharges. Instead, PPN neurons fire tonically for stimuli that are effective in causing rises of the EOD frequency (Fig. 18).

INTRACELLULAR PHYSIOLOGY: THE TEMPORAL DECODING QUESTION

In both the auditory and electrosensory systems, fluctuations in the amplitude of sensory signals over time are represented in the temporal patterns of discharges of primary afferents. In individual afferents, however, this representation in real time is crude; when constructing a histogram of the spike activity over time, many repetitions of the stimulus are required before the amplitude structure of the stimulus can be estimated with confidence. The nervous system, however, cannot average over multiple trials and must process information in real time. This raises the question of whether the time-varying pattern of action potentials can be "read" to recover an analog representation of the amplitude envelope of the stimulus. Filtering algorithms for such processing have been proposed (Bialek et al., 1991), but how might they be implemented by nervous systems? Further, from a functional standpoint, what might follow or accompany the recovery of an analog representation of the amplitude envelope?

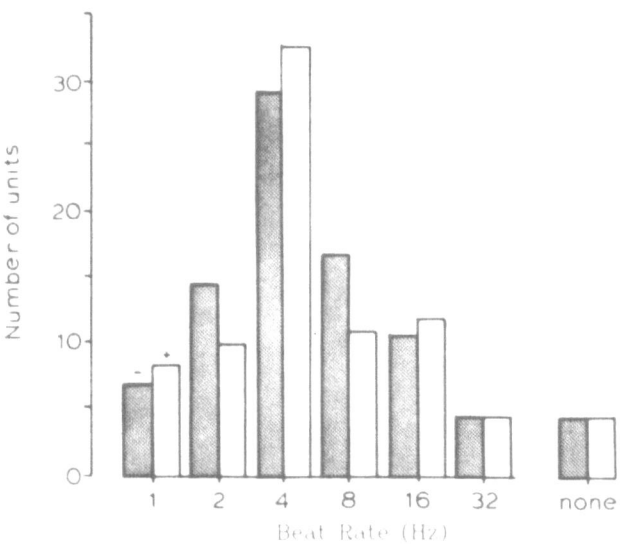

Figure 16. Summary histogram to show the distribution of most effective beat rates for neurons in the torus of *Eigenmannia*. Shaded bars represent most effective beat rate for jamming signals of frequencies lower then that of the fish's EOD; open bars for jamming signals higher than the frequency of the EOD mimic (from Partridge et al., 1981).

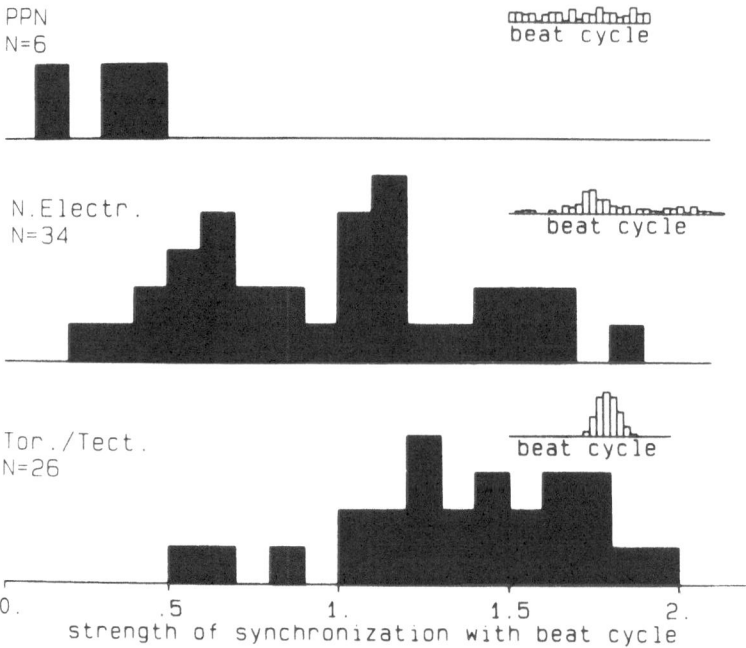

Figure 17. Strength of synchronization between spike rate and beat cycle is weakest for neurons in the PPN and strongest for neurons in the torus semicircularis and optic tectum. An intermediate degree of response synchronization is found for neurons in the nucleus electrosensorius. To derive synchronization coefficients, a Fourier analysis was performed on the spike rate histograms, and the amplitudes at the fundamental and second harmonic frequency were added. The length of the beat cycle was approximately 150-300 ms in each case. Each histogram differs significantly from the others (p<.0005, Mann Whitney U-test). Representative beat cycle histograms are shown in the insets at the upper right of each graph.

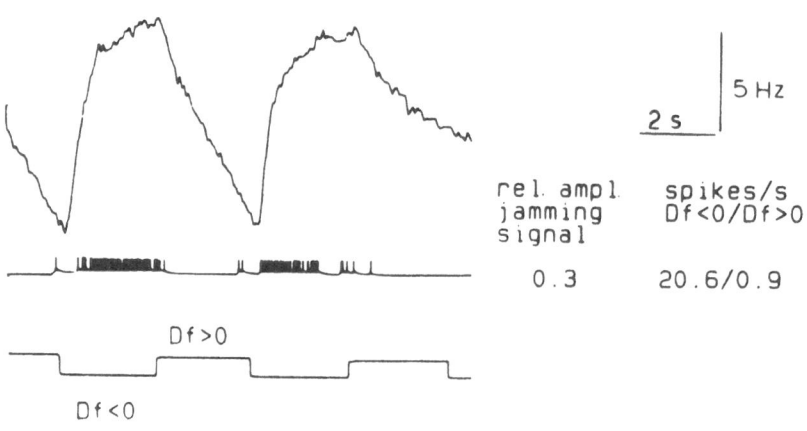

Figure 18. Spike occurrence (middle trace) of a prepacemaker (PPN) neuron and pacemaker/EOD frequency (top trace) as frequency of a jamming signal was changed periodically from 3-4 Hz above (Df>0) to 3-4 Hz below (Df<0) the frequency of the EOD-substitute signal.

To determine whether an analog representation of the amplitude envelope of a stimulus exists in the fluctuations of the membrane potential of individual neurons, intracellular recordings were made from neurons in the torus semicircularis of *Eigenmannia* (Rose and Call, 1992; 1993). In some neurons, individual, fast postsynaptic potentials (psps) were evident and their number fluctuated over time in approximate correspondence with stimulus amplitude; the amplitude envelope of the stimulus, however, was poorly represented in the membrane potential of these cells. In other neurons, however, the membrane potential fluctuated smoothly over time, nicely reflecting the amplitude envelope of the stimulus. Further, the magnitude of these stimulus-related potentials declined at higher beat rates (Fig. 19).

These recordings demonstrated that an analog representation of the amplitude envelope of the stimulus can be recovered and raised questions of what mechanisms might underlie this process and what the functional significance of this representation might be.

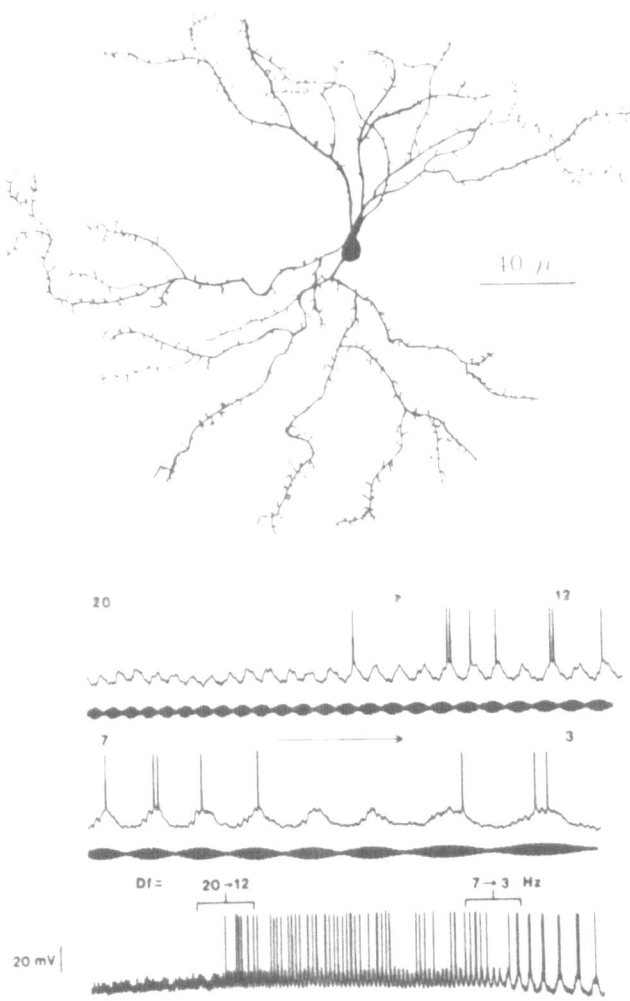

Figure 19. Neuron in lamina 4 of the torus semicircularis of *Eigenmannia*. The bottom trace shows intracellular recordings taken from this neuron while the stimulus beat rate was varied from 20 to 3 Hz. The portions of this trace indicated by brackets are expanded in the upper and middle traces. Beats were produced by the addition of two sinusoidal signals. The resulting signal was presented through a pair of electrodes, one in the mouth, the other at the tail.

Mechanisms for Temporal Filtering

The recovery of an analog representation of the beat envelope is essentially a filtering phenomenon. In this process, individual inputs to a neuron, represented by the conductances associated with the arrival of each spike at each synapse onto the cell, contribute little to the depolarization of the soma. Instead, depolarization of the soma appears to reflect the sustained aggregate level of synaptic activity within the neuron's dendrites. That is, the time constant governing the temporal integration of inputs appears large.

How might such temporal integration be achieved? Neurons that show high-fidelity representation of the amplitude envelope of the stimulus are richly endowed with dendritic spines. These neurons also have the most pronounced low-pass temporal filtering characteristics in that they respond well to low beat rates and poorly to high beat rates (Fig. 20).

These findings are consistent with the hypothesis that spines serve as a source of high impedance for inputs, and thereby increase the effective time constant of the cell. The specific features of the spines that could represent high impedances include the narrow, long stems and

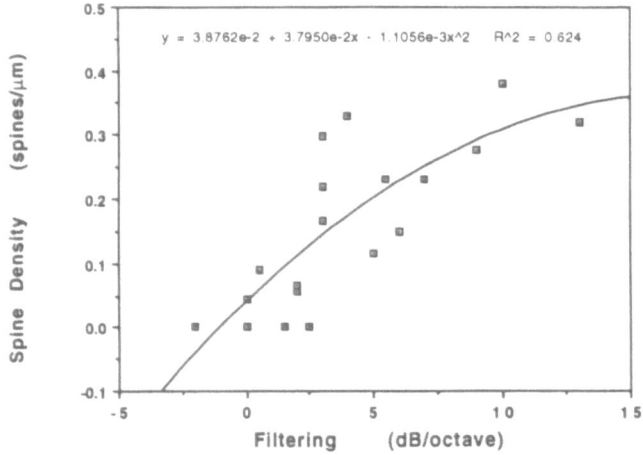

Figure 20. Relation between spine density (spines per μm length of dendrite) and low-pass filtering properties of toral neurons. Each neuron was recorded from a separate fish. Low-pass filtering (dB/octave) was quantified by performing a Fourier analysis on the intracellular recordings at various beat rates and measuring the height of the peak in the power spectrum at the frequency equal to the stimulus beat rate. A 6 dB/octave slope would result if, for example, postsynaptic potentials (psps) to a 20 Hz beat rate were half the amplitude of psps to a 10 Hz beat stimulus. Negative filtering values indicate that psp amplitude increased with beat rate.

the conductance channels associated with the synapses at the spine head. While these data support the notion that spines play a role in the temporal filtering properties of central neurons, they do not rule out other mechanisms.

Specifically, low-pass filtering, as measured by the decline in psp amplitude with increasing beat rate, may be due, in part, to "network" types of mechanisms. One potential type of network mechanism involves inhibition from a local-circuit high-pass neuron. This mechanism can be tested by superimposing a fast pattern of AM upon a slow one; the fast AM

18

should recruit the putative inhibitory interneuron and thereby diminish the response to the slower modulation pattern. Recordings from toral neurons that respond well over a wide range of AM rates demonstrate that the superimposed fast AM pattern is effective in recruiting toral afferents (Fig. 21). The fast AM does not, however, attenuate the responses of low-pass neurons to the slow AM pattern (Fig. 22).

These data also suggest that the decrease in psp size at high beat rates is not mediated by a voltage and time-dependent conductance mechanism where ligand-gated channels open only after experiencing maintained depolarization. If this were the case, one would expect the responses to fast AM to be facilitated when combined with the slow modulations.

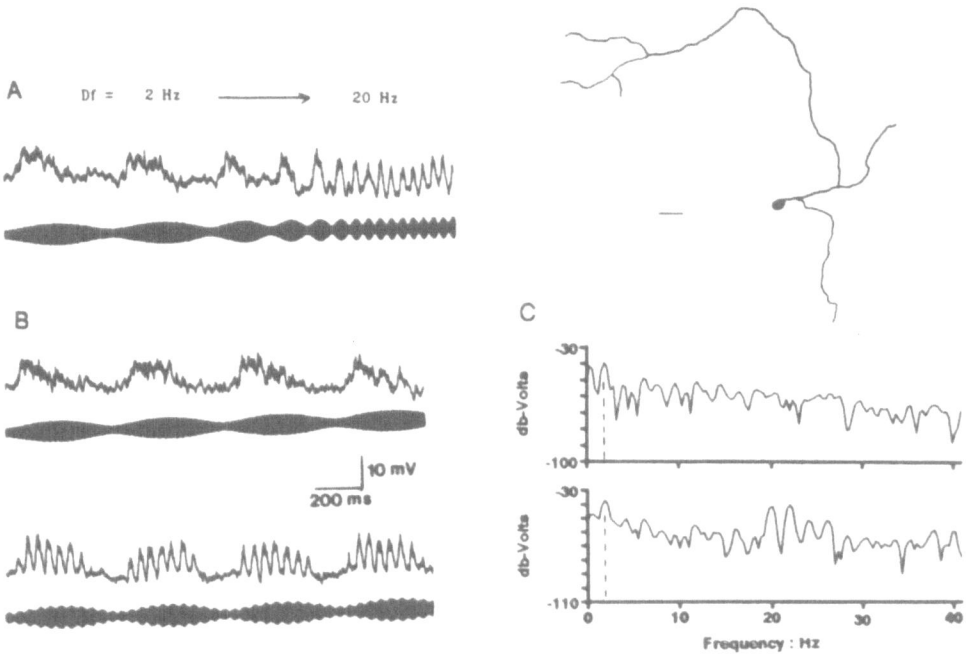

Figure 21. A: Responses of a temporally nonselective neuron in the torus semicircularis of *Eigenmannia* to a stimulus whose beat pattern was varied from 2 Hz to 20 Hz. B: Responses of the same neuron to a 2 Hz stimulus alone (upper traces) and stimulus in which a 20 Hz modulation pattern was superimposed onto a 2 Hz pattern (lower traces). C: Power spectra of the recordings shown in (B); dotted line identifies peaks at 2 Hz. Scale bar is 40 μm.

In many cases, psps related to the fast modulation pattern are largest for the first few cycles that accompany the depolarization resulting from the slower AM, then decline in magnitude (adapt) throughout the remainder of the slow AM cycle. This response profile is most apparent in band-pass neurons. Such adaptation clearly contributes in many cases to the strong decline in psp amplitude as the beat rate is increased. How is this adaptation process mediated? One likely substrate is the descending feedback network (Bastian, 1986a,b) that inhibits ELL neurons via projections from nucleus praeeminentialis to the cerebellum (Fig. 15). This descending, negative-feedback system strongly influences the adaptation properties of amplitude-sensitive neurons in the ELL.

CONCLUSIONS AND FUTURE DIRECTIONS

The transformation that occurs in the neural representation of changes in stimulus amplitude over time (amplitude modulations, AM) is very similar in the auditory system of anurans and in the electrosensory system of *Eigenmannia*. In both sensory systems, temporal patterns of AM are represented by a periodicity code in the peripheral nervous system and by a temporal filter ensemble in the midbrain. The fact that these two sensory systems have

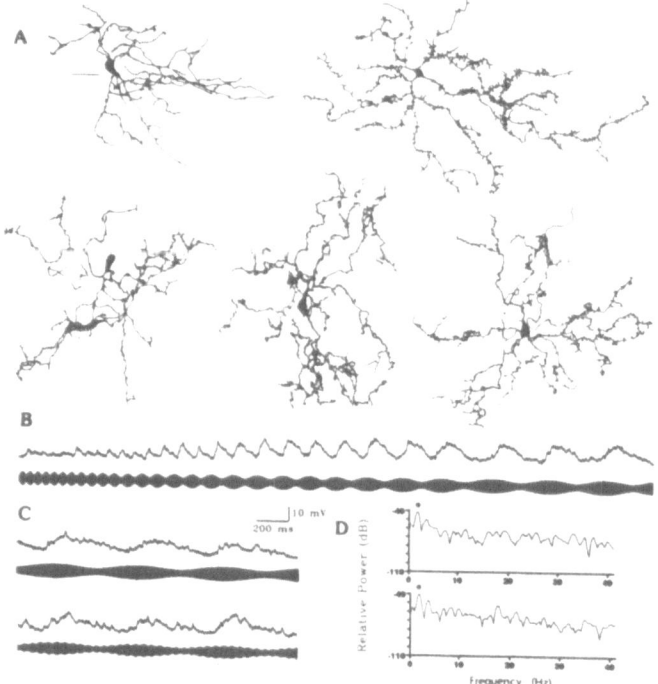

Figure 22. A: Drawings of neurons in laminae 4 and 5 of the torus semicircularis of *Eigenmannia* that exhibited low-pass temporal selectivity. Top row: two type-b neurons of lamina 5. High background staining precluded drawing all structural details of the neuron on the left. Bottom row: from left to right, octopus cell, vertical neuropil cell and pyramidal cell. Scale bar is 40 µm. **B**: Intracellular recordings from the neuron shown in the upper left corner in response to a beat stimulus that changed in rate from approximately 20 to 2 Hz. **C**: Responses of the same neuron to a 2 Hz beat pattern (upper traces) and to the complex modulation pattern described in Fig. 21 (lower traces). **D**: Power spectra of the intracellular recordings shown in (C).

converged, in general, on strikingly similar solutions to a similar problem, strongly argues for the functional importance of this transformation. It is particularly fascinating that the representation of AM rate in the periodicity of discharges is lost in both systems. The loss of periodicity coding in the central auditory system of anurans is, by itself, perhaps not entirely surprising since neurons in this system are tuned to moderately high AM rates. Such periodicity information may not be preserved accurately across many synapses. In the electrosensory system, however, AM rates below 10 Hz are biologically relevant; temporal precision of synaptic transmission is more than adequate to preserve information about the pattern of slow AMs (e.g. 4 Hz) in the temporal pattern of discharges, even in neurons 3-4 synapses away from the periphery. The loss of periodicity representation is not required for generating temporal filtering, since strong filtering is present in the torus of *Eigenmannia*, where neurons still show robust periodicity coding. It would seem more likely, therefore, that the loss of periodicity coding is functionally important, but to what end? One possibility is that this transformation is important at the sensorimotor interface. Converting from a periodic to continuous discharge pattern would provide a tonic signal to motor or "motivational" areas. For example, a smooth rise in the EOD frequency can be produced in response to a stimulus that is periodic. Further work is needed to determine the mechanism underlying this transformation.

Is there evidence of similar transformations in the neural representation of amplitude modulations in other species? Langner (1992) has recently reviewed periodicity coding in the auditory nervous systems of vertebrates. Rather than re-review this extensive literature, I will present some general conclusions that pertain to this question. The same classes of AM-selective neurons present in anurans (Rose and Capranica, 1983; 1984; Rose et al., 1985; Walkowiak, 1984) are also present in the midbrain of the guinea fowl (Albert et al., 1989) and in the cricket (Shildberger, 1984). In mammals, there is a loss of periodicity coding at the midbrain, and the temporal code appears transformed into a spike rate code (Langner and Schreiner, 1988; Rees and Palmer, 1989). The processing of temporal patterns of amplitude modulations by an ensemble of central temporal filters appears to be a widespread phenomenon. Neurons that respond in a band-pass fashion to AM rates (in terms of spike rate) have been found at several levels of the central auditory systems of several mammals (Schreiner and Urbas ,1986; 1988; Langner and Schreiner, 1988; Rees and Palmer, 1989), birds (Albert et al., 1989) and crickets (Schildberger, 1984).

While the mechanisms underlying the temporal filtering properties of central neurons are still uncertain, both electrical filtering and adaptation processes appear to play a role. In the electrosensory system, descending negative feedback from nucleus praeeminentialis to the ELL provides a powerful inhibitory input to pyramidal cells and results in adaptation to a maintained stimulus (Bastian, 1986a,b). Adaptation progressively attenuates the overall firing rate as the beat rate is decreased, so that these neurons exhibit high-pass filtering. The finding that toral neurons are sensitive to the rise time of tone bursts (Gooler and Feng, 1992) strongly suggests that a similar form of adaptation plays a role in the temporal filtering properties of neurons in the torus of anurans. Adaptation might also contribute to the decline of response at high beat rates if integrating elements in the feedback circuit are able to sum over several modulation cycles of the stimulus and provide an inhibitory influence. The moderate band-pass and low-pass properties of some ELL neurons may result from such a process. In the electrosensory system, further work is needed to determine whether other adaptation mechanisms exist at the level of the torus.

The brainstem mechanisms underlying adaptation in the anuran auditory system remain to be explored, but clearly this process is well developed in many neurons in the dorsal medullary nucleus, the first central auditory region in anurans (Hall and Feng, 1991).

In the electrosensory system of *Eigenmannia*, spiny toral neurons faithfully represent the amplitude envelope of beat stimuli in the smooth fluctuations of their membrane potential. The magnitude of these psps declines with AM rate. It appears, therefore, that individual inputs to these cells have little effect on the membrane potential. Rather, the membrane potential reflects aggregate, sustained input to the cell. Aspiny neurons, however, poorly represent the amplitude envelope of beat stimuli in their membrane potential. These neurons show fast psps and code high temporal (beat) frequencies.

It is presently unclear as to exactly how dendritic spines might limit the inward current flow associated with activity at each synapse. Potential sites of high resistance to current flow include the synaptic channels and the long, thin stems of the spines. It is clear from light microscopic examination that the stems of spines on toral neurons are quite long (2-4 µm is common), but EM is required to determine the dimensions of the conducting core of the stem. An important extension of these considerations is the notion that the time constant governing the transmission of input to a spiny neuron, through the spines, is likely to differ significantly from that measured by passing current into the soma of a neuron. Simulation studies, employing models based on measured structural and electrical features of particular neurons, are needed to further examine the role of dendritic spines in the low-pass filtering process. For these studies, it will be important to determine whether spiny neurons receive all inputs onto their spines, as appears to be the case in other spiny neurons (Gray, 1959).

If adaptational processes are present in the torus, they may be associated with the spines. For example, maintained activity at a synapse might result in elevated calcium levels at the spine head, and if calcium-sensitive potassium channels were present, synaptic current could be shunted before reaching the parent dendrite. These ideas remain quite speculative at present, but high levels of calcium have been observed in some spines of hippocampal cells (Muller and Connor, 1991), consistent with the "chemical compartmentalization" model of spine function (Koch et al., 1992; Koch and Zador, 1993). The role of dendritic spines in creation of temporal filter characteristics is likely to be a fruitful area of future investigation.

The temporal selectivity of neurons in the auditory and electrosensory systems result primarily from central processes. Will a map of AM tuning be found in species that have more than a single "AM channel"? The Pacific treefrog has been shown to have at least two AM channels (Brenowitz and Rose, 1993) and would be an excellent species in which to explore this question. Results may help elucidate the functional significance of computational maps in the CNS.

There is some behavioral evidence indicating sensitivity to the shape as well as rate of pulses within a call. Is pulse shape sensitivity based on the same mechanisms that underlie AM rate selectivity? Is there, as a consequence, a tradeoff between pulse rate and shape. Toral neurons recorded thus far show only small differences in firing rate for natural vs sinusoidal patterns of AM.

The sharpness of tuning to AM rate varies considerably among the central neurons that have been recorded. Are sharply tuned neurons and broadly tuned neurons segregated in different areas of the torus? Mechanistically, what gives rise to the sharper temporal tuning? Since the most sharply tuned neurons in anurans also exhibit little, if any, periodicity coding and have long latencies, an integration mechanism involving long channel open times (as seen in the case of NMDA forms of glutamate channels) may be present.

The evolution of AM selectivity remains an unexplored and interesting question. Do more primitive anurans exhibit the same sharpness of temporal tuning as more derived species? Are all classes of temporally selective neurons that are found in highly derived anurans also present in more primitive ones? The ancestral condition may have simply involved adaptation mechanisms, with integration mechanisms, as seen in band-suppression neurons, being incorporated later.

SUMMARY

A similar transformation in the neural representation of changes in stimulus amplitude over time occurs in the auditory system of anurans and in the electrosensory system of *Eigenmannia*. In both sensory systems, temporal patterns of amplitude modulation (AM) are represented by a periodicity code in the peripheral nervous system and by a temporal filter ensemble in the midbrain. In the latter representation, each band-pass neuron in the midbrain discharges maximally (is tuned) to a particular rate of AM. AM-tuned neurons that appear to be closest to the sensory-motor interface fail to code the rate of AM in their periodicity of discharges. The fact that these two sensory systems have converged, in general, on strikingly similar solutions to a similar problem, strongly argues for the functional importance of this mechanism. The loss of periodicity coding may play an important role in the translation of sensory information to motor commands.

In the electrosensory system, the generation of AM tuning is accompanied by representation of the amplitude envelope of the stimulus in the smooth fluctuations of the membrane potential of spiny neurons. These observations suggest the hypothesis that dendritic spines serve as a source of high impedance for inputs, and thereby increase the effective time constant of the cell. As a consequence of this electrical filtering process, fast fluctuations in stimulus amplitude would be less effective than slower modulations in depolarizing the soma membrane.

REFERENCES

Albert, M., Hose B., and Langner G., 1989, Modulation transfer functions in the auditory midbrain (MLD) of the guinea fowl (*Numida meleagris*), in: "Dynamics and Plasticity in Neuronal Systems", N. Elsner and W. Singer, eds, Thieme Verlag, Stuttgart.

Bastian, J., 1981a, Electrolocation I. How the electroreceptors of *Apteronotus albifrons* code for moving objects and other electrical stimuli, *J. Comp. Physiol. A*, 144:465.

Bastian, J., 1981b, Electrolocation II. The effects of moving objects and other electrical stimuli on the activities of two categories of posterior lateral line lobe cells in *Apteronotus albifrons, J. Comp. Physiol. A*, 144:481.

Bastian, J., 1986a, Gain control in the electrosensory system mediated by descending inputs to the electrosensory lateral line lobe, *J. Neurosci.*, 6:553.

Bastian, J., 1986b, Gain control in the electrosensory system: A role for the descending projections to the electrosensory lateral line lobe, *J. Comp. Physiol. A*, 158:505.

Bastian, J., and Yuthas, J., 1984, The jamming avoidance response of *Eigenmannia*: Properties of a diencephalic link between sensory processing and motor output, *J. Comp. Physiol. A*, 154: 895.

Bialek, W., Rieke, F., De Ruyter van Steveninck, R.R., and Warland, D., 1991, Reading a neural code, *Science*, 252:1854

Brenowitz, E.A., and Rose, G.J., 1994, Behavioural plasticity mediates aggression in choruses of the Pacific treefrog, *Anim. Behav.*, 47:633.

Brenowitz, E.A., Rose, G.J., and Capranica, R.R., 1985, Species specificity and temperature dependency of temporal processing by the auditory midbrain of two species of treefrogs, *J. Comp. Physiol. A*, 157:763.

Bullock, T.H., 1982, Electroreception, *Ann. Rev. Neurosci.*, 5:121.

Bullock, T.H., Hamstra, R.H., and Scheich, H., 1972, The jamming avoidance response of high-frequency electric fish, *J. Comp. Physiol. A*, 77:1.

Emlen, S.T., 1972, An experimental analysis of the parameters of bird song eliciting species recognition, *Behavior*, 41:130.

Gerhardt, H.C., 1978, Temperature coupling in the vocal communication system of the gray treefrog, *Hyla versicolor, Science*, 199:992.

Gerhardt, H.C., 1982, Sound pattern recognition in some North American treefrogs (*Anura:Hylidae*): Implications for mate choice, *Am. Zool.*, 22:585.

Gerhardt, H.C., 1988, Acoustic properties used in call recognition by frogs and toads. in: "The Evolution of the Anuran Auditory System", Fritzsch, B., Ryan M., Wilczynski, W., Hetherington, T., Walkowiak, W., eds, John Wiley and Sons, New York, NY.

Gerhardt, H.C., and Dougherty, J.A., 1988, Acoustic communication in the gray treefrog, *Hyla versicolor*: Evolutionary and neurobiological implications, *J. Comp. Physiol. A*, 162:261.

Gooler, M., and Feng, A.S., 1992, Temporal coding in the frog auditory midbrain: Influence of duration and rise-fall time on the processing of complex amplitude-modulated stimuli, *J. Neurophysiol.*, 67:1.

Gray, E.G., 1959, Axo-somatic and axo-dendritic synapses of the cerebral cortex: an electron-microscopic study, *J. Anat,.* 93:420.

Hall, J.C., and Feng, A.S.,1991, Temporal processing in the dorsal medullary nucleus of the northern leopard frog (*Rana pipiens pipiens*), *J. Neurophysiol.*, 66:955.

Heiligenberg, W., 1977, Principles of electrolocation and jamming avoidance in electric fish. A neuroethological approach. in: "Studies of Brain Function, Vol. 1," Braitenberg, V., ed, Springer, Berlin.

Heiligenberg, W., 1989, Central processing of electrosensory information in gymnotiform fish, *J. Exp. Biol.*, 146:255.

Heiligenberg, W., 1991, The neural basis of behavior: A neuroethological view, *Ann. Rev. Neurosci.*, 14:247.

Heiligenberg, W., 1991, "Neural Nets in Electric Fish", MIT press, Cambridge, MA.

Heiligenberg, W., Baker, C., and Matsubara, J., 1978, The jamming avoidance response in *Eigenmannia* revisited: The structure of a neuronal democracy, *J. Comp. Physiol. A*, 127: 267.

Kay, R.H., 1982, Hearing modulations in sound, *Physiol. Rev.*, 62:894.

Keller, C., 1988, Stimulus discrimination in the diencephalon of *Eigenmannia*: The emergence and sharpening of a sensory filter, *J. Comp. Physiol. A*, 162:747.

Koch, C., and Zador, A., 1993, The function of dendritic spines: Devices subserving biochemical rather than electrical compartmentalization, *J. Neurosci.*, 13:413.

Koch, C., Zador, A., and Brown, T.H., 1992, Dendritic spines: Convergence of theory and experiment, *Science*, 256:973.

Langner, G., 1992, Periodicity coding in the auditory system, *Hearing Res.*, 60:115.

Langner, G., and Schreiner, C.E., 1988, Periodicity coding in the inferior colliculus of the cat. I. Neuronal mechanisms, *J. Neurophysiol.*, 60:1799.

Matsubara, J., and Heiligenberg, W., 1978, How well do electric fish electrolocate under jamming? *J. Comp. Physiol. A*, 149:339.

Muller, W., and Connor, J.A., 1991, Dendritic spines as individual neuronal compartments for synaptic Ca^{2+} responses, *Nature*, 354:73.

Partridge, B.L., Heiligenberg, W., and Matsubara, J., 1981, The neural basis for a sensory filter in the jamming avoidance response: No grandmother cells in sight., *J. Comp. Physiol. A*, 145:153.

Rees, A., and Palmer, A.R.,1989, Neuronal responses to amplitude-modulated and pure-tone stimuli in the guinea pig inferior colliculus, and their modification by broad-band noise, *J. Acoust. Soc. Am.*, 85:1978.

Rose, G.J., 1986, A temporal processing mechanism for all species? *Brain Behav. Evol.*, 28: 134.

Rose, G.J., and Capranica, R.R., 1983, Temporal processing in the central auditory system of the leopard frog (*Rana pipiens*), *Science*, 219:1087.

Rose, G.J., and Capranica, R.R., 1984, Processing amplitude-modulated sounds by the auditory midbrain of two species of toads: Matched temporal filters, *J. Comp. Physiol. A*, 154: 211.

Rose, G.J., and Capranica, R.R., 1985, Sensitivity to amplitude modulated sounds in the anuran auditory system, *J. Neurophysiol.*, 53:446.

Rose, G.J., and Heiligenberg, W., 1986, Neural coding of difference frequencies in the midbrain of the electric fish *Eigenmannia*: Reading the sense of rotation in an amplitude-phase plane, *J. Comp. Physiol. A*, 158:613.

Rose, G.J., and Call, S.J., 1992, Evidence for the role of dendritic spines in the temporal filtering properties of neurons: The decoding question and beyond, *Proc. Natl. Acad. Sci. USA*, 89:9662.

Rose, G.J., and Call, S.J., 1993, Temporal filtering properties of midbrain neurons in an electric fish: Implications for the function of dendritic spines, *J. Neurosci.*, 13:1178.

Rose, G.J., Brenowitz, E.A., and Capranica, R.R., 1985, Species specificity and temperature dependency of temporal processing by the auditory midbrain of two species of treefrogs, *J. Comp. Physiol. A*, 157:763.

Rose, G.J., Zelick, R., and Rand, A.S., 1988, Auditory processing of temporal information in a neotropical frog is independent of signal intensity, *Ethology*, 77:330.

Rose, G.J., Kawasaki, M., and Heiligenberg, W., 1988, "Recognition units" at the top of a neuronal hierarchy? *J. Comp. Physiol. A*, 162:759.

Ryan, M.J., 1985), "The Tungara Frog", University of Chicago Press, Chicago, IL.

Schildberger, K., 1984, Temporal selectivity of identified auditory neurons in the cricket brain, *J. Comp. Physiol. A*, 155:171.

Schreiner, C.E., and Urbas, J.V., 1986, Representation of amplitude modulation in the auditory cortex of the cat. I. The anterior auditory field (AAF), *Hearing Res.*, 21:227..

Schreiner, C.E., and Urbas, J.V., 1988, Representation of amplitude modulation in the auditory cortex of the cat. II. Comparison between cortical fields, *Hearing Res.*, 32:49.

Walkowiak, W., 1984, Neuronal correlates of the recognition of pulsed sound signals in the grass frog, *J. Comp. Physiol. A*, 155:57.

Wells, K.D., 1977, The social behavior of anuran amphibians, *Animal Behav.*, 25:666.

MECHANISMS FOR ANALYSIS OF AUDITORY TEMPORAL PATTERNS IN THE BRAINSTEM OF ECHOLOCATING BATS

John H. Casseday and Ellen Covey

Department of Neurobiology
Duke University Medical Center
Durham, NC 27710
USA

INTRODUCTION

Temporal Patterns in Biologically Important Sounds

A vital function of the auditory system in all vertebrates is to identify sounds that are important for social interactions, predation and predator avoidance. Examples of these behaviorally important sounds are communication signals of conspecifics, noises made by movements of other animals and highly specialized species-specific sounds such as the biosonar signals used by echolocating bats. Identification of many behaviorally important sounds, especially those made by prey or predators, must occur rapidly to activate other neural systems that produce a motor response. Many biologically important sounds are characterized by simple temporal features, such as duration of the sound or its components, direction of a frequency sweep, or the rate of modulation in sounds that periodically change in frequency or amplitude. Many sounds are further characterized by complex sequences of elements that follow a specific order over time.

Echolocation Signals and Auditory Processing

For echolocating bats, success in foraging depends on detection of echoes of the sounds that the bat emits. Therefore, it is easy to determine what aspects of sound are biologically important to the bat. The echolocation signals of all bats have specific patterns of temporal and spectral characteristics that are related to specific stages and strategies of foraging behavior. The echolocation sounds of bats that use frequency modulated (FM) sounds are particularly rich in variation (Kalko and Schnitzler, 1989; Schnitzler et al., 1987; Neuweiler, 1990). For example, Figure 1 shows the echolocation calls of a big brown bat during a typical hunting sequence. When the bat is searching for an insect it emits a relatively long (10-20 ms) signal that has most of its energy concentrated within a shallow FM sweep that changes only from ~28 kHz to ~23 kHz. The repetition rate of this signal is about 3/s.

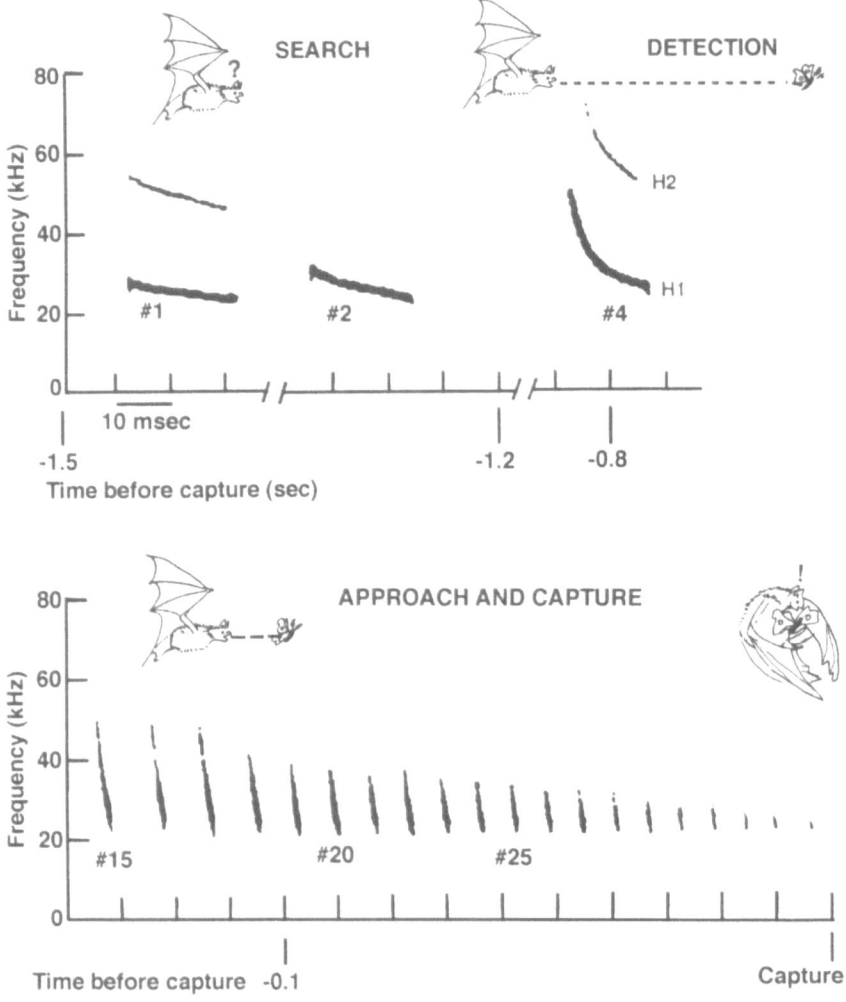

Figure 1. Echolocation sounds: excerpts from a typical hunting sequence. During the searching phase (top left), *Eptesicus* emits a shallow frequency sweep, the main harmonic of which changes in frequency from about 28 kHz to about 23 kHz. This signal is called the quasi-constant frequency (QCF) signal. Sometimes it has a second harmonic. In a later stage, presumably after the bat has detected a target (top right), the sweep starts at about 50 kHz, becomes slightly shorter although a short QCF tail may remain. During the approach stage of hunting (lower figure), the echolocation sounds are brief frequency sweeps with 8 to 10 ms between sweeps. Only the fundamental harmonic of the signals is shown in the lower figure. The numerals indicate the sequential position of the echolocation sounds as they were recorded in the hunting sequence. (Adapted from Simmons, 1989).

    When pursuing an insect the bat emits short (<1 to 5 ms) FM sounds that are spectrally broad (~80 kHz to ~20 kHz), and the repetition rate increases up to as much as 150/s (Simmons, 1989). Long duration signals with narrow bandwidth are used to detect the presence of prey, while short duration, broad band signals are used to determine the distance of targets (Neuweiler, 1990). In addition, most bats vary their signal designs according to the acoustical conditions of the environment. For example, the narrow-band signal of the searching phase is changed to a wider band signal under conditions of background clutter of nearby foliage (von der Emde and Schnitzler, 1986; Kalko and Schnitzler, 1989).

The processing of these signals begins in the auditory structures of the lower brainstem. Certain parts of the auditory brainstem in bats and dolphins have obvious anatomical specializations that appear to play a role in analyzing the temporal structure of echolocation sounds (Poljak, 1926; Baron, 1974; Schweizer, 1981; Zook and Casseday, 1982a; Covey and Casseday, 1986; 1991; Zook et al., 1988).

Why Study Auditory Temporal Processing in the Brainstem?

There are two reasons why we would expect processing in the lower brainstem to play an important role in the analysis of temporal patterns. First, "real-time" processing of events must be accomplished at an early stage in the system before timing precision is lost due to the jitter introduced by transmission across multiple synapses. Thus the fine temporal precision that exists initially may be translated into a non-temporal code such as neural place, or it may be translated into a different format of temporal pattern that is resistant to degradation across synapses. The best known example of a transformation of time to place is the encoding of binaural time differences into neural place in the medial superior olive (Jeffress 1948; Goldberg and Brown, 1969; Yin and Chan, 1990). In this chapter, we propose that for the same reasons, sound onset is encoded into neural place in the ventral nucleus of the lateral lemniscus. Little is known about translations from one temporal code to another.

The second reason to suspect that a significant amount of temporal processing occurs in the brainstem has to do with the fact that rapid sensory processing is needed for some kinds of behavior. For behaviors such as predator avoidance or prey capture, the initial analysis of sound pattern must be accomplished very quickly. The hunting behavior of echolocating bats is based on the analysis of trains of emitted calls and reflected echoes that occur at rates up to 150/s. To achieve the required rapidity of processing, some basic analysis of sound patterns must occur at an early level in the system, and this level must have some fairly close connections with motor systems (Grinnell, 1963b; Casseday and Covey, 1995). The auditory midbrain, or inferior colliculus (IC), seems to satisfy these requirements.

In this chapter, we use the echolocating bat as a model to examine how parallel pathways below the IC contribute to the analysis of temporal patterns of sound. The analysis consists of multiple transformations of the neural input to the IC. These transformations include changes from excitatory to inhibitory pathways, changes in the temporal patterns of response, and construction of delay lines. We then show how inputs with various properties are integrated at the IC to create tuning for temporal features of sound. The consequences of this integrative process are creation of tuning for biologically important parameters of sound and a reduction in the speed of processing at the IC, possibly to match the rate of sensory input to the pace of behavior (Casseday and Covey, 1995).

PARALLEL PATHWAYS TO THE MIDBRAIN

In all vertebrates, the central auditory system below the midbrain consists of a complex network of parallel pathways. These lower brainstem pathways consist of a number of separate processing centers, each dedicated to extracting and emphasizing a particular class of information. These centers can be viewed as a set of "preprocessors" that transform auditory signals in various ways for use at the next stage, the IC, where the lower pathways converge. The IC is the target of ten or more pathways, each of which is anatomically and functionally distinct (e.g., Zook and Casseday, 1982b; 1985; 1987; Ross and Pollak, 1989; Covey and Casseday, 1986; Casseday and Covey, 1992; Pollak and Casseday, 1989; Covey and Casseday,

1995). From this evidence it is reasonable to hypothesize that some of these pathways converge at the cellular level (Vater et al., 1995).

The parallel pathways of the lower brainstem can be grouped into two broad classes, a *binaural system* that receives input from both ears, and a *monaural system* that receives input only from the contralateral ear (Fig. 2). The structure and function of the binaural pathways of the brainstem have been studied in sufficient detail to show that they perform the initial computations for sound localization (Jeffress, 1948; Goldberg and Brown, 1969; Kuwada and Yin, 1987; Yin and Chan, 1990). Here we will consider mainly the monaural pathways, because they are especially highly developed in echolocating bats and seem to play an important role in the initial stages of processing temporal patterns of sound (Schweizer, 1981; Zook and Casseday, 1982a; Covey and Casseday, 1986; 1991; 1995).

The monaural system in echolocating bats includes direct pathways from the divisions of the cochlear nucleus to the midbrain and indirect pathways via the intermediate nucleus of the lateral lemniscus (INLL) and two divisions of the ventral nucleus of the lateral lemniscus (VNLL) (Fig. 2). Of these nuclei, the most structurally distinct is the columnar division of VNLL (VNLLc), a specialized group of cells that is most highly developed in echolocating bats and dolphins (Covey and Casseday, 1986; Zook et al., 1988).

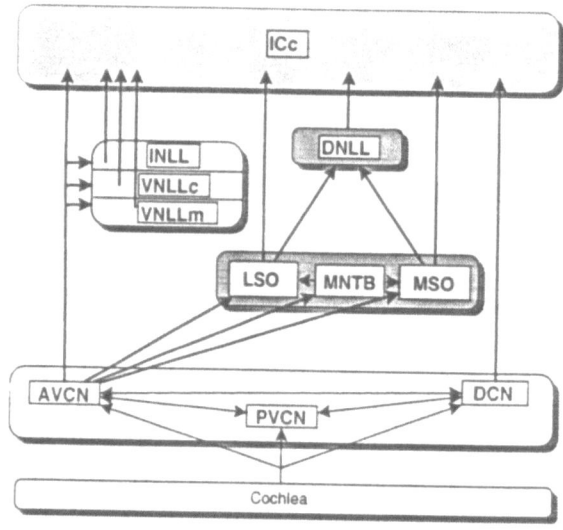

Figure 2. Schematic diagram to show that the IC integrates information that arises directly from the cochlear nuclei with information that undergoes one or more transformations at intermediate processing stages. The light stippled boxes indicate monaural systems, and the dark stippled boxes indicate binaural pathways. The first central station in the pathway is comprised of the anteroventral cochlear nucleus (AVCN), the posteroventral cochlear nucleus (PVCN) and the dorsal cochlear nucleus (DCN). The DCN and the AVCN project directly to the central nucleus of the inferior colliculus (ICc). The AVCN is a major source of projections to the intermediate, columnar and multipolar nuclei of the lateral lemniscus (INLL, VNLLc and VNLLm). The AVCN is also the major source of input to the first stage for binaural processing, comprised of the medial superior olive (MSO) and the lateral superior olive (LSO). The medial nucleus of the trapezoid body (MNTB), is itself monaural but contributes to the binaural connections of LSO and MSO. The LSO and MSO send binaural information to the ICc directly and indirectly via the dorsal nucleus of the lateral lemniscus (DNLL). The monaural processing centers in the lateral lemniscus (INLL, VNLLc, VNLLm) receive input from the AVCN as shown, and each of these centers projects to the IC. For simplification, monaural inputs from PVCN and MNTB to the lateral lemniscus are not shown.

SIGNAL PROCESSING IN THE LOWER BRAINSTEM

The processing of signals in the monaural auditory pathways seems to accomplish three transformations: a reversal in the sign of the neural signal from excitation to inhibition in some pathways, changes in the temporal pattern of neural discharge, and an increase in the range of latencies.

Change in Sign from Excitation to Inhibition

Although the projections from the cochlea to the central nervous system are excitatory, there is much evidence that as early as the level of the cochlear nucleus inhibition plays an important role. One line of evidence for excitatory and inhibitory pathways comes from immunocytochemical staining for the two inhibitory neurotransmitters GABA and glycine. Neurons in some brainstem pathways to the IC are known to contain GABA, while neurons in other pathways contain glycine (Glendenning and Baker, 1988; Godfrey et al., 1978; Wenthold et al., 1986; 1987; Roberts and Ribak, 1987; Saint-Marie et al., 1989). Many neurons in the INLL and VNLL stain for glycine, indicating that a high proportion of these neurons provide inhibitory input to the IC. Neuropharmacological experiments in which the inhibitory inputs to IC neurons are blocked show that both GABA and glycine affect the response properties of IC neurons (Faingold et al., 1991; Park and Pollak, 1993; Pollak and Park, 1993; Vater et al., 1992; Johnson, 1993). Much of the inhibitory input from monaural pathways undoubtedly arises from INLL and VNLL. To see how these inputs might participate in tuning for temporal features of sound, we now examine the different temporal patterns of neural discharge in the monaural nuclei of the lateral lemniscus.

Changes in Discharge Pattern

The basic discharge patterns of neurons in the INLL and VNLL have been studied using electrophysiological methods for extracellular recording of action potentials evoked by sound (Covey and Casseday, 1991). Some examples of typical discharge patterns are shown in Figure 3. Neurons in the VNLLc have the most distinctive response properties. These cells have no spontaneous activity and respond to a sound with one and only one spike. Furthermore, an individual neuron responds over an exceptionally broad range of sound frequencies. For bursts of pure tones or noise, the timing of the spike is precisely correlated with the time of sound onset. Under constant stimulus conditions, the trial-to-trial variability of the spike time is only a few tens of microseconds; moreover, the spike times remain virtually constant over a wide range of stimulus amplitudes and frequencies. Figure 4 shows examples of constant latency neurons in the VNLLc and compares them with neurons in the cochlear nucleus. For an echolocating animal such as a bat, constant latency neurons might be especially important for marking the time of an outgoing high intensity biosonar signal, information that would be used at a higher level to measure target distance by calculating the time between the emission of a signal and the return of its echo.

Transient responses consisting of a single spike are not present in the auditory nerve; therefore, they must arise through neural processing at higher levels. Figure 5 shows how the transient responses of neurons in the medial superior olive are created through convergence of two sustained inputs, one excitatory and one inhibitory, slightly offset in time. The result is a transient onset response if excitation leads, a transient offset response if inhibition leads. Although the mechanism for producing level-tolerant constant-latency responses in the VNLLc is not known with certainty, it may also depend on interaction of excitation and inhibition. Alternatively, it could be created through some intrinsic biophysical properties of the neurons themselves. If the basis for this property can be discovered, it will represent an important breakthrough in understanding the early stages of auditory processing.

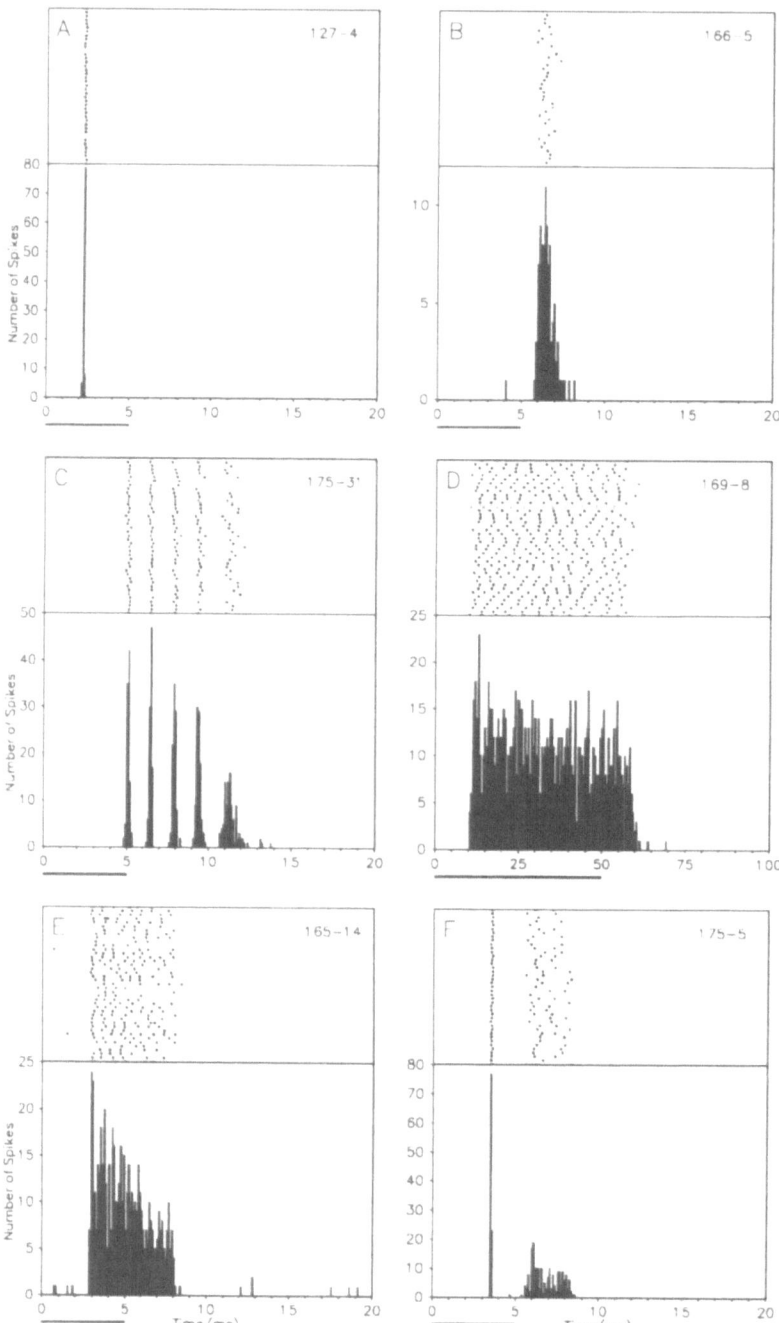

Figure 3. Basic discharge patterns of neurons in the INLL and VNLL as seen in dot rasters (upper portion of each panel) and post stimulus time histograms (PSTHs, lower portion of each panel). Each dot raster illustrates the neuron's response to 40 stimulus presentations, with individual trials stacked from top to bottom. Each dot represents the time when an action potential occurred. In the PSTHs, each bar represents the total number of spikes per time bin summed over all the trials. **A.** A transient constant latency neuron in VNLLc. **B.** A transient variable latency neuron in INLL. **C.** A chopper neuron in VNLLm. **D.** A sustained, nonadapting neuron in INLL. **E.** A sustained, adapting neuron in INLL. **F.** A pauser in VNLLm. The *bar* below each panel indicates the stimulus duration. Note the long duration signal in D. Numerals in the upper right of the graphs indicate animal and unit numbers. (From Covey and Casseday, 1991)

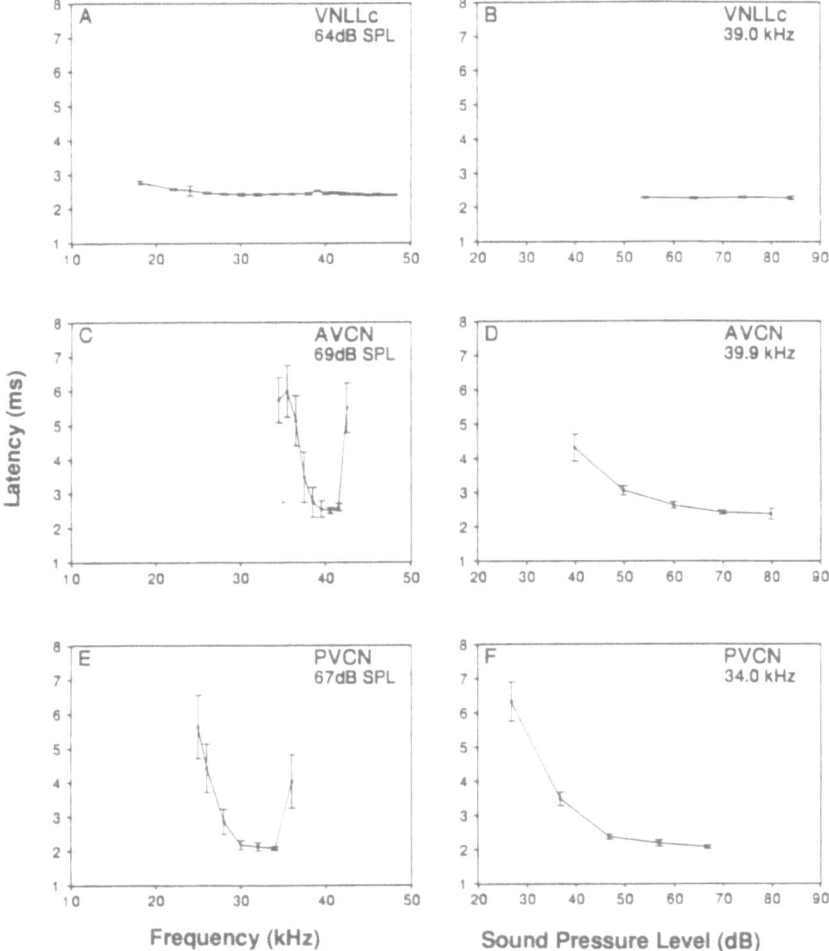

Figure 4. Trial to trial variability in first spike latency as a function of frequency or sound pressure level for units in the cochlear nucleus vs. units in VNLLc. **A.** A neuron in VNLLc shows little change in latency across a wide range of frequencies. **B.** The latency of the same neuron shown in **A** is also tolerant to wide changes in sound pressure level. **C.** A neuron in AVCN shows large changes in latency as frequency is varied, and the variability of the first spike increases on either side of best frequency. **D.** The latency of the same neuron shown in **C** decreases systematically as sound pressure level is increased. **E** and **F.** A neuron in PVCN shows similar response characteristics as the neuron in AVCN. (From Haplea et al., 1994)

The predominant response pattern in INLL is a sustained discharge that continues at a relatively steady rate throughout the duration of a pure tone. Sustained responses provide a real-time measure of stimulus duration for discrete stimuli; for modulated stimuli they could provide a real-time measure of the period during which the stimulus is within the neuron's amplitude-frequency area of sensitivity.

The most common response type in the multipolar cell division of VNLL is a sustained chopper response with spikes occurring at very regular intervals. The chopper response, like the sustained response, continues throughout the duration of a non-modulated stimulus. The origin and function of the chopper response is still obscure. In a later section we propose a hypothesis for integration of different chopping inputs at the IC.

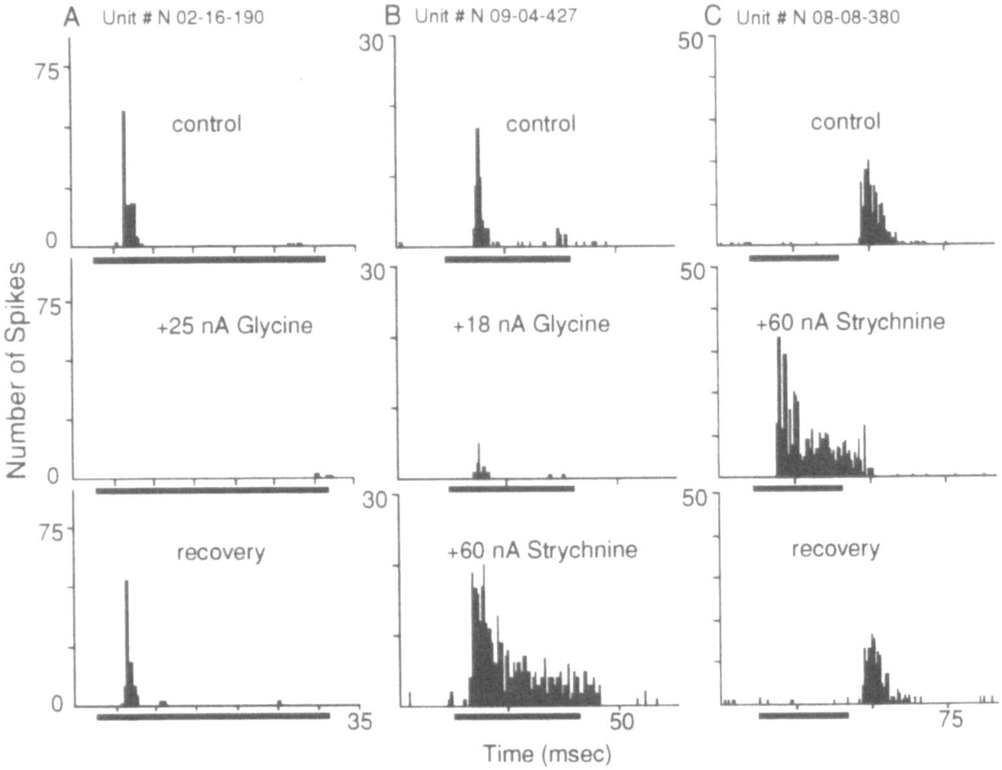

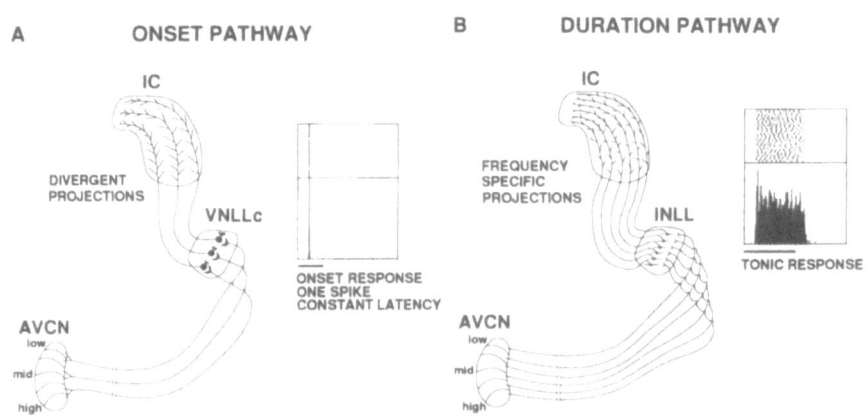

A Unit # N 02-16-190   B Unit # N 09-04-427   C Unit # N 08-08-380

control

control

control

75

30

50

0

+25 nA Glycine

+18 nA Glycine

+60 nA Strychnine

75

30

50

0

recovery

+60 nA Strychnine

recovery

75

30

50

0

35   50   75

Number of Spikes

Time (msec)

Figure 5. Temporal response properties of three neurons in the MSO of *Pteronotus parnellii* before and after application of the inhibitory transmitter glycine or its antagonist, strychnine. The PSTHs show the responses to 100 presentations of a pure tone with a duration of 30 ms, indicated by the bar below each PSTH. **A.** Glycine abolishes the transient response. **B.** Glycine abolishes most of the response, but strychnine produces a sustained response. **C.** Strychnine changes the transient off response to a sustained on response. (From Grothe et al., 1992)

A   ONSET PATHWAY   B   DURATION PATHWAY

IC

IC

DIVERGENT
PROJECTIONS

VNLLc

FREQUENCY
SPECIFIC
PROJECTIONS

INLL

ONSET RESPONSE
ONE SPIKE
CONSTANT LATENCY

TONIC RESPONSE

AVCN
low
mid
high

AVCN
low
mid
high

Figure 6. Schematic illustration of two kinds of temporal inputs from the nuclei of the lateral lemniscus to the IC. **A.** One pathway from the VNLLc is broadly tuned and transmits information about the onset of sound. **B.** The other pathway, from the INLL, transmits information that corresponds to the real-time duration of sound.

On the basis of the response types just described, the monaural pathways from the brainstem to the IC can be divided into two streams of processing, one of which transmits information about stimulus onset and the other of which transmits information about stimulus duration (Fig. 6). In both of these streams, information corresponds to real time events of the modulation. As will be shown in a later section, combining excitatory and inhibitory inputs with these different patterns provides a means of tuning for simple temporal patterns of sound. However, for this mechanism to be effective, the inputs must also be offset from one another in time. This offset requires delay lines.

Delay Lines in the Nuclei of the Lateral Lemniscus

When the first-spike latencies of neurons in the nuclei of the lateral lemniscus are compared with those in the cochlear nucleus and IC (Fig. 7), it is clear that there is a large

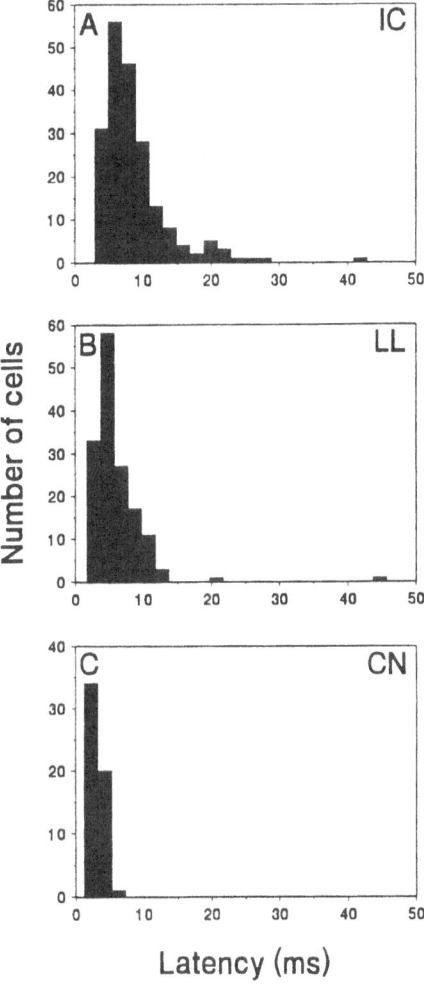

Figure 7. First spike latencies at three levels of the auditory pathway. **A.** The widest spread of latencies, 3 ms to more than 30 ms, is found in the IC. The largest range of latencies is for neurons with best frequencies between 20 and 35 kHz. **B.** Neurons in the nuclei of the lateral lemniscus contribute to the broadening of latencies. Latencies here range from <2 ms to ~ 20 ms. **C.** In AVCN and PVCN only a very small range of latencies, from 1.3 to 5.7 ms, are found. Measurements were obtained using pure tones at best frequency, 10 dB above threshold. (From Haplea et al., 1994).

increase in latency at levels above the cochlear nucleus. The cochlear nucleus is the first central stage of the auditory system and, as would be expected, response latencies of neurons in the cochlear nucleus are short, ranging from approximately 1 ms to 6 ms. The major source of excitatory input to the nuclei of the lateral lemniscus is the cochlear nucleus. In the nuclei of the lateral lemniscus, however, the range of latencies recorded is considerably greater than would be expected from synaptic delays alone, approximately 2 ms to 20 ms. Similarly, the latency range in the IC is about double that, approximately 3 ms to 40 ms. Because the range of latencies can not be accounted for by synaptic delays across excitatory synapses, these increases in delay at each stage must be at least partly created through synaptic inhibition.

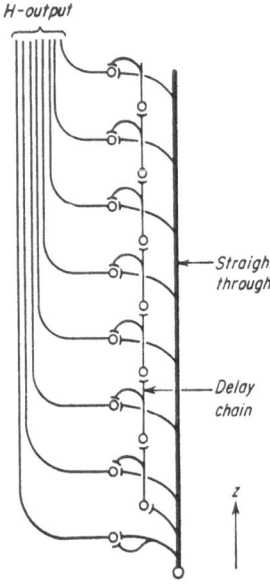

Figure 8. Licklider's neuronal autocorrelator. The caption for this figure stated in part: "The straight through neuron is one of the ascending neurons [from a lower level delay line]. With each straight through neuron is associated a number of delay chain neurons and a number of H-output...neurons. Several or many such networks operate simultaneously and in parallel on inputs that differ only in 'microscopic' detail. From the assumption that, for an H-input neuron to fire, both fibers impinging upon it must fire within a short interval of temporal integration, it follows that the level of neural flux, regarded as a function jointly of time t and of the spatial coordinate z (which represents time delay τ), is approximately the running autocorrelation function of the macroscopic input." (Licklider, 1959, p. 102)

We conclude that part of the purpose of the brainstem circuitry is to create a system of delay lines. The structure of the monaural brainstem pathways is remarkably like that illustrated by Licklider (1951; 1959; Fig. 8) who proposed a neural autocorrelator for pitch discrimination. It is useful to reexamine Licklider's circuit in the light of what we now know about the excitatory and inhibitory transmitters used in different pathways, as well as what we know about the different discharge patterns that are present. Figure 9 shows the circuitry proposed by Licklider (1959), modified to reflect the connectivity in the nuclei of the lateral lemniscus. One set of rows, e.g., the A cells, is virtually identical to Licklider's model. In the lateral lemniscus of bats, there appear to be several duplications of this basic plan, as illustrated by the multiple rows A, B and C. At the time Licklider proposed this model, the role that inhibition might play was not appreciated. We now know that inhibitory components add an important temporal dimension by establishing a mechanism for coincidence detection of

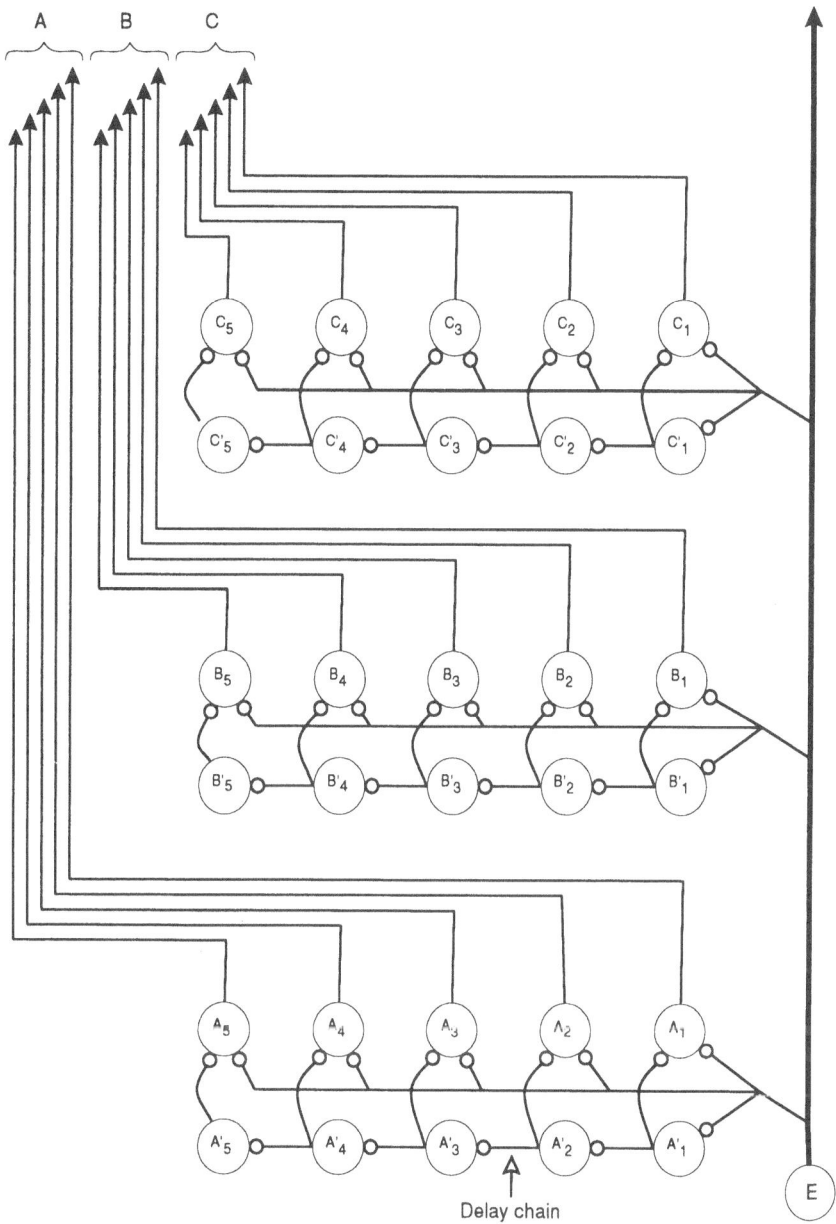

Figure 9. Set of delay lines modified from Licklider's (1959) illustration. Three sets of delay lines A, B and C, are shown to correspond to the connectional architecture of the nuclei of the lateral lemniscus. In each set the output neurons (A_i, B_i, C_i) receive direct excitatory input from one "straight through" neuron (E) and indirect input from a set of delay neurons (A'_i, B'_i, C'_i). A periodic stimulus will excite the A neurons directly from the E neuron, with a synaptic delay from the corresponding A' neuron. The A neuron will fire if the delayed input from its corresponding A' neuron is equal to the period of the stimulus. The A neurons farther down the chain have longer delayed inputs from their corresponding A' neurons and will respond to stimuli with longer periods. Due to axonal conduction time of the E neuron, the neurons in the B and C rows will respond at delays very slightly offset from neurons in the A row.

events on a much longer time scale than the timing of phase differences in low frequency sounds. Inhibition provides the basis for coincidence detection of events related to the sound envelope, that is, onset, duration and offset of sounds. Because the intrinsic and extrinsic sources of inhibition have not been completely identified, it is not clear exactly how inhibitory components should be added to the circuit in Figure 9. Later we will show one way in which inhibition can be incorporated with the delay line-coincidence mechanism of Figure 9 to provide windows of coincidence with time frames extending from milliseconds to tens of milliseconds, and how this mechanism in turn can produce tuning for different sound durations.

CONVERGENCE AND INTEGRATION AT THE INFERIOR COLLICULUS

Tuning for at least some temporal features of sound is achieved by the level of the IC. Examples of some of these features are tuning for sound duration, sweep direction and modulation rate.

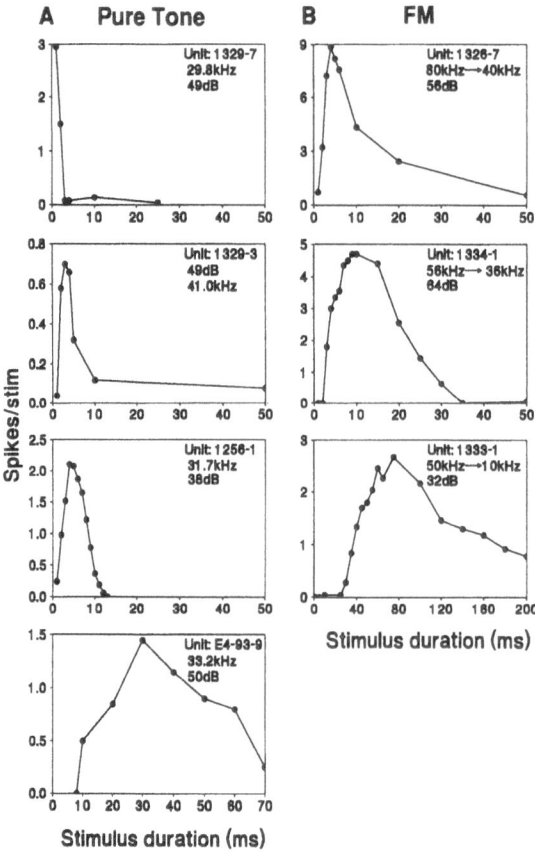

Figure 10. Examples of IC neurons that are tuned to sound duration. Spikes per stimulus presentation are shown as a function of stimulus duration. **A.** Neurons tuned to duration of a pure tone. The range of best durations was from less than 1 ms (top) to about 30 ms (bottom). Most neurons responded to durations of 5 ms or less. **B.** Some duration tuned neurons responded only to frequency sweeps (FM); most of these neurons responded best to sweep durations of 9 ms (top) to 75 ms (bottom). (Reprinted with permission from Casseday et al., *Science*, 264:847-850. Copyright 1994, American Association for the Advancement of Science.)

Tuning for the Duration of Sound

Duration of sound is a biologically important feature of signals as diverse as echolocation calls, animal communication sounds, music and speech. However, unlike frequency, there is no peripheral mechanism for encoding sound duration into neural place. Therefore, if there is a filtering mechanism in which different neurons are tuned to different sound durations, it must be in the central nervous system. Studies of the midbrain of frogs led to the first evidence that neurons are tuned to the duration of sound (Potter, 1965; Narins and Capranica, 1980). Duration tuning first emerges at the level of the midbrain, as it is not found at lower levels (Feng et al., 1990; Hall and Feng, 1991; Condon et al., 1991). Duration tuned neurons in the midbrain of frogs seem to have a biological corollary; the durations to which these neurons are tuned are the same as the durations of sound pulses in conspecific vocalizations (Gooler and Feng, 1992).

Echolocation is another system in which sound duration has biological importance. Neural tuning to the duration of sound would serve the bat by providing an additional filter besides those for frequency and intensity. Neurons in the IC of the big brown bat (Pinheiro et al., 1991; Casseday et al., 1994) and the pallid bat (Fuzessery, 1994) are tuned to sound duration. Figure 10 shows the results from an experiment in which we systematically varied sound duration while recording the responses of single neurons in the IC of the big brown bat.

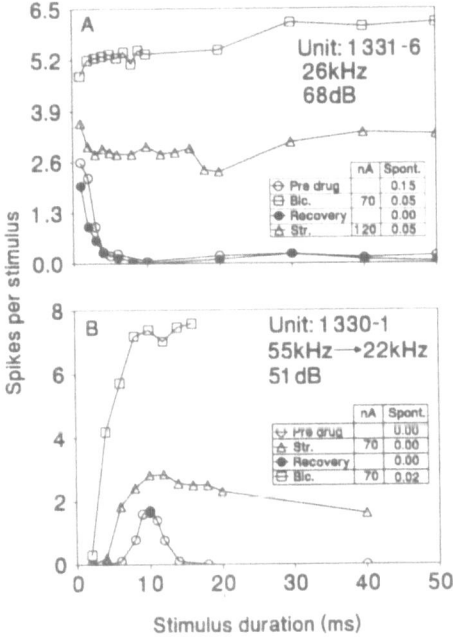

Figure 11. Effects of application of antagonists of GABA and glycine on duration tuning of two neurons. Open circles show spike counts before application of the drug (Pre-drug) and closed circles show spike counts after application (Recovery). Open squares indicate spike count during application of bicuculline (Bic); open triangles indicate spike counts during application of strychnine (Str). The tables show the amount of current (nA) used for iontophoretic application of drug and the spontaneous activity (spont) in spikes per trial when no stimulus was presented. **A.** A neuron that responded best to a 1 ms tone prior to drug application, responded to sounds of all tested durations after application of bicuculline or strychnine. **B.** A neuron that responded best to 10 ms downward FM sweeps before drug application, responded well to all tested stimulus durations after application of bicuculline or strychnine. The drug applications had little or no effect on spontaneous responses. (Reprinted with permission from Casseday et al., *Science*, 264:847-850. Copyright 1994, American Association for the Advancement of Science.)

Somewhat more than 30% of IC neurons show an increase of 50% or more in the response to specific durations. These neurons are maximally sensitive to different durations of sound ranging from one ms to 10 ms or more. The durations of the echolocation sounds used by this species of bat match the range of duration tuning of its IC neurons. One way to address the question of whether duration tuning is constructed in the IC is to block the action of inhibitory inputs by applying substances that act as antagonists of GABA or glycine, the two inhibitory transmitters in the IC. These antagonists can be applied directly to duration tuned neurons while recording their activity in response to sound (Havey and Caspary, 1980). For most duration tuned neurons in the big brown bat, blocking inhibitory input completely or nearly completely eliminates their duration tuning (Casseday et al., 1994). Figure 11 shows the results of experiments on two duration tuned neurons for which application of bicuculline, a GABA antagonist, eliminated duration tuning. Application of strychnine, a glycine antagonist, almost completely eliminated duration tuning; however, there was still a slightly stronger response to the control best duration. The fact that application of either antagonist eliminated

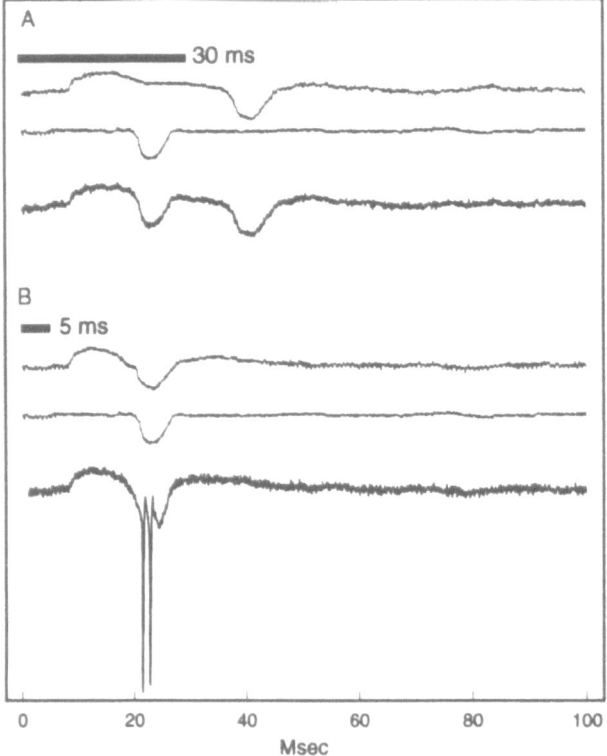

Figure 12. Model of the mechanism for duration tuning. In this example the neuron does not respond to a 30 ms tone (A) but responds well to a 5 ms tone (B). The traces are modeled after whole-cell patch-clamp recordings. Upward deflections are outward current, reflecting inhibition, and downward deflections are inward current, reflecting excitation. Two kinds of input are postulated. One input (top traces in A and B) is short latency inhibition, sustained for the stimulus duration and followed by transient excitation that is locked to stimulus offset. The offset excitation alone is not sufficient to generate a spike. The second input (middle trace in A and B) is excitation that is delayed with respect to the inhibition, is transient and is locked to stimulus onset. The response of the cell (lower trace in bold lines in A and B) is the sum of the two inputs. A. For a long stimulus, inhibition over-rides the delayed excitatory input, preventing a spike. B. For a brief stimulus, the inhibition is brief, and the excitation at offset coincides with the delayed, transient onset excitation, generating a spike.

or greatly reduced duration tuning suggests first that both GABAergic and glycinergic inputs contribute to this tuning and, second, that duration tuning is the result of several convergent excitatory and inhibitory inputs.

Excitatory-Inhibitory Interactions as the Mechanism for Duration Tuning

A second way to study the origin of duration tuning is through the examination of sound-evoked synaptic potentials or synaptic currents in individual IC neurons. Recently it has proven possible to apply the technique of whole-cell patch-clamp recording, usually used to study cells in tissue culture or brain slices, to record synaptic currents of IC neurons in intact, awake bats (Casseday et al., 1994). In this technique, the tip of a glass micropipette is brought into contact with the cell membrane of a neuron, where it forms a tight seal. Gentle suction is then used to rupture the disk of membrane under the pipette tip, providing continuity between the cell's interior and the recording electrode. The resulting recordings provide information about the magnitude and time course of subthreshold excitatory input, seen as inward current or depolarization, and inhibitory inputs to the cell, seen as outward current or hyperpolarization. This information about subthreshold synaptic events provides important clues about how the different inputs to a neuron interact to determine its response properties.

The results of intracellular recording from IC neurons, using whole-cell patch clamp recording, suggest a model for the computation of sound duration (Casseday et al., 1994; Fig. 12). Inhibitory input arrives first and is sustained for the duration of the stimulus. At stimulus offset, there is either excitatory input or "rebound" from inhibition; however this offset depolarization in itself is insufficient to produce a spike. Excitatory input is transient, and it is delayed relative to the stimulus onset.

When the sound is so long that the sustained inhibition overlaps the transient excitation, the excitation is insufficient to produce a spike. However, when the duration of the sound is such that the delayed excitation coincides with the offset depolarization, a spike occurs. To construct tuning to longer or shorter sounds, it is only necessary to have longer or shorter delays in the excitatory inputs. Figure 7 shows that the range of latencies of neurons in the lateral lemniscus is adequate to account for most of this range. Latencies in the IC can account for an even longer range of delays.

Figure 13 shows how the delay-line concept can be modified to produce rows of neurons in which each neuron in a row responds to a different duration. The neurons nearest the "straight through" pathways respond to the shortest durations, while neurons farther to the left respond to progressively longer durations. The neurons in one row would respond to slightly shifted ranges of durations compared to the neurons in another row. This shift would be due to the delay in transmission of input from the straight through neurons.

Tuning for Frequency Sweeps

Neurons that respond only to downward FM sweeps and not to other test sounds such as pure tones, noise or clicks have been found consistently in the IC of FM bats (Suga, 1965; 1968; 1969; Casseday and Covey, 1992; Casseday et al., 1994; Fuzessery, 1994). Suga used the term "FM specialists" to describe this class of neurons. A reasonable hypothesis is that this specialization is created in the IC, because FM specialists have not been found at lower levels. Suga (1969; 1972) and Fuzessery (1994) have described several possible mechanisms for creating FM direction specialization. Some of these models share features of the model for duration tuning in that they combine excitatory and inhibitory inputs. However, they have not incorporated features of different discharge patterns and different latencies.

The model for duration tuning suggests a simple model for FM sweep direction specialization. To adapt the model of duration tuning to produce sweep direction specialization, it is only necessary to assign different frequencies to the excitatory and

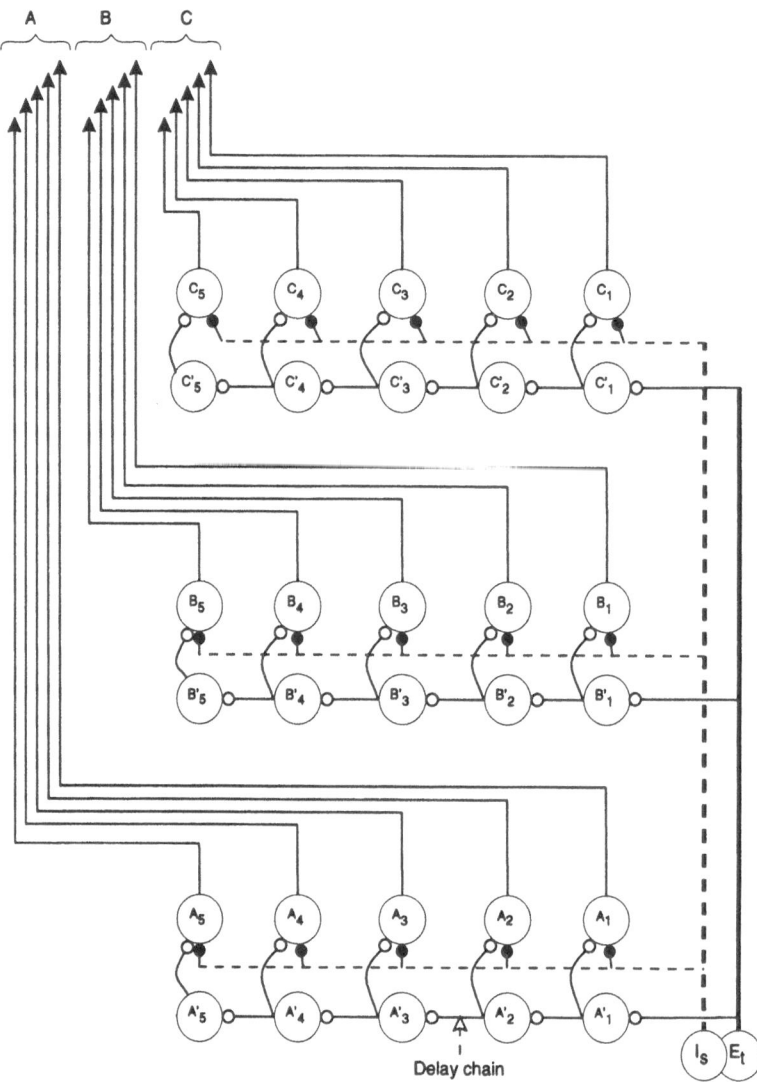

Figure 13. Set of delay lines with inhibitory pathways and specification of the temporal patterns of inputs added to the basic plan proposed by Licklider (1959). This modification provides a set of duration tuned neurons with each neuron in a row tuned to different durations, expanding on the model shown in Figure 9. Three sets of delay lines A, B and C, are shown, possibly corresponding to different isofrequency contours in the IC. In each set the output neurons (A_i, B_i, C_i) receive direct input from a neuron (I_s) that is inhibitory and has a sustained response pattern. As in the original model, there is a delay chain via neurons (A'_i, B'_i, C'_i). The input to the delay chain is excitatory and transient (E_t). The inhibitory input from I_s will prevent the A neurons from firing for a period of time equivalent to the duration of the stimulus. Excitatory "rebound" from inhibition is assumed to occur at the offset of inhibition, but this excitation is insufficient to create an action potential. Transient excitation arrives with a synaptic delay from the A' neurons. If the transient excitation arrives at the same time that the inhibitory input terminates, then A cell will fire. With this scheme, there is a topographical localization according to the best duration of the cells; the cells further to the left respond to longer duration sounds and those further to the right respond to shorter duration sounds.

inhibitory inputs. To create a neuron specialized to downward FM sweeps, inhibition from high frequency input would arrive first, and would be sustained for the period during which the FM sweep remained within the frequency response areas of the input neurons. The inhibition would be followed by a rebound as the sweep exits the input neuron's range of frequency tuning. Excitation from low frequency inputs would be delayed relative to the inhibitory input, and the excitation would be transient. The neuron would fire when the rebound coincided in time with the excitatory input. The neuron could not respond to upward FM sweeps because the timing of inhibitory rebound and delayed excitation would not coincide. The neuron would not respond to pure tones of low frequency because this excitatory input alone is too weak to initiate an action potential. It would not respond to pure tones of high frequency because this excitatory input is normally counterbalanced by an inhibitory input with the same latency, temporal pattern and frequency tuning characteristics. The latter assumption is supported by evidence that the thresholds of many IC neurons are raised by inhibitory inputs with the same frequency tuning characteristics as the excitatory inputs (Faingold et al, 1991; Park and Pollak; 1993; Pollak and Park, 1993; Johnson, 1993). However for the purposes of describing the model, it is sufficient to assume early inhibition from inputs tuned to high frequencies and delayed transient excitation that is in itself insufficient to drive the cell.

Because of the dependence on the timing of the excitatory and inhibitory inputs, FM specialized neurons should also be tuned to the duration of the sweep and to the rate of the sweep. The attraction of this model is that it incorporates the known biological features of the IC circuitry including inputs that differ in three ways: 1) sign of input - excitatory vs. inhibitory, 2) timing of input - delayed vs. undelayed, 3) pattern of input - sustained vs. transient.

The main components of this model were initially proposed by Suga for two different purposes. The combination of excitatory-inhibitory inputs were described in a model to account for FM specialized neurons (Suga and Schlegel, 1973). In that model, the FM direction tuning was explained by the sound sweeping from an excitatory region to an inhibitory region, just the opposite direction of our model. A system of delay lines, including inhibitory and excitatory inputs, forms the basis of Suga's (1990) model to account for tuning of thalamic and cortical neurons to pulse-echo delays. The present model builds on these models by utilizing experimental evidence that synaptic delay lines exist in the lateral lemniscus, that many of the delayed and undelayed inputs to the IC are inhibitory, and that both types of inputs are useful in constructing coincidence detectors. It seems likely that the mechanisms that produce tuning to FM sweeps and other simple temporal patterns take advantage of all of these properties of the system.

Tuning for Modulation Rate

Another type of tuning to temporal features that may be constructed in the IC is tuning to specific rates of modulation, especially those that occur in communication sounds or echo-location sounds. For example, in the torus semicircularis, the auditory midbrain of amphibians, single neurons are tuned to specific AM rates (Rose and Capranica, 1984; 1985; Gooler and Feng, 1992; Rose, this volume). Most are tuned to rates that resemble the AM rates of the mating calls of conspecifics. In the IC of the big brown bat, many units respond only to sinusoidally frequency modulated (SFM) tones. These units do not respond to pure tones or noise, and if they respond to single presentations of a frequency sweep, the response is much less than the response to SFM. Figure 14 shows that these cells are tuned to both modulation rate and modulation depth. It is possible that this type of cell is really specialized for a sequence of FM signals such as occur in the final stages of insect pursuit (Fig. 1) and that the modulation rate of the SFM stimulus simply resembles this natural sound. In any case, these neurons are clearly tuned to very specific temporal patterns of sounds.

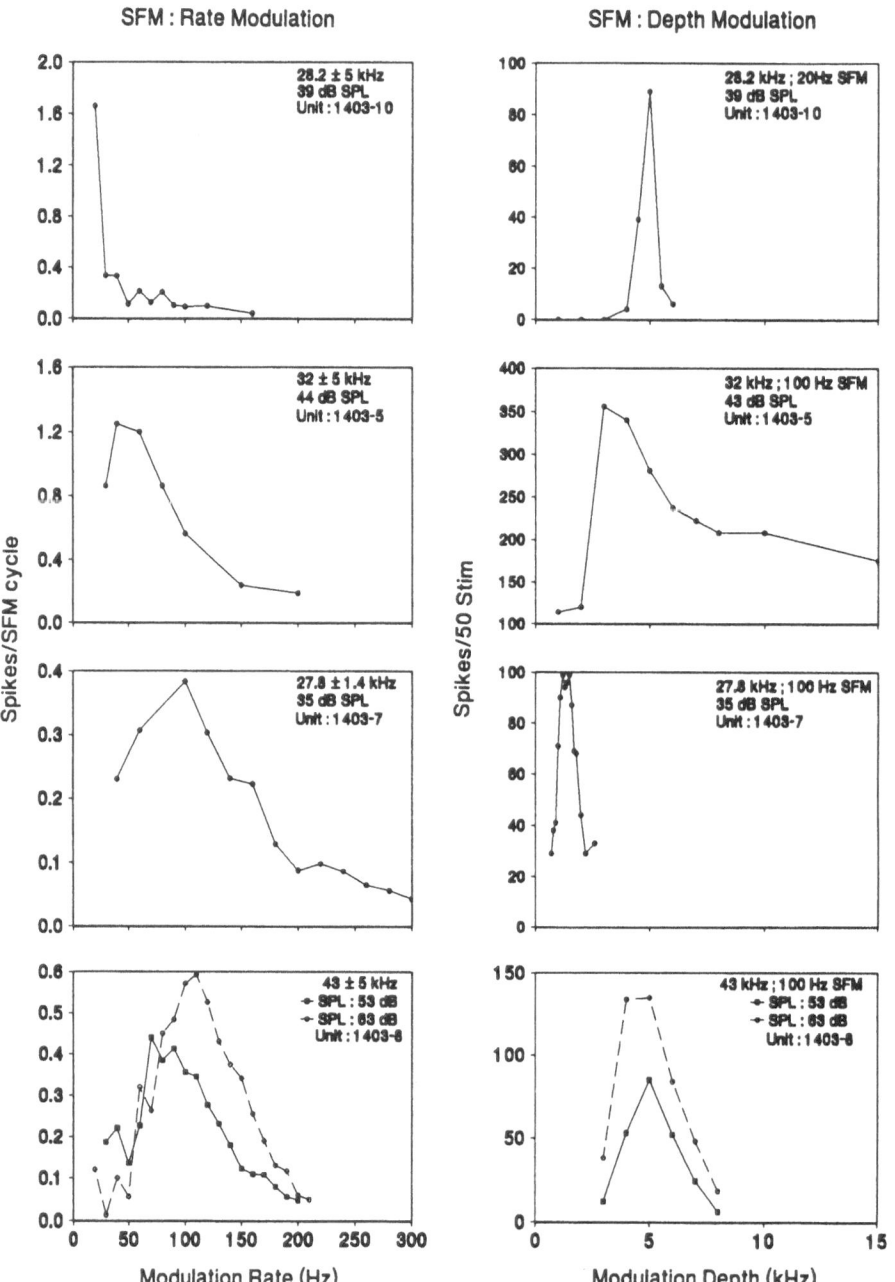

Figure 14. Examples of IC units tuned to sinusoidally frequency modulated sounds. The graphs in the left column show the response (spikes per SFM cycle) as a function of modulation rate, at best modulation depth. Center frequency and modulation depth are indicated at upper right of the graphs. The graphs in the right column show the response (spikes to 50 stimulus repetitions) as a function of modulation depth, at best modulation rate. Center frequency and modulation rate are indicated at upper right of these graphs. The left and right pairs of graphs show data from the same neuron. These neurons did not respond to pure tones. All units were tested at 10 dB above threshold. The unit in the lower pair of graphs was also tested at 20 dB above threshold.

If specialized tuning to SFM stimuli arises in the IC, what is the mechanism of its origin? Here we propose a hypothesis based on interaction of chopper inputs, from cells with regularly spaced sustained discharges, to the IC. In the lateral lemniscus, different chopper neurons have different chopping rates, or interspike intervals. Most of the intervals fall within the range of 2.5 ms to 0.5 ms, i.e., chopping rates of 400 Hz to 2000 Hz. The chopping interval is not correlated with best frequency but is correlated with latency (Covey and Casseday, 1991). The hypothesis is that neurons with different chopping rates converge at an IC neuron. The converging inputs produce membrane potentials that oscillate at a frequency corresponding to the beat frequency of the inputs. The idea is a straightforward analogy with "beat frequencies" in sound. The frequency at which two frequencies, f_1 and f_2 'beat' is the difference between them. Thus if the two inputs had chopping intervals of 2.5 ms and 2.0 ms, corresponding to frequencies of 400 and 500 Hz, then the membrane beat frequency would be 100 Hz. We next assume that the membrane beat oscillations are subthreshold, or possibly inhibitory. The consequence to the IC neuron would be that its membrane potential would oscillate at 100 Hz. This membrane beat potential is essentially a filter centered at 100 Hz. A third excitatory input is postulated that follows the SFM stimuli in a frequency dependent manner but in itself is not sufficient to make the IC neuron fire. If the auditory signal is then modulated at 100 Hz, the signal will match the neural filter, and spikes will have a greater probability of occurring than when the signal is modulated at some other frequency.

PRINCIPLES OF OPERATION OF THE INFERIOR COLLICULUS

It seems likely that the temporal processing mechanisms described in the previous section are examples of general principles. The first principle is that inhibitory and excitatory inputs, which themselves have different temporal properties, interact to produce filters for specific temporal features of sounds. The second principle is that the filters are mainly for biologically important sounds. Many biologically important sounds are sounds produced by other animals. At one level of analysis, these sounds consist of the very rapid frequency components of the sounds. Auditory nerve fibers can follow, or "phase lock", to sounds up to about 3 kHz. Phase-locking becomes degraded at levels above the superior olive, and it is likely that processing in the IC is not concerned with the rapid events related to the waveform of sounds. At another level of analysis, biologically important sounds have a slower rate that is described in terms of the sound envelope, the duration, the envelope modulation rate or in terms of frequency modulation. These temporal aspects of biologically important sounds are limited by the rate of movement of the skeletal and muscular elements that produce them during running, flying or vocalizing. The rate of change of these sounds is slow relative to the rate at which neurons can fire action potentials. For example, the most rapidly produced echolocation sounds of the big brown bat have a rate of about 150/s. Wing beat frequencies of flying insects that the bat hunts are much less. If it is the job of the IC to filter for these kinds of sounds, then its operation must become slower to accommodate the filtering. Duration tuning is a good example: in order to measure the duration of sound, the neurons that do so can not respond until the end of the sound. Thus a by-product of filtering for biologically important sounds is that inhibitory mechanisms reduce the rate of temporal operations in the IC to match the rate at which the sounds can be analyzed. The consequence of the processes that occur in the IC is a temporal window or multiple windows during which the neuron can or can not fire. We examine the evidence for this statement next.

Windows for Excitation

From the time of the earliest electrophysiological studies of neurons in the IC, it has been a common observation that there are many more neurons with transient responses than there are at lower levels (Rose et al., 1963; Suga, 1964a). Further, there is a wide range of latencies, and it is difficult to drive neurons at a high rate of stimulation. Current evidence shows that transient responses are produced by excitation that occurs within a time frame flanked by early and late inhibition. The resistance to high repetition rates of stimulation seems to be produced by a process that lasts even longer than the late inhibition.

Onset Inhibition. For a large number of neurons in the IC, the initial input appears to be dominated by inhibition. Direct evidence for this assertion comes from two sources. First, blocking of inhibitory inputs decreases the response latency to sound for many neurons (Park and Pollak, 1993; Johnson, 1993). Figure 15 shows changes in latency for a population of IC cells after application of bicuculline, a $GABA_A$ antagonist, and after application of strychnine, a glycine antagonist. The result is that either antagonist decreases the response latency for most neurons in the IC. The second way to show this role of inhibition is by observing the time course of excitatory and inhibitory events within the cell. Whole-cell patch-clamp recordings indicate that inhibition dominates the early part of the response (Covey et al., 1993; Casseday et al., 1994). Thus for many IC neurons there is an early unresponsive period correlated with the onset of the stimulus.

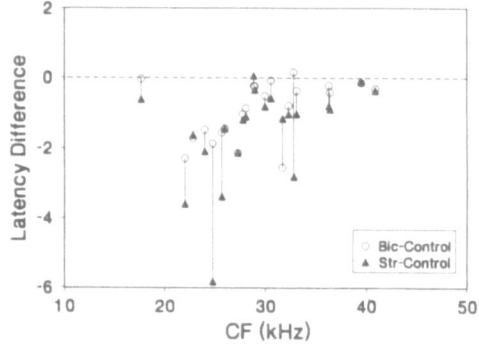

Figure 15. Latency shifts for a set of neurons after GABA was antagonized with bicuculline and after glycine was antagonized with strychnine. Open circles indicate the latency change between control latency and latency after application of bicuculline (Bic-Control). Closed triangles indicate the latency change between control and strychnine application (Str-Control). The vertical lines connecting the symbols indicate that data are from the same neuron. Each neuron was tested under one condition, allowed to recover, and retested under the other condition. For most cells, application of strychnine resulted in a greater decrease in latency than did application of bicuculline. (From data in Johnson, 1993)

Late Inhibition. The studies just cited also show that antagonists of inhibitory transmitters produce an increase in the time frame during which the neuron fires. Evidence that inhibition also curtails the duration of response to make it transient comes from neuropharmacological experiments to block inhibitory transmitters in the IC. When bicuculline is applied to block GABA inhibition while recording the response of an IC neuron to sound, the duration of the response greatly increases (Faingold et al., 1991; Park and Pollak, 1993) (Fig. 16). This increase is only partly accounted for by the decreased latency. These results mean that excitatory input lasts much longer than would be indicated from the spikes normally generated by a sound.

Long-Lasting Post-Excitatory Depression. Electrophysiological studies of the neural correlates of temporal parameters of echolocation sounds support the idea that neurons in the IC have windows of facilitation or inhibition following the occurrence of a sound. Two early reports show evidence that some units in the IC have periods during which the response to the second sound of a pair is either facilitated or suppressed (Grinnell, 1963b; Suga, 1964b). Behavioral studies also reveal the presence of temporal windows. Following the emission of an echolocation pulse, there is an optimal time interval for processing of echoes (Roverud and Grinnell, 1985). All of these studies indicate that these windows, even when evoked by very short duration stimuli, can last tens of milliseconds. The absence of further physiological evidence of this phenomenon in the IC may be due to the suspicion that the early results were influenced by anesthetic, which is known to affect processing in the IC (Kuwada et al., 1989).

We repeated these recovery-time experiments in unanesthetized bats. We presented two 5 ms sounds, either pure tones or frequency sweeps, depending on which type of sound was most effective in driving the neuron. We varied the interval between the two sounds and counted the number of spikes fired in response to the first and to the second sound. The results

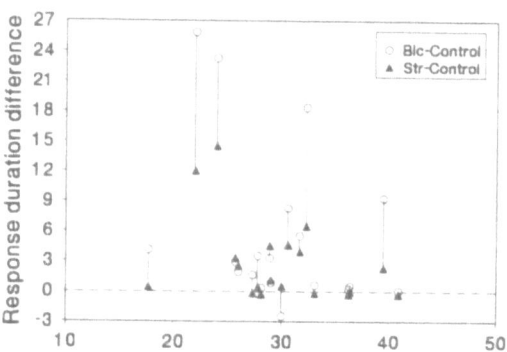

Figure 16. Changes in response duration for a set of neurons after GABA was antagonized with bicuculline and after glycine was antagonized with strychnine. Open circles indicate the change in response duration between control response and response after application of bicuculline (Bic-Control). Closed triangles indicate the change in response duration between control response and response after strychnine application (Str-Control). The vertical lines connecting the symbols indicate that data are from the same neuron. Each neuron was tested under one condition, allowed to recover, and retested under the other condition. For most cells, application of bicuculline resulted in a greater increase in response duration than did application of strychnine. (From data in Johnson, 1993)

are shown in Figure 17 as the ratio of the second response to the first. Although some neurons show a facilitation of the response to the second sound, the surprising result, in an echolocating bat, is that this was not the usual case. For most neurons, there was some period of time during which the response to the second sound was suppressed. Out of a sample of 48 units, 44 (92%) had inhibition of the response to the second sound. Our statistics indicate that this is an extremely robust phenomenon, with 96% of a sample of neurons in the IC showing either suppressive or facilitatory periods following the response to a sound. Using 50% recovery as a criterion, the suppressed periods ranged from 0.7 ms to 50.2 ms. Using 90% recovery as a criterion, the suppressed periods ranged from 5.6 ms to 68.7 ms. Half the neurons required more than 5 ms for 50% recovery, and two-thirds of the neurons required more than 10 ms for 90% recovery. In contrast, out of 37 units we have tested with the two-pulse paradigm in the cochlear nucleus, one was facilitated, and only 3 had a measurable period of suppression; these were all less than 2 msec.

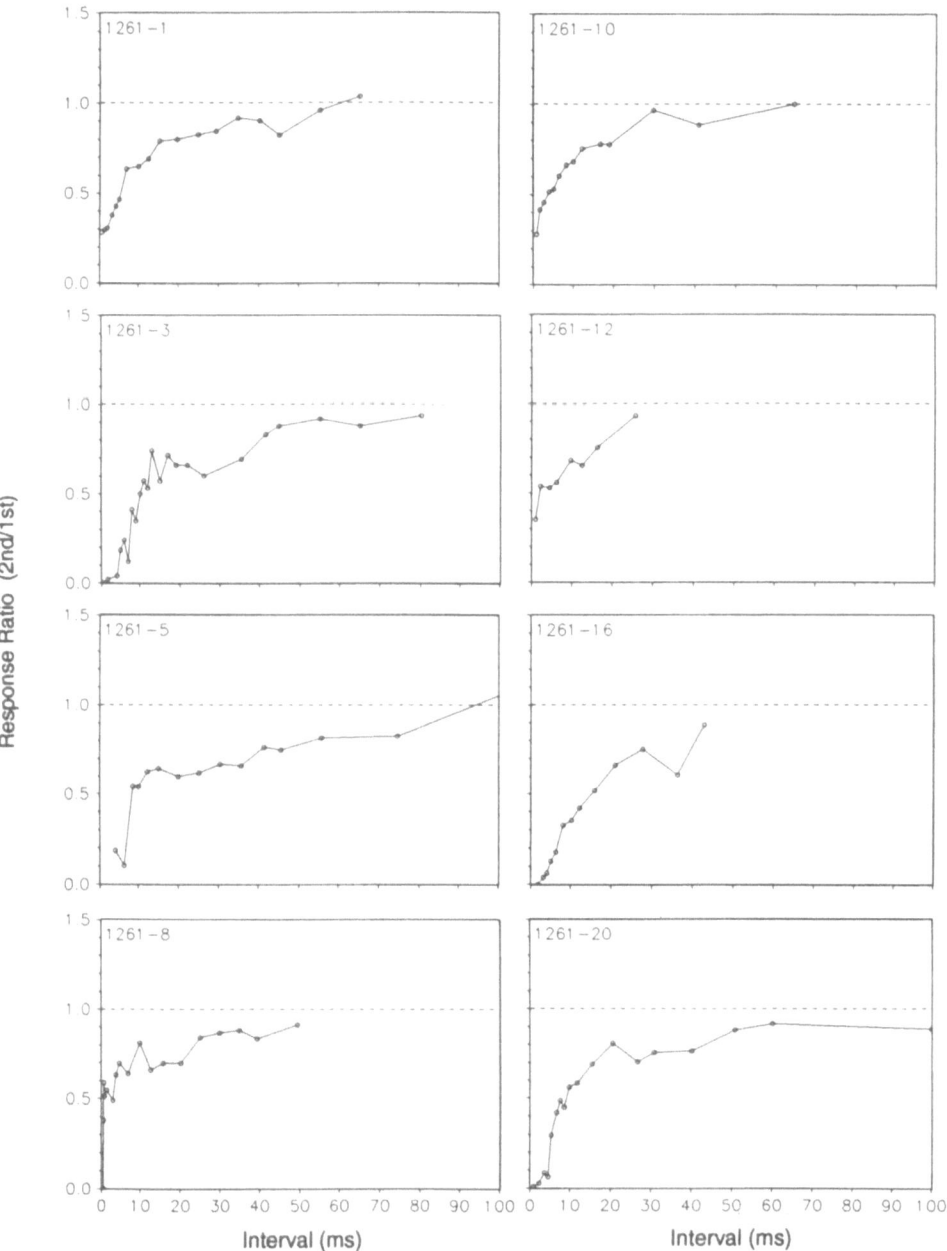

Figure 17. Recovery cycles for a set of neurons in one electrode penetration in the IC of the big brown bat. The curves indicate the ratio of spikes to two sequential sounds that were otherwise identical. One sound (2nd) followed the other (1st) at the interval shown on the X-axis. Animal and unit number are shown in the upper left of each graph.

Excitatory Windows and Temporal Filters

The above evidence raises the hypothesis that a long lasting sequence of subthreshold excitatory and inhibitory events following presentation of a sound is a fundamental mechanism in signal processing in the IC. The windows of excitation created by this mechanism could be the functional basis for temporal filters. The idea that the IC has temporal filters suggests a basis for tuning to amplitude modulation (AM) rates. Neurons with the same frequency tuning have different sensitivities to rate of AM (Rees and Møller, 1983; Schreiner and Langner, 1988; Pinheiro et al., 1991). It is not difficult to see how temporal filters could be generated through delay lines that produce sequential windows of facilitation and suppressions that repeat at intervals.

In summary, these studies indicate that inhibitory mechanisms are crucial for the integrative properties of the IC, especially those involving temporal processing. The ubiquitous presence and robust nature of windows of excitation opens a new avenue of investigation into the mechanisms for the neural analysis of complex sounds.

Short-Term Memory and Analysis of Sound Sequences

One component of the excitatory window is early inhibition which increases response latency. These inhibitory mechanisms can last for varying time periods to create a broad range of latencies among a population of neurons. A wide range of latencies can provide a mechanism for short-term memory.

An illustration of the utility of a broad latency range is given in the following model, in which the spread of latencies for neurons with the same best duration could be the first stage in a delay-line mechanism for auditory depth perception. Echoes returning from multiple targets all at the same distance (same Δt) would activate duration-tuned neurons at their "normal" latencies. However, echoes returning from multiple targets at different distances (different Δt) would alter the temporal firing pattern; some short latency neurons would respond to echoes from a distant target at the same time that long latency neurons responded to echoes from a nearby target. Because the durations of the bat's pulses change over time, a different set of neurons would respond to echoes from pulses of different durations, thus keeping the pulses and echoes "sorted" in time. Bats encounter echoes from multiple targets whenever they approach the foliage of trees. The bat's response to these "clutter" echoes is to change the structure of its echolocation pulse (Kalko and Schnitzler, 1989). It is likely that the population of duration-tuned units would have a different pattern of firing to the cluttered than to the uncluttered echo situations, and this information would be the input necessary for the motor system to alter the echolocation pulse design.

CONCLUSIONS AND FUTURE DIRECTIONS

Transformations in the brainstem are important for setting up mechanisms for tuning to biologically important parameters of sound. These transformations result in a slowing of the rate of neural processing in the IC. The brainstem transformations include change in sign from excitatory to inhibitory, change in discharge pattern, and creation of delay lines. These transformations are segregated into separate pathways below the IC. Convergence of these pathways at the IC produces tuning for temporal features of sound. The bat's brainstem circuitry provides a good general model of neural mechanisms for analysis of temporal patterns. The analyses presented here yield a number of questions that can be answered by further experimental studies or by modeling.

Do temporal response patterns at one level, e.g., the nuclei of the lateral lemniscus, produce temporal modulations of membrane potential at another level that in turn are the basis for tuning to specific temporal patterns or features ? The example we gave earlier concerned two chopping inputs converging at a third neuron, but other patterns could be added to produce selectivity to more complex temporal sound patterns. As a step in testing this idea, it would be useful to construct models of synaptic events that result from convergence of different patterned inputs.

Have we discovered the extent to which temporal filters operate in the IC? Our view is that temporal processing, in the range of milliseconds to tens of milliseconds, may be the main function of the IC and that we are only beginning to appreciate, from an experimental view, this aspect of midbrain function. If we are correct that the IC is a filter for biologically important sounds, then the use of such sounds in physiological studies should provide further evidence for temporal filters in the IC.

What drives the temporal transformation at the IC? Is the transformation the consequence of sensory processing, or is it related to some behavioral process such as the rate of motor output? There are good reasons to consider both ideas. We have proposed elsewhere (Casseday and Covey, 1995) that the IC operates at a rate that coincides with the rate of motor output, such as running, flying, vocalizing. Briefly the idea is as follows. Because the rate of motor output is limited by the speed at which muscle systems can produce coordinated movements, the motor rate is considerably slower than the rate at which action potentials can travel in an auditory fiber ascending to the IC. The rate at which IC neurons operate is closer to the rate of motor actions. Because the IC has outputs to motor systems, it is reasonable to suggest that IC processing has a direct effect on some behaviors, particularly those behaviors in which speed of response to sound is critical, such as escape and hunting behaviors. Whatever the "purpose" of the slowed pace of IC processing, its immediate cause is almost certainly due to the action of inhibitory neurotransmitters. The sources of inhibitory influence not only include neurons within the auditory brainstem, but also neurons in the substantia nigra, part of the motor system. Therefore, the factors that govern the pace of IC output may be both sensory and motor.

Is a temporal response pattern an encoding alternative to a neural place code (Middlebrooks et al 1994; Buonomano and Merzenich, 1995)? Here we raise the idea that the temporal patterning of a stimulus, after multiple transformations, results in a different temporal pattern, or set of patterns across multiple neurons, that is immediately used in behavior. In this scheme, topographical representation is only necessary at the point at which the neural pattern connects to the appropriate motor response. If, for example, two different neural patterns result in two different responses via the same common motor output, then the two pattern representations have the same topographical representation. The perception or response to a temporal pattern is itself a temporal pattern or set of temporal patterns.

ACKNOWLEDGEMENTS

The authors thank Boma Fubara for help in preparing illustrations, Neil Das Gupta for contributing to the experiments illustrated in Figure 17, obtained when he was an undergraduate student at Duke University, and Drs. Gerhard Neuweiler and Harold Hawkins for their useful comments on earlier versions of this manuscript. Research supported by grants from NIH (DC-00287 and DC-00607) and NSF (IBN-9210299).

REFERENCES

Baron, G., 1974, Differential phylogenetic development of the acoustic nuclei among chiroptera, *Brain Behav. Evol.*, 9:7.

Buonomano and Merzenich, 1995, Temporal information transformed into a spatial code by a neural network with realistic properties, *Science*, 267:1028.

Casseday, J.H. and Covey, E., 1992, Frequency tuning properties of neurons in the inferior colliculus of an FM bat, *J. Comp. Neurol.*, 319: 34.

Casseday, J.H. and Covey, E., 1995, A neuroethological theory of the operation of the inferior colliculus, *Brain Behav. Evol.*, in press.

Casseday, J.H., Ehrlich, D., and Covey, E., 1994, Neural tuning for sound duration: role of inhibitory mechanisms in the inferior colliculus, *Science*, 264:847.

Condon, C.J., Chang, S.H., and Feng, A.S., 1991, Processing of behaviorally relevant temporal parameters of acoustic stimuli by single neurons in the superior olivary nucleus of the leopard frog, *J. Comp. Physiol.*, 168:709.

Covey, E. and Casseday, J.H., 1986, Connectional basis for frequency representation in the nuclei of the lateral lemniscus of the bat, *Eptesicus fuscus*, *J. Neurosci.*, 6:2926.

Covey, E. and Casseday, J.H., 1991, The monaural nuclei of the lateral lemniscus in an echolocating bat: parallel pathways for analyzing temporal features of sound, *J. Neurosci.*, 11:3456.

Covey, E. and Casseday, J.H., 1995, The lower brainstem auditory pathways, *in*: "Springer Handbook of Auditory Research, vol 5, Hearing by Bats", A.N. Popper, and R.R. Fay, eds., Springer-Verlag, New York, NY.

Covey, E., Johnson, B.R., Ehrlich, D., and Casseday, J.H., 1993, Neural representation of the temporal features of sound undergoes transformation at the auditory midbrain: Evidence from extracellular recording, application of pharmacological agents and *in vivo* whole cell patch clamp recording, *Neurosci. Abstr.* 19:535.

Faingold, C.L., Boersma Anderson, C.A., and Caspary, D.M., 1991, Involvement of GABA in acoustically-evoked inhibition in inferior colliculus neurons, *Hear. Res.*, 52:201.

Feng, A.S., Hall, J.C., and Gooler, D.M., 1990, Neural basis of sound pattern recognition in anurans, *Prog. Neurobiol.*, 34:313.

Fuzessery, Z.M., 1994, Response selectivity for multiple dimensions of frequency sweeps in the pallid bat inferior colliculus, *J. Neurophysiol.*, 72:1061.

Glendenning, K.K. and Baker, B.N., 1988, Neuroanatomical distribution of receptors for three potential inhibitory neurotransmitters in the brainstem auditory nuclei of the cat, *J. Comp. Neurol.*, 275:288.

Godfrey, D.A., Carter, J.A., Lowry, O.H., and Matchinsky, F.M., 1978, Distribution of gama-aminobutyric acid, glycine, glutamate and aspartate in the cochlear nucleus of the rat, *J. Histochem. Cytochem.*, 26:118.

Goldberg, J.M. and Brown, P.B., 1969, Response of binaural neurons of dog superior olivary complex to dichotic tone stimuli: Some physiological mechanisms of sound localization, *J. Neurophysiol.*, 32:613.

Gooler, D.M. and Feng, A.S., 1992, Temporal coding in the frog auditory midbrain: the influence of duration and rise-fall time on the processing of complex amplitude-modulated stimuli, *J. Neurophysiol.*, 67:1.

Grinnell, A.D., 1963a, The neurophysiology of audition in bats: intensity and frequency parameters, *J. Physiol.*, 167:38.

Grinnell, A.D., 1963b, The neurophysiology of audition in bats: temporal parameters, *J. Physiol.*, 167:67.

Grothe, B., Vater, M., Casseday, J. H., and Covey, E., 1992, Monaural interaction of excitation and inhibition in the medial superior olive of the mustached bat: an adaptation for biosonar, *Proc. Nat. Acad. Sci.*, 89:5108.

Hall, J.C. and Feng, A.S., 1991, Temporal processing in the dorsal medullary nucleus of the northern leopard frog (*Rana pipiens pipiens*), *J. Neurophysiol.*, 66:955.

Haplea. S., Covey, E., and Casseday, J.H., 1994. Frequency tuning and response latencies at three levels in the brainstem of the echolocating bat, *Eptesicus fuscus*, *J. Comp. Physiol. A*, 174:671.

Havey, D.C., and Caspary, D.M., 1980, A simple technique for constructing "piggy-back" multibarrel microelectrodes. *Electroencephalogr. Clin. Neurophysiol.*, 48:249-251.

Jeffress, L.A., 1948, A place theory of sound localization, *J. Comp. Physiol. Psychol.*, 41:35.

Johnson, B. R., 1993, GABAergic and Glycinergic Inhibition in the Central Nuclues of the Inferior Colliculus of the Big Brown Bat, Ph.D. dissertation, Duke University.

Kalko, E.K.V. and Schnitzler, H.-U., 1989, The echolocation and hunting behavior of Daubenton's bat, *Myotis daubentoni*, *Behav. Ecol. Sociobiol.*, 24:225-238.

Kuwada, S., Batra, R., and Stanford, T.R., 1989, Monaural and binaural response properties of neurons in the inferior colliculus of the rabbit: Effects of sodium pentobarbital, *J. Neurophysiol.*, 61:269.

Kuwada, S. and Yin, T.C.T., 1987, Physiological studies of directional hearing, *in*: "Directional Hearing", W.A. Yost and G. Gourevitch, eds., Springer, New York, NY.

Licklider, J.C.R., 1951, A duplex theory of pitch perception, *Experentia*, 7:128.

Licklider, J.C.R., 1959, Three auditory theories, *in*: "Psychology: A Study of a Science, Study I. Conceptual and Systematic, Vol. 1. Sensory, Perceptual, and Physiological Formulations", S. Koch, ed., McGraw-Hill, New York, NY.

Middlebrooks, J.C., Clock, A.E., Xu, L, and Green, D.M., 1994, A panoramic code for sound location by cortical neurons, *Science*, 264:842.

Narins, P.M. and Capranica, R.R., 1980, Neural adaptations for processing the two-note call of the Puerto Rican treefrog, *Eleutherodactylus coqui, Brain Behav. Evol.*,17:48.

Neuweiler, G., 1990, Auditory adaptations for prey capture in echolocating bats, *Physiol. Rev.*, 70:615.

Park, T.J. and Pollak, G.D., 1993, GABA shapes a topographic organization of response latency in the mustache bat's inferior colliculus, *J. Neurosci.*, 13:5172.

Pinheiro, A.D., Wu, M., and Jen, P.H.S., 1991, Encoding repetition rate and duration in the inferior colliculus of the big brown bat, *Eptesicus fuscus, J. Comp. Physiol. A*, 169:69.

Poljak, S., 1926, Untersuchungen am Oktavussystem der Säugetiere und an den mit diesem koordinierten motorischen Apparaten des Hirnstammes, *J. Psychol. Neurol.*, 32:170.

Pollak, G.D. and Casseday, J.H., 1989, "The Neural Basis of Echolocation in Bats", Springer-Verlag, Berlin.

Pollak, G.D. and Park, T.J., 1993, The effects of GABAergic inhibition on monaural response properties of neurons in the mustache bat's inferior colliculus, *Hear. Res.*, 65:99.

Potter, H.D., 1965, Patterns of acoustically evoked discharges of neurons in the mesencephalon of the bullfrog, *J. Neurophysiol.*, 28:1155.

Rees, A. and Møller, A.R., 1983, Responses of neurons in the inferior colliculus of the rat to AM and FM tones, *Hear. Res.*, 10:301.

Roberts, R.C. and Ribak, C.E., 1987, GABAergic neurons and axon terminals in the brainstem auditory nuclei of the gerbil, *J. Comp. Neurol.*, 258:267.

Rose, G.J. and Capranica, R.R., 1984, Processing amplitude-modulated sounds by the auditory midbrain of two species of toads: matched temporal filters, *J. Comp. Physiol. A*, 154:211.

Rose, G.J. and Capranica, R.R., 1985, Sensitivity to amplitude modulated sounds in the anuran auditory nervous system, *J. Neurophysiol.*, 53:446.

Rose, J.E., Greenwood, D.D., Goldberg, J.M., and Hind, J.E., 1963, Some discharge characteristics of single neurons in the inferior colliculus of the cat. I. tonotopical organization, relation of spike-counts to tone intensity, and firing patterns of single elements, *J. Neurophysiol.*, 26:294.

Ross, L.S. and Pollak, G.D., 1989, Differential ascending projections to aural regions in the 60 kHz contour of the mustache bat's inferior colliculus, *J. Neurosci.*, 9:2819.

Roverud, R.C. and Grinnell, A.D., 1985, Echolocation sound features processed to provide distance information in the CF/FM bat, *Noctilio albiventris*: evidence for a gated time window utilizing both CF and FM components, *J. Comp. Physiol. A*, (1985)156:457.

Saint Marie, R., Ostapoff, E.M., Morest, D.K., and Wenthold, R.J., 1989, Glycine- immunoreactive projection of the cat lateral superior olive: possible role in midbrain ear dominance, *J. Comp. Neurol.*, 279:382.

Schnitzler, H.-U., Kalko, E., Miller, L., and Surlykke, A., 1987, The echolocation and hunting behavior of the bat, *Pipistrellus kuhli, J. Comp. Physiol. A*, 161:267.

Schreiner, C.E. and Langner, G., 1988, Periodicity coding in the inferior colliculus of the cat. II. Topographical organization, *J. Neurophysiol.*, 60:1823.

Schweizer, H., 1981, The connections of the inferior colliculus and the organization of the brainstem auditory system in the greater horseshoe bat (*Rhinolophus ferrumequinum*), *J. Comp. Neurol.*, 201:25.

Simmons, J.A., 1989, A view of the world through the bat's ear: the formation of acoustic images in echolocation, *Cognition*, 33:155.

Suga, N., 1964a, Single unit activity in cochlear nucleus and inferior colliculus of echo-locating bats, *J. Physiol.*, 172:449.

Suga, N., 1964b, Recovery cycles and responses to frequency modulated tone pulses in auditory neurones of echo-locating bats, *J. Physiol.*, 175:50.

Suga, N., 1965, Analysis of frequency modulated sounds by neurons of echolocating bats, *J. Physiol. (Lond.)*, 179:26.

Suga, N., 1968, Analysis of frequency modulated and complex sounds by single auditory neurons of bats, *J. Physiol. (Lond.)*, 198:51.

Suga, N., 1969, Classification of inferior collicular neurons of bats in terms of responses to pure tones, FM sounds and noise bursts, *J. Physiol. (Lond.)*, 200:555.

Suga, N., 1972, Analysis of information bearing elements in complex sounds by auditory neurons of bats, *Audiology*, 11:58.

Suga, N., 1973, Feature extraction in the auditory system of bats, *in*: "Basic Mechanisms in Hearing", A.R. Møller, ed., Academic Press, New York, NY.

Suga, N., 1990, Cortical computational maps for auditory imaging. *Neural Networks*, 3:3.

Suga, N., and Schlegel, P., 1973, Coding and processing in the auditory systems of FM-signal-producing bats. *J. Acoust. Soc. Am.*, 54:174.

Vater, M., Casseday, J.H., and Covey, E., 1995, Convergence and divergence of ascending binaural and monaural pathways from the superior olives of the mustached bat, *J. Comp. Neurol.*, 351:632.

Vater, M., Habbicht, H., Kössl, M., and Grothe, B., 1992, The functional role of GABA and glycine in monaural and binaural processing in the inferior colliculus of horseshoe bats, *J. Comp. Physiol. A*, 171:541.

Von der Emde, G. and Schnitzler, H.-U., 1986, Fluttering target detection in Hipposiderid bats, *J. Comp. Physiol.*, 159:765.

Wenthold, R.J., Zempel, J.M., Parakkal, M.H., Reeks, K.A., and Altschuler, R.A., 1986, Immunocytochemical localization of GABA in the cochlear nucleus of the guinea pig, *Brain Res.*, 380:7.

Yin, T.C.T. and Chan, J.C.K., 1990, Interaural time sensitivity in medial superior olive of cat, *J. Neurophysiol.*, 64:465.

Zook, J. M. and Casseday, J.H., 1982a, Cytoarchitecture of auditory system in lower brainstem of the mustache bat, *Pteronotus parnellii*, *J. Comp. Neurol.*, 207:1.

Zook, J. M. and Casseday, J.H., 1982b, Origin of ascending projections to inferior colliculus in the mustache bat, *Pteronotus parnellii*, *J. Comp. Neurol.*, 207:14.

Zook, J. M. and Casseday, J.H., 1985, Projections from the cochlear nuclei in the mustache bat, *Pteronotus parnellii*, *J. Comp. Neurol.*, 237:307.

Zook, J. M. and Casseday, J.H., 1987, Convergence of ascending pathways at the inferior colliculus in the mustache bat, *Pteronotus parnellii*, *J. Comp. Neurol.*, 261:347.

Zook, J.M., Jacobs, M.S., Glezer, I., and Morgane, P.J., 1988, Some comparative aspects of auditory brainstem cytoarchitecture in echolocating mammals: Speculations on the morphological basis of time-domain signal processing, *in*: "Animal Sonar: Processes and Performance", P.E. Nachtigall and P.W.B. Moore, eds., Plenum, New York, NY.

AN OSCILLATORY CORRELATION THEORY OF TEMPORAL PATTERN SEGMENTATION

De Liang Wang

Laboratory for AI Research
Department of Computer and Information Science
and Center for Cognitive Science,
The Ohio State University
Columbus, OH 43210-1277
USA

INTRODUCTION

Acoustic energy from many different sources is present in the environment at all times. In order for a listener to recognize and understand the auditory environment, it is necessary to disentangle the acoustic wave form and analyze each separate event. This process is referred to as *temporal pattern segmentation*, or *auditory scene analysis* (Bregman, 1990). Its task is to break down an auditory scene, or total acoustic input, into a number of coherent segments, each of which has a high probability of coming from the same source. Temporal pattern segmentation (temporal segmentation for short) is a remarkable achievement of the auditory system, playing a fundamental role in auditory perception. It has much in common with visual segmentation of a scene into different objects. Segmentation can be based on either current input or prior knowledge. Segmentation based on current input relies on the similarities of local qualities within the input pattern itself, such as frequency, timing, or amplitude. On the other hand, segmentation based on prior knowledge relies on patterns stored in memory to segregate the auditory input. These two processes occur simultaneously in auditory scene analysis.

Temporal pattern segmentation as addressed here is the process of segregating auditory input into (potentially simultaneous) multiple segments, which should be distinguished from the process of separating a single auditory stream into different sequential components. The former is probably a precursor to the latter.

This chapter examines temporal segmentation from the neural network perspective. The remaining part of this introduction provides a review of psychological evidence for auditory stream segregation as well as a review of pertinent computational studies on temporal segmentation. We then present an oscillatory correlation theory of temporal pattern segmentation, linking auditory segregation with neural oscillations in the auditory system. The theory is quantified by a neural architecture of a laterally coupled two-dimensional neural oscillator network with a global inhibitor. Computer simulation results show that this architecture can group auditory features into a segment by phase synchrony and segregate different segments by desynchronization, both in real time.

Psychological Evidence

Psychophysical investigation of temporal pattern segmentation was first reported by Miller and Heise (1950) who noted that listeners split a signal composed of two alternating sine wave tones into two segments. Temporal segmentation was obtained with as little as a 15% difference in frequency and was observed throughout the discernible range from about 150 Hz to 7000 Hz. Bregman and his collaborators have carried out a series of studies on this subject. In one of the early studies (Bregman and Campbell, 1971), subjects were asked to judge the order of six different tones in a sequence. Three tones were in the high frequency range, and the other three in the low frequency range. Figure 1 illustrates the situation in a simplified form. The results showed that at high rates of presentation, subjects perceived two separate sequences corresponding to the high and the low frequency tones respectively, and were able to report the temporal order of the tones within each sequence, but not across the two sequences. This phenomenon is called *stream segregation*. The loss of order information across different streams was observed earlier by Warren et al. (1969). In stream segregation, there is a trade-off between frequency separation and presentation rate, so that the higher the presentation rate, the smaller the frequency difference required to bring about stream segregation (van Noorden, 1975).

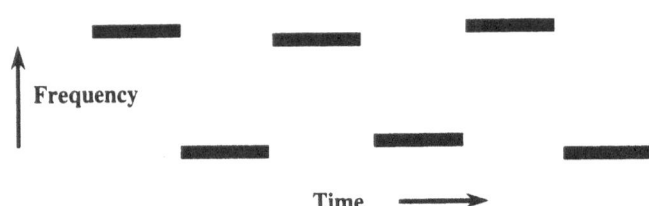

Figure 1. Six alternating high and low frequency tones as displayed in a spectrogram.

The phenomenon of stream segregation has been verified repeatedly in different contexts (see among others van Noorden, 1975; Jones, 1976; Bregman, 1978; Bregman et al., 1985; Bregman, 1990; Roberts and Bregman, 1991; Jones and Yee, 1993). While stream segregation concerns itself primarily with tones presented sequentially, the basic phenomenon also occurs for simultaneously presented tones. For example, two overlapping tones can be fused into the same segment if either their onset times or offset times are the same. On the other hand, two overlapping tones are likely to be segregated into two streams if the tones have different onset times as well as different offset times (Rasch, 1978). In general, if the auditory patterns are displayed on a spectrogram, the results are consistent with the Gestalt laws of grouping (the reverse process of segmentation) that have been expressed in the visual domain (see also Handel, 1989). The following is a list of several important determinants of temporal segmentation:

1. Frequency (pitch) proximity. The closer the frequencies of two tones, the more likely they are to be grouped into the same segment. Frequency proximity is considered the prime factor in temporal segmentation.

2. Presentation rate. The faster the presentation rate, the more readily stream segregation occurs. Since a faster presentation rate corresponds to a smaller time separation

between successive tones of similar frequency, the role of presentation rate in stream segregation may be viewed as the Gestalt proximity rule in the time domain.

3. Common modulation. Simultaneous tones that undergo the same change at the same time tend to be bound together into a common segment. This principle applies to both frequency modulation and amplitude modulation.

4. Onset or offset synchrony. Listeners tend to perceive two or more tones as a segment if they have the same onset or offset time. This is a major factor for integrating frequency partials (harmonics) into a complex sound.

5. Smooth transition. Tones tend to be grouped into the same segment if there is a gradual and systematic transition from one tone to another (e.g., ascending frequency progression).

Jones and her colleagues have also argued that rhythmic structure between successive tones can promote their grouping into the same segment (Jones et al., 1981; Jones et al., 1994). For example, they observed that if a sequence of tones forms a simple rhythm, i.e., the duration between the onset times of successive tones in the sequence is a constant, the sequence tends to be integrated into the same segment.

In speech perception, temporal segmentation seems to contribute to a listener's ability to separate utterances from different speakers into different streams. This observation leads to the question of what prevents the listener from segmenting speech sounds such as syllables produced by a single speaker into competing streams, since the rate of speech production is very high and articulation by the same speaker varies greatly. Roughly speaking, the production of a syllable involves a set of transitions from several frequency partials associated with the consonant to the same number of frequency partials associated with the following vowel. These frequency transitions are called formant transitions. Two acoustic factors seem to be responsible for the coherence of speech sounds uttered by the same speaker (Handel, 1989; Bregman, 1990). The first factor is the relative degree of onset or offset synchrony within the set of formant transitions generated during the production of a syllable. A second factor leading to coherence is the existence of a continuous frequency transition within each one of the formant transitions. The continuous transition results from co-articulation, which refers to the phenomenon of simultaneous production of multiple sounds due to overlapping articulatory movements of the vocal tract (Cole and Scott, 1973). These two factors are consistent with the properties of temporal segmentation.

Music, like speech, is subject to temporal segmentation. If the notes of two melodies whose pitch ranges do not overlap are interleaved in time so that successive tones come from the different melodies, the resulting sequence of tones is perceptually divided into two seemingly co-occurring segments that correspond to the two melodies (Dowling, 1973; Dowling et al., 1987). If the two melodies overlap in the pitch domain, they can no longer be identified by listeners who are familiar with each of the melodies. In fact, composers of music have made use of pitch difference to permit the perception of interleaved melodies (Dowling, 1973). On the other hand, if listeners know in advance what melody to listen for, they can attend to it even if it is intermixed in pitch with the other notes by making use of attentional expectancy, a "cocktail party effect" (Cherry, 1953) in music perception. Although the above experiments were performed with interleaved melodies, other observations have shown that the essential properties of temporal segmentation are also true for simultaneously presented melodies (see Chapter 5 in Bregman, 1990).

Computational Studies

The technology for temporal pattern recognition, particularly speech recognition, has advanced rapidly in recent years. It has been proven that neural networks can make a significant contribution to this technology (Morgan and Scofield, 1991; Wang, 1995). However, segmentation of interleaving or simultaneous temporal signals remains a tremendous challenge, which has hardly been addressed at all. Almost all speech recognition systems assume pre-segmentation, and many can perform well only if the input is from a single stream (Rabiner and Juang, 1986; Morgan and Scofield, 1991). Obviously, temporal segmentation is essential in order for any speech recognizer to work in a realistic auditory environment. Thus, a successful method of temporal segmentation would be a breakthrough in making speech recognition technology reach the real world.

Parsons (1976) developed a program to separate two speakers on the basis of different fundamental frequencies. The system uses Fourier analysis to extract frequency partials, and then an algorithm to compute the first fundamental frequency. The first fundamental frequency is later used to cancel those partials which fall in the harmonics of the fundamental. The remaining partials are used to compute the second fundamental frequency. The system is programmed to separate just two voices, and it cannot determine how many voices are there to be separated. Using a similar strategy of separating fundamental frequencies, Weintraub (1986) proposed a different model for separating two simultaneous speakers. The voices were provided by a male speaker and a female speaker. The input signal was first processed by a cochlear model. The model uses a dynamic programming algorithm to compute the periods of two fundamentals corresponding to the two speakers, and then a Markov model is used to identify certain characteristics of the voices and thus separate the two talkers. Because many other factors contributing to speech separation were not considered, the success of both models is quite limited even just for two input sources. It is not clear how the models could be extended to handle sound separation for more than two speakers producing voiced sounds.

More recently, Beauvois and Meddis (1991) proposed a computational model to simulate stream segregation. The model uses a bandpass-filter bank to extract frequency components of an auditory input, and assumes competition between different filter channels. The activity of the winning channel does not experience decay while the activities of the other channels do. Their model is targeted to successive high and low tones, and streaming is assumed to occur if the overall system consistently shows a higher response to one of the two alternating tones. This criterion of judging whether streaming occurs is somewhat arbitrary, and cannot explain the basic fact that the subject hears two streams when streaming occurs (Bregman and Campbell, 1971). Also, the model cannot explain temporal segmentation of simultaneously presented streams.

Perhaps the only neural network model that really addresses the problem of temporal segmentation is one proposed by von der Malsburg and Schneider (1986). They proposed the idea of using the phases of neural oscillations for expressing different segments, a form of the temporal correlation theory proposed earlier by von der Malsburg (1981). Since an oscillatory pattern has an extra degree of freedom, its *phase*, it can be used to elegantly represent synchronization and desynchronization among a group of oscillators. In this representation, a set of auditory features forms a segment if the corresponding oscillators oscillate in phase with no phase-lag. Oscillators representing different segments oscillate out of phase, that is, they are desynchronized. Similar ideas for representing different segments in a general context were suggested earlier (Milner, 1974; see also Ellias and Grossberg, 1975). Using these ideas as the basis, von der Malsburg and Schneider constructed a network of oscillators, each representing a specific auditory quality. Each oscillator is connected to all the others in the network, and a global inhibitory oscillator is introduced to desynchronize different segments. They demonstrated that rapid modulation of connection strengths led to segmentation based on onset synchrony, i.e., simultaneously triggered oscillators synchronize with each other, and they desynchronize with oscillators representing another segment presented at a different time.

Although the idea of using oscillators for segmentation has been around for quite some time, it has not been very successful in helping solve the problem of temporal segmentation. For example, the model proposed by von der Malsburg and Schneider cannot simulate the basic phenomenon of stream segregation shown in Fig. 1. One reason for this lack of success is that temporal patterns are represented as sets of qualities which have no geometrical relationships. Stream segregation in contrast depends on specific relationships among tones in the time/frequency domain. In other words, temporal segmentation that is best explained by Gestalt laws of grouping requires a representational framework with specific geometrical, i.e. spectral, relations among different tones, such as proximity or continuity.

The following section presents a neurocomputational theory of temporal pattern segmentation. The theory is based on the idea of using synchronization to represent a segment and desynchronization to represent different segments. Then we present the computational elements of the theory. Both the building block, the single oscillator model, and the neural architecture are fundamentally different from those used by von der Malsburg and Schneider. Simulations presented later show that the computational model is capable of replicating the basic phenomenon of stream segregation. Some further discussions are provided in the conclusion section. The model proposed here promises to explain a variety of experimental observations and provide an effective computational approach to temporal segmentation.

A NEUROCOMPUTATIONAL THEORY

One of the dominant theories, Bregman's auditory scene analysis (this term is overloaded since it is used to refer to both the theory and the process of auditory segmentation), attempts to explain experimental phenomena of temporal segmentation by extending Gestalt principles of grouping to the auditory domain (Bregman, 1990). As illustrated in Fig. 1, once time is treated as a separate dimension, this theory can explain a variety of experimental data in terms of proximity in the frequency and time dimensions, smooth (continuous) transitions, and common modulation. On the other hand, Jones' theory of rhythmic attention emphasizes the uniqueness of auditory processing (Jones, 1976; Jones et al., 1981). This theory states that listeners group tones that follow a certain rhythmic structure and anticipate the onset of the next tone using the acquired rhythm. Both theories, however, make no reference to the underlying neural processes.

We present the following theory to explain temporal pattern segmentation. The theory is based on the idea of using synchronization to represent a segment (stream) and desynchronization to represent different segments. The basic building block of the theory is a nonlinear oscillator which exhibits approximately periodic behavior. This kind of dynamics is called limit cycle dynamics (Guckenheimer and Homes, 1983), which differs fundamentally from the commonly used limit point dynamics. The latter always approaches stationary behavior. The fact that temporal segmentation is a time-varying process and depends on the rate of presentation calls for a representation of time in any system that deals with temporal segmentation. In the theory described here, time is treated as a separate dimension. To simplify the discussions, we consider only time and frequency in this model. The network is composed of two dimensional oscillators that are laterally coupled, with a global inhibitor. One dimension represents time, and another one represents frequency. The architecture of the network is depicted in Fig. 2. We assume that lateral connections are all excitatory, and the connection strength between two oscillators decreases as a function of distance. The global inhibitor receives excitation from every oscillator in the network, and sends inhibition back to it. There is an input end to the network, and this input end consists of units representing distinct frequencies (called input channels). Each input channel connects to a corresponding row of oscillators representing the same frequency through a set of delay lines with different delays. These delay lines are arranged so that delays increase systematically from left to right. Thus, each oscillator senses input with a specific frequency at a specific time relative to the present time.

Given this network, when a tone is presented at some time it triggers certain frequency channels in the input end corresponding to the frequency of the tone. When the tone is a pure tone, it can trigger only one frequency channel. Since a tone has a certain duration, when the

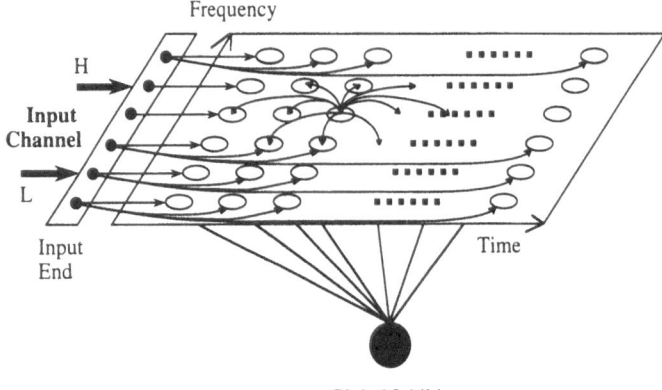

Frequency

H

Input
Channel

L

Input
End

Time

Global Inhibitor

Figure 2. Two dimensional time-frequency matrix for temporal segmentation. External input is always applied to the left end, called the input end. Each input channel in the input end projects to a row of isofrequency oscillators with systematically increasing delays from left to right. Also shown in the figure are the lateral connections of a typical oscillator. The connections from other oscillators are omitted for clarity. The global inhibitor receives an excitatory input from and sends an inhibitory output to every oscillator in the matrix. As in the following figures, H is used to indicate an auditory input with a high frequency and L to indicate an auditory input with a low frequency.

later part of the tone is still being presented, the earlier part is already triggering appropriate oscillators along the frequency axis. One can see that a continuous tone triggers a neighboring set of oscillators, but the set is transient and keeps shifting from left to right as time progresses. Also, the duration of the tone is transformed into the length of the set. Computationally, we want to synchronize a group of oscillators representing different tones in the same segment, and desynchronize different groups representing different segments. This desired property is schematically shown in Fig. 3, where two segments are illustrated. This idea can be easily extended to handle more than two segments, at least conceptually. How this computational property is achieved will be discussed in the next two sections.

Now let us see how such a network of oscillators with the above computational property might explain the phenomenon of auditory stream segregation. A fundamental effect of stream segregation is the loss of information about the order of different tones presented sequentially. Take the classic example of a sequence of alternating pure tones $HLHLHL$, where H denotes a pure tone of high frequency and L denotes a pure tone of low frequency. All H or L tones do not have to have the same frequency. When stream segregation occurs, the subject perceives two streams, one for H tones and another for L tones. In the meantime, the subject cannot identify relative order across the two streams, e.g. the second L tone and the third H tone. This phenomenon can be explained on the basis of two assumptions. First, we assume that *attention is paid to a segment when its constituent oscillators reach the peak of their oscillations*, a natural assumption for temporal segmentation based on oscillations. So in our computational model, attention quickly switches between the different segments (see Fig. 3).

Since each segment can be composed of multiple tones, this assumption also implies that attention can be simultaneously paid to multiple tones. Second, we assume that the temporal order of two items (tones) can be perceived only if (1) they have no temporal overlap in their presentations; (2) they belong to the same stream or they constitute two different streams, one of which comes before (or after) the other. While the first requirement is a natural one since the order would be undefined otherwise, the second one merits further discussion. If the two items belong to the same segment (stream), they will be attended to simultaneously according to our assumption. However, since they do not overlap in time, the subject should be able to tell their order of occurrence by recency (which may be coded in short-term memory). If, on the other hand, the two items constitute two separate segments having a clear temporal order with respect to which one comes first, the clear sequential order of the segments should enable the subject to perceive the relative order of the two items. We refer to the theory thus formed as the *oscillatory correlation theory*. The computational issues involved will be discussed in detail in the next section, so for the time being let us assume that the computational properties described in the remainder of this section can be achieved. The

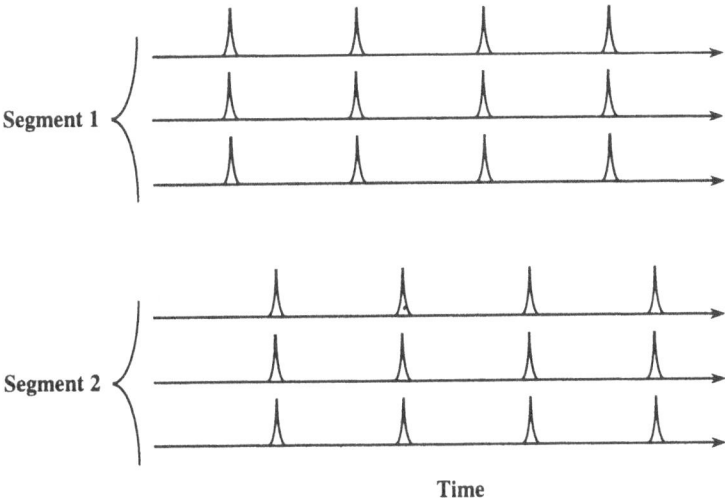

Time

Figure 3. Schematic representation of segmentation by synchrony and desynchrony. Two segments are shown, each corresponding to a set of oscillators oscillating in synchrony. The oscillations between the two segments, however, are desynchronized.

length of the silences between successive tones docs not seem to affect streaming (Dannenbring and Bregman, 1976), so for simplicity it is assumed that there are no silences between successive tones.

The oscillatory correlation theory can simulate properties of stream segregation observed in psychophysical experiments under different conditions of presentation rate. For fast presentation rates (short duration for each tone), all H tones should be grouped into a single segment by phase locking of the constituent oscillators due to the factor that the sets of oscillators triggered by successive H tones are near each other and thus coupled strongly by lateral excitatory connections (Fig. 2). The same is true for all L tones. Due to global inhibition exerted by the global inhibitor, the oscillators representing different segments should be out of phase (anti-phase locking). According to the second assumption, the relative order between the tones within each segment is perceived because these tones are attended to simultaneously (see Fig. 3). However, the relative order of tones across the two segments is not perceived because they are not attended to simultaneously and they are in the two segments that interweave in time.

For slow presentation rates, a number of situations are possible. If the frequency separation between the tones is not large, all of the tones may form a single segment. In this case, the temporal order of the tones is perceived according to the basic assumptions. If the frequency separation is very large, each tone forms its own segment. In this case, the order is still perceived because the segments do not overlap in time. For intermediate frequency separation, neighboring pairs of H and L tones may form a single segment. In this case the order of tones is again perceived, because segments do not overlap in time. We thus conclude that at slow presentation rates, order is fully perceived.

For intermediate rates, successive pairs of tones with similar frequencies may form a single segment, together with segments that are made of neighboring H and L tones. This is because the distance between the sets of oscillators representing successive tones with similar frequencies is neither small enough nor large enough to give rise to a decisive outcome of synchronization or desynchronization. Because this situation is in some ways similar to each of the two previous cases, the model predicts that the subject should be able to tell the order across different frequencies to a certain extent. In other words, the subject's performance on recalling the order of tones should be somewhere between total confusion and full perception of the order. This intermediate level of stream segregation seems to happen experimentally (Bregman, 1990, pp. 143-165).

The main point of the oscillatory correlation theory is that attention is directed to segments which are formed by phase synchrony of their constituent oscillators, and the interweaving of different segments in time is responsible for the subject's inability to report the temporal order of the tones across different segments. The latter prediction can be verified by psychological experiments. The theory is also useful for explaining psychological data on temporal pattern segmentation phenomena other than streaming. For instance, if common modulation can help synchronize a group of oscillators, the theory will be able to explain the psychophysical observation that common modulation, either in frequency or in amplitude, promotes perceptual grouping of a set of auditory stimuli. To realize this computationally, the simple architecture of Fig. 2 would have to be extended to incorporate detection of common modulation, but the basic idea of oscillatory correlation seems to be applicable.

How can the oscillatory correlation theory be related to processing in neural systems? Biologically, an oscillator could be interpreted as a single neuron since a neuron generates action potentials from time to time. However, in this model an oscillator is interpreted as the collective behavior of a local group of neurons. More specifically, an oscillator corresponds to the result of interaction between a set of excitatory cells and a set of inhibitory cells. The excitatory cells have recurrent excitatory connections among themselves and also excite the inhibitory cells. The inhibitory cells inhibit the excitatory cells. It has been shown that a network of excitatory and inhibitory binary neurons with the kind of interactions just described exhibit oscillatory neural activity (Sporns et al., 1989; Buhmann, 1989). Thus, a single oscillator could be replaced by a local neuron group. The advantage of using oscillators directly lies in the high computational efficiency.

There is ample evidence for the existence of neural oscillations in the brain. It has been known for quite some time that the central olfactory system exhibits oscillatory activity in response to olfactory stimulation (Freeman, 1991). More recently, it has been observed that local field potentials in the visual cortex and the sensorimotor cortex show stimulus-driven oscillations (Eckhorn et al., 1988; Gray et al., 1989; Murthy and Fetz, 1992). The range of the frequencies of all these oscillations is between 20 and 80 Hz, often referred to as 40 Hz oscillations. In addition to stimulus-driven oscillations, neural oscillations seem to exhibit synchronization across remote regions of the visual system in response to highly correlated stimulus. In the auditory domain, 40 Hz oscillations of auditory evoked potentials by a tone were first observed in humans by Galambos et al. (1981). This oscillation can last for several cycles after the stimulus presentation is over. The observation was later confirmed by Madler and Pöppel (1987) who further found that these characteristic oscillations of auditory evoked potentials were absent in patients who were deeply anesthetized and by Mäkelä and Hari (1987) who showed evidence suggesting that the oscillations are generated within the auditory cortex.

Perhaps the most direct support of the oscillatory correlation theory comes from the recent work by Ribary et al. (1991). They used a noninvasive imaging technique called magnetic field tomography to record three-dimensional human brain activity during auditory processing. They found 40-Hz activity in localized brain regions both at the cortical level and at the thalamic level of the auditory system. Furthermore, the oscillations are synchronized over cortical areas during auditory processing, and the synchronized oscillations can be elicited by both rhythmic and transient sound stimuli. In a later report, Llinás and Ribary (1993) also described 40 Hz oscillations triggered by frequency modulated tones.

An important element of the architecture of the model is the use of an array of delay lines (Fig. 2). It has been argued that this architecture may be neurally plausible (Hopfield and Tank, 1989), and similar models have been used as a basis for temporal pattern recognition (see Tank and Hopfield, 1987; Waibel et al., 1989). The range of latencies of neuronal responses increases greatly at higher levels of the auditory pathway. For instance, in the cat auditory cortex, electrophysiological recordings identify up to 1.6 second delays in response to a sequence of different tones (Hocherman and Gilat, 1981; McKenna et al., 1989). In the auditory cortex of echolocating bats, Dear et al. (1993) have proposed that cells with different systematic response latencies together encode multiple objects located at different distances from the bat so that the echoes arriving at different times can concurrently trigger corresponding cortical cells. In this theory, the delay lines serve to provide some form of short-term memory (STM) that can make simultaneously available a recent history of external stimulation. Towards this end, Gottlieb et al. (1989) reported that most neurons in the monkey auditory cortex exhibit STM characteristics during a delay task.

Other structural characteristics of the model include tonotopic organization and lateral connections. Tonotopic organization is a wide-spread feature of the auditory system, including the auditory cortex (Popper and Fay, 1992). There exist intricate local connections within the auditory cortex (for a recent review see Winer, 1992). The local projections of cortical cells can span up to 3 mm in the auditory cortex.

The global inhibitor exerts global control over the entire oscillator network in order to desynchronize multiple segments that are simultaneously active. Since attention is supposed to shift between different streams, it is reasonable to assume that the global inhibitor is involved in attentional control. Crick (1984, 1994) has suggested that part of the thalamus, the thalamic reticular complex in particular, may be involved in selective attention. The thalamus sends projections to and receives input from almost the entire cortex. In light of this suggestion and the reciprocal nature of the connections between the thalamus and cortex, the global inhibitor could be thought of as a neuronal group in the thalamus. In any case, the global inhibitor is assumed to represent the collective behavior of a large group of neurons, not necessarily confined to a single location in the brain.

Although very preliminary, it is tempting to suggest that the two dimensional oscillator network as shown in Fig. 2 describes an aspect of the functions of the auditory cortex. As argued above, this suggestion seems to be consistent with the neurobiology of the auditory cortex, at least structurally. In other words, as a working hypothesis, we suggest that auditory segmentation is achieved in the auditory cortex.

It is interesting to compare the neural pathways underlying vision and audition. Roughly 130 million light receptors comprise the retina of each human eye. The receptors converge onto roughly one million fibers of the visual nerve, which in turn stimulate 100 million neurons in each side of the visual cortex. In contrast, roughly 15,000 receptors (hair cells) in each cochlea converge on roughly 30,000 fibers of the auditory nerve. The pathway to the auditory cortex has many relay nuclei, each having an increased number of auditory neurons so that there are about 100 million auditory cells in each side of the auditory cortex (Handel, 1989; Kandel et al., 1991). Therefore, the number of visual and auditory cortical cells is roughly identical. Why does the number of auditory cells significantly increase along the auditory pathway? The time delay network of Fig. 2 provides a possible explanation for this phenomenon. The representation of auditory stimuli in the cochlea is in real time. The numerous serial and parallel relay stations along the auditory pathway progressively expand the time dimension so that by the level of the auditory cortex a broad range of latencies (delays) are present. The increased ranges of latencies may accompany the significant increase in the cell populations along ascending stations of the auditory pathway. But ever increasing latencies cannot stretch endlessly. One possible solution to the problem would be for STM to provide a sliding representation that is short-lived, i.e. items in STM would disappear after a short time period.

To distinguish from other theories and models that make no reference to the underlying neural processes, we call the oscillatory correlation theory a *neurocomputational* theory. The following section addresses the computational issues involved in modeling temporal segmentation.

NEURAL ARCHITECTURE

The building block of an oscillator network model is a single oscillator, which is defined in the simplest form as a feedback loop between an excitatory unit whose activity is represented by x_i and an inhibitory unit y_i (Fig. 4)

$$\frac{dx_i}{dt} = -x_i + g_x(x_i - \beta y_i + S_i + I_i + \rho) \tag{1a}$$

$$\frac{dy_i}{dt} = -\lambda y_i + g_y(\alpha x_i) \tag{1b}$$

$$g_r(v) = \frac{1}{1 + \exp[-(v-\theta_r)/T]}, \quad r \in \{x, y\} \tag{1c}$$

where α and β are the coupling parameters (connection weights) between the two units. S_i represents overall input from other oscillators of the network and I_i represents stimulation external to the network. λ is the decay parameter, and ρ denotes the amplitude of a Gaussian noise term. $g_r(v)$ is a sigmoid gain function with threshold θ_r and parameter T, the latter used to determine the shape of the sigmoid function. Eq. 1 is essentially a simplification of the oscillator model defined by Wilson and Cowan (1972), who have shown that this type of feedback loop produces oscillations within a wide range of parameters. Consistent with the neural oscillations in the brain, the oscillations in this model are stimulus-driven, i.e., the oscillator stays inactive unless it receives excitatory inputs from either an external source (I_i) or other oscillators in the network (S_i).

The synchronization properties of a network of the oscillators defined in (1) have recently been investigated (Wang, 1993a; Wang, in press). Inspired by von der Malsburg's idea of dynamic links (von der Malsburg, 1981; von der Malsburg and Schneider, 1986), we introduced a mechanism called *dynamic normalization*. In this scheme, the connection from oscillator j to oscillator i is assigned two weights, one permanent weight (T_{ij}), and a dynamic weight (J_{ij}). Permanent connections represent the hardwired structure of a network, while dynamic connections quickly change their strengths from time to time, depending on the current state of the network. More specifically, dynamic normalization is a two-step procedure. First, dynamic connections are updated and then they are normalized to a constant parameter γ

$$\Delta J_{ij} = \eta\, T_{ij}\, h(x_i)\, h(x_j) \tag{2a}$$

$$J_{ij} = \gamma (J_{ij} + \Delta J_{ij}) / [\varepsilon + \sum_k (J_{ik} + \Delta J_{ik})] \tag{2b}$$

where parameter η controls the speed of dynamic modification. Parameter γ represents the overall strength of dynamic links converging on a single active oscillator at any given time, i.e., every active oscillator receives the same amount of dynamic connections. The small constant ε is introduced to prevent division by 0. Function $h(x)$ measures whether the oscillator whose excitatory unit is x (Eq. 1a) is active. It is here defined as $h(x) = 1$ if $<x>$ is greater than a specified constant (0.01 in the following simulations) and $h(x) = 0$ otherwise, where $<x>$ stands for temporal averaging of the activity x over a most recent time period roughly corresponding to the period of the oscillation as defined in (1). All J_{ij}'s are initialized to 0. Thus, if any given oscillator i is not active, i.e. $h(x_i) = 0$, then $J_{ij} = 0$ for any j.

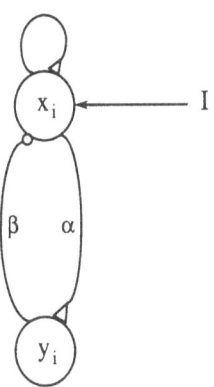

Figure 4. Basic oscillator model formed by a feedback loop between an excitatory unit x_i and an inhibitory unit y_i. α and β are mutual connection strengths. The external input I_i is always applied to x_i.

Once dynamic normalization has been performed, we have shown that a network of oscillators with local connections as defined in (1) and (2) exhibits emergent (long-range) synchrony across the whole network when triggered by a stimulus (Wang, 1993a; Wang, in press). Note that oscillations and synchronization are present only when the network is stimulated. Otherwise, all the oscillators remain silent. Independent of our study, Somers and Kopell recently showed that a network of so-called relaxation oscillators, when locally coupled, can also demonstrate global phase synchrony (Somers and Kopell, 1993).

Emergent synchrony across a stimulated oscillator network on the basis of local connections eliminates the use of either long-range all-to-all connections (von der Malsburg and Schneider, 1986; Sompolinsky et al., 1990) or a global phase coordinator (Kammen et al., 1989), the two mechanisms commonly used for reaching global phase synchrony. From the perspective of scene segmentation, this is a significant result since global mechanisms would lead to indiscriminate synchronization. Imagine that there are two objects, say a desk lamp and a cup, on an image. We can easily separate them visually if they are not connected in space. Even this trivial task could not be performed by an oscillator network that relies on global mechanisms. This is because critical information about geometrical (spatial) relations between the objects is lost in a globally connected network, and thus the oscillator group representing the lamp and the group representing the cup are connected together just as the oscillators within the same group are connected together (for more details, see Wang, 1993b).

We now come back to the oscillator network of Fig. 2 for temporal segmentation. Each oscillator in the matrix is as defined in (1). The global inhibitor whose activity is y is defined as:

$$\frac{dy}{dt} = -\lambda y + g_y\left(\frac{1}{N_x}\sum_i x_i\right) \qquad (3)$$

where λ is the decay parameter of the inhibitor as already defined in (1b). $N_x = \sum h(x_i)$ is the number of active oscillators. The inhibitor itself does not generate oscillations, i.e. it approaches a stationary point if its input is constant. But since its inputs are oscillatory signals, y also generally exhibits oscillations. As explained in the last section, the role of the global inhibitor is to prevent the oscillator groups representing different segments (streams) from reaching synchrony. Intuitively, when the oscillators representing one segment reach the peak value of their oscillations, they cause the inhibitor to fire strongly, which in turn exerts strong inhibition on other oscillators representing different segments, and thus prevents them from reaching their peaks of activities. Consequently, the oscillator groups representing different segments are desynchronized from each other by the global inhibitor.

The excitatory lateral connections on the oscillator network represent permanent connections (T_{ij} in Eq. 2a). The strengths of these permanent connections, except the self connection, are assumed to take on a two dimensional Gaussian distribution: the connection strength between any pair of oscillators on the matrix of Fig. 2 falls off exponentially with the distance between the two oscillators. More specifically, assume that the two dimensional indices of oscillator i are (t_i, f_i), representing the time and frequency coordinates of the oscillator respectively. Oscillator i connects to oscillator j with strength

$$T_{ij} = Exp[\frac{-(t_j-t_i)^2}{\sigma_t^2} - \frac{-(f_j-f_i)^2}{\sigma_f^2}] \qquad (4)$$

where the parameters σ_t and σ_f determine the widths along the time axis and the frequency axis of the Gaussian distribution, respectively. The self connectivity $T_{ii} = 0$. With this definition, it is easy to see that $T_{ij} = T_{ji}$, or that connections are symmetrical. Figure 5 illustrates a two dimensional Gaussian distribution, which shows the distribution of the strengths with which the center oscillator connects with others, namely those with higher (indicated by a positive number) or lower (indicated by a negative number) frequency coordinates and with higher or lower time coordinates on the network. The two dimensional Gaussian distribution characterizes exponential decay in connection strength with respect to distance, and it is often used to describe the lateral connection pattern in the brain in the theoretical community.

The excitatory lateral connections as defined in (4) across both time and frequency dimensions are a fundamental requirement to synchronize oscillators representing the same segment, because they express the cooperation between the components of a segment. The nearness in frequency and time is characterized by stronger mutual connections in such a connectivity pattern. Stronger mutual connections promote mutual synchrony of the corresponding oscillators.

Once T_{ij}'s are defined, J_{ij}'s are updated according to (2). In summary, the input S_i to oscillator i is defined as (cf. Eq. 1a):

$$S_i = \sum_j J_{ij} x_i - \mu y \tag{5}$$

where the first term on the right-hand side represents *the effective input* that oscillator i receives from other oscillators in the network. The second term is the inhibition that i receives from the global inhibitor, with the positive parameter μ representing the absolute weight of inhibition. By the effective input we mean the input that comes through the dynamic connections J_{ij}'s that change rapidly.

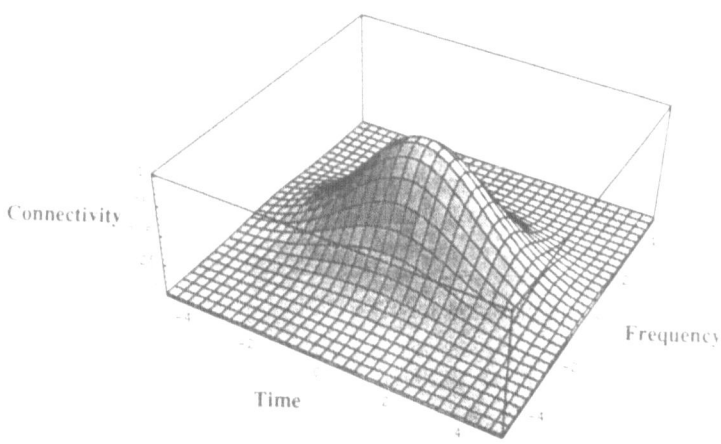

Figure 5. Two-dimensional Gaussian distribution of the strengths of the connections from the center oscillator according to (4) in the text. The parameter values for this distribution are $t_i = 0, f_i = 0, \sigma_t = 2.75$, and $\sigma_f = 1.8$.

Let us see how the oscillator network functions in general terms. From the study of the general oscillator network without the global inhibitor (Wang 1993a; Wang, in press), we know that a pair of oscillators synchronize if they both are externally stimulated and they have sufficient mutual excitation. With the global inhibitor taken into account, the mutual excitation has to be strong enough to overcome the global inhibition. Otherwise, the two oscillators would be anti-phase locked. From (2), one can see that in order to form a dynamic connection between two oscillators, they both have to be active and there have to be non-zero permanent connections between them. If a dynamic connection can be formed, its strength will be proportional to the strength of the corresponding permanent connection. The earlier result on emergent synchronization on the basis of local connections allows synchronization to be transitive: if separately oscillator i can synchronize with oscillator j and oscillator j can synchronize with oscillator k, then all three can be synchronized together. This property of synchronization plus Gaussian distribution of permanent connections should promote overall grouping (synchronization) of a sequence of successive tones which have similar frequencies (proximity in frequency) and/or high presentation rates (proximity in time). At the same time, tones that cannot be grouped will be segregated due to the global inhibition.

SIMULATION RESULTS

The above architecture for temporal segmentation has been simulated using a matrix of oscillators with six rows representing different frequency channels, as shown in Fig. 2 (see Wang, 1994 for an earlier version of the simulations). A simulated sequence of alternating tones *HLHLHL* was used as input. The sequence was presented to the network in *real time* by triggering the input end with each tone successively (see Fig. 1 for an example of a stimulus sequence). For simplicity, *H* tones are assumed to all trigger the same frequency channel and so are *L* tones. The two triggered channels were three rows apart (see Fig. 2). An oscillator in the network was triggered by a stimulus if the stimulus had the same frequency as represented by the oscillator and an appropriate delay had elapsed after the stimulus was presented to the input end. When an oscillator in the leftmost column (shortest latency) was triggered, a random phase was generated for it. The sequence was repeatedly presented to the network, as in the psychological experiments.

The system of equations (1)-(5) was solved using a numerical differential equation solver (see Press et al., 1992, for details on the Runge-Kutta method used to solve the differential equations). In order to relate to real time, the basic time unit in the simulation was set at 0.05 ms. The basic delay interval, i.e. the difference in delay between two neighboring oscillators in the time dimension, was set at 80 ms. This value would be unrealistic if it were taken to be a single synaptic delay. However, as explained in Section 2, an oscillator is assumed to correspond to a local group of neurons. In the time-frequency domain, the duration of a tone presentation corresponds to the number of oscillators occupied by the tone along the time axis. We conducted three groups of simulations with tone durations of 160 ms, 240 ms, and 320 ms, corresponding to fast, medium, and slow presentation rates, respectively. Thus, for the fast presentation rate, each tone occupied two oscillators, for the medium presentation rate three oscillators, and for the slow presentation rate four oscillators.

When the presentation rate was fast (2 oscillators per tone), a set of 12 isofrequency oscillators were simulated for each of the six frequency channels. This network can simultaneously encode up to 6 tones. Figure 6 displays the response of the two channels triggered by the tones. Since the other four frequency channels were not stimulated, the oscillators in those rows were inactive, and hence omitted in the display. Each trace displays the activity of the excitatory unit of one oscillator. The top 12 traces represent the 12 isofrequency oscillators with progressively increasing response delays (latencies) in the second frequency channel (*H* tone), the top one corresponding to the leftmost oscillator of channel 2 in Fig. 2. Similarly, the bottom 12 traces represent the 12 oscillators with increasing delays in channel 5 (*L* tone). We refer to frequency channel 2 as the high frequency channel, and channel 5 as the low frequency channel. The time axis (horizontal) is divided into 18 time increments, each of which equals 80 ms, the basic delay interval. Thus the figure represents a total time of 1440 ms, which equals one and a half presentation cycles of the entire sequence. At each step of the numerical simulation, the values of x_i and y_i were computed based on the their own previous values (Eq. 1), as well as that of the global inhibitor and x_j's of other oscillators (Eq. 5). Then, the dynamic link J_{ij} was computed on the basis of T_{ij} using Eq. 2. As can be seen from the figure, except for a brief period at the beginning of each time increment, all of the active oscillators in the high frequency channel quickly reached synchronization, and so did the oscillators in the low frequency channel. Except for the first few time increments (e.g., column 6 of Fig. 6), the active oscillators of one channel (high or low) desynchronized with those of the other channel. In other words, the oscillators triggered by *H* tones were synchronized and so were those triggered by *L* tones, but those triggered by *H* tones desynchronized with those triggered by *L* tones. This pattern was particularly stable

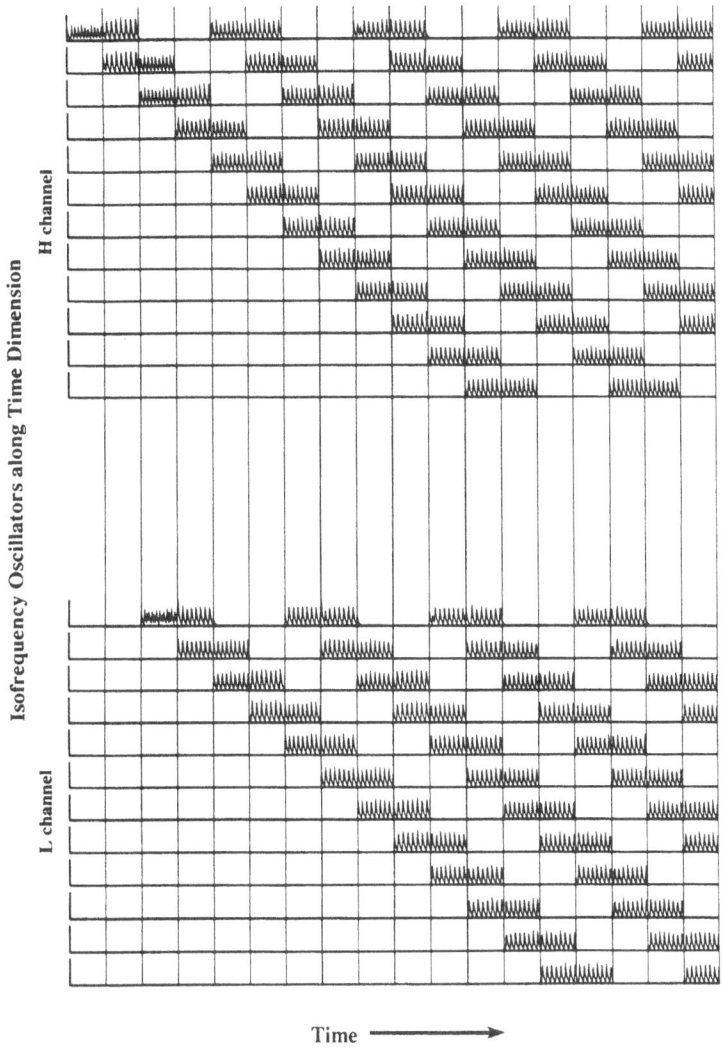

Time →

Figure 6. Responses of two frequency channels to fast presentation of alternating H and L tones (2 oscillators per tone). Each isofrequency channel (H or L) consists of 12 oscillators with increasing delays from top to bottom. Each activity trace represents the value of the excitatory unit of an oscillator. The vertical lines are used to divide the time axis into increments, each corresponding to a basic delay interval. The parameters $\alpha = 0.6$, $\beta = 2.5$, $\gamma = 1.6$, $\rho = 0.01$, $\lambda = 1.0$, $\theta_x = 0.6$, $\theta_y = 0.15$, $T = 0.025$, $\eta = 10.0$, $\sigma_i = 2.75$, $\sigma_f = 1.8$, $\mu = 0.5$, $I_i = 0.7$ if oscillator i is externally stimulated, and $I_i = 0.0$ otherwise.

after the first cycle of tone sequence presentation was finished, i.e. after the first 12 time increments. According to the oscillatory correlation theory, this pattern would cause all H tones to be grouped into one segment, and all L tones to be grouped into another segment.

Relating back to psychophysical experiments, stream segregation occurred in the simulations for this rate of presentation, and two streams were segmented apart in real time on the basis of differing frequencies between the streams.

Figure 7 shows an enlarged version of the last two time increments of Fig. 6, together with the activity of the global inhibitor during this period, to better illustrate the phase relations of the oscillators. The frequency of the global inhibitor is double that of an active oscillator in the oscillator matrix. This shows that there are two distinct oscillation phases in the network, corresponding to the two segments, because each phase causes the inhibitor to oscillate with the same frequency. Figure 8 shows the combined x values (see Eq. 1) of all oscillators of one frequency channel during one time increment, the second one of Fig. 7. The top panel of the figure depicts the stimulation pattern, showing that one tone occupies two oscillators, while the middle and the bottom panels show the combined activities of the high

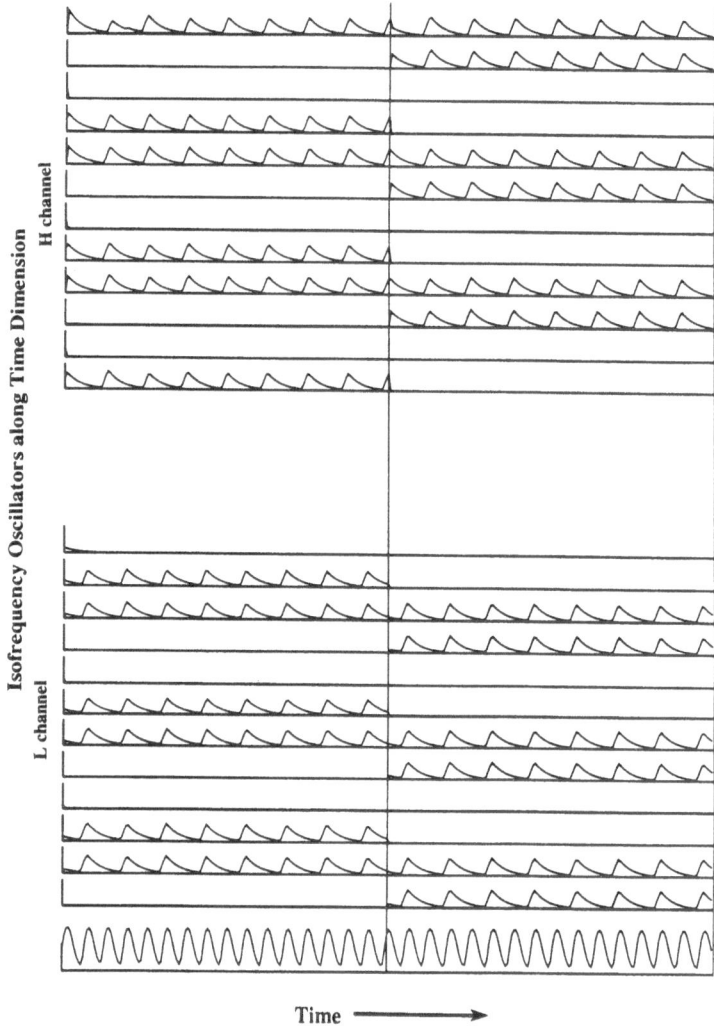

Figure 7. Enlarged display of the last two time increments of Fig. 6 plus the activity of the global inhibitor (bottom trace) during the same time period.

and low frequency channels respectively. Synchronization within the same frequency channels and desynchronization across the two channels are clearly illustrated in this form of display, which is also used for the following two simulation conditions discussed below.

It should be emphasized that synchrony and desynchrony in the above simulations are the emergent properties of the oscillator network, and not induced by the input. The input, or tone stimulation, to a particular input channel was constant (see Fig. 6 caption). The rhythmic relationship between the sequence of H (or L) tones, as shown in the top panel of Fig. 8, has nothing to do with the oscillations generated in the oscillator network. One can easily check that their frequencies are very different. Moreover, to generate synchrony and desynchrony as shown in Fig. 6 does not require a rhythmic input sequence of tones, although Jones et al. (1993) have suggested that rhythm in a tone sequence helps in grouping it into a single segment.

When the presentation rate was medium (3 oscillators per tone), a network of 6x18 oscillators was simulated. Otherwise the simulation procedures were similar to those for the

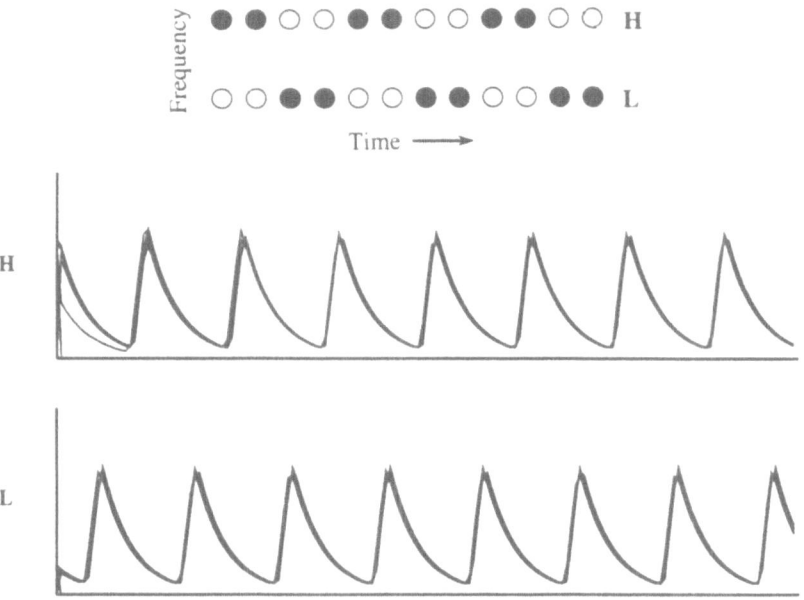

Figure 8. Top panel: the stimulus pattern at a specific time in the two frequency channels that are stimulated. Middle and bottom panels: The combined activities of all the oscillators in the high and the low frequency channels, respectively. Only the last time increment of Fig. 6 is shown.

fast presentation rate. Figure 9 shows the corresponding results. The top panel of the figure illustrates the stimulation condition. The middle two panels show the combined oscillatory activities for the high frequency and the low frequency channels, respectively, for one typical delay interval after a full cycle of presentation of six tones was completed. As shown in the figure, the oscillators within each channel did not synchronize, but instead exhibited two phases after an initial irregular transient. Closer inspection of the detailed simulation results (not shown) reveals that two successive H tones out of three formed one segment, two successive L tones formed another segment, and one remaining H tone and its neighboring L tone formed yet another segment (see the top panel of Fig. 9). Together, there were three distinct phases on the oscillator network, and each frequency channel exhibited two distinct phases. The bottom panel of Fig. 9, which shows the activities of a set of the three neighboring oscillators representing one single tone, illustrates that the three oscillators representing the same tone were synchronized in the simulation. In sum, for the medium presentation rate, phase synchrony was not reached across the same frequency channel. Rather, partial stream segregation, e.g. among two consecutive H or L tones, was exhibited in the simulation.

Finally, when the presentation rate was further slowed to 4 oscillators per tone, a network of 6x24 oscillators was simulated, so that the entire sequence of six tones could be represented in the simulation. Figure 10 shows the results of the simulation in a way similar to Figs. 8 and 9. Again, the top panel shows the stimulation pattern. From the combined oscillations in the middle two panels, one can observe that there are three overlapping waveforms corresponding to three distinct phases within each frequency channel. Each phase corresponds to a single tone. In other words, there was no grouping at all between different tones of the same frequency (H or L). Instead, one tone was grouped into a segment with a neighboring tone of the other frequency. This can be seen by comparing the two middle panels with respect to time. The waveforms between the panels align with each other. As in Fig. 9, the bottom panel

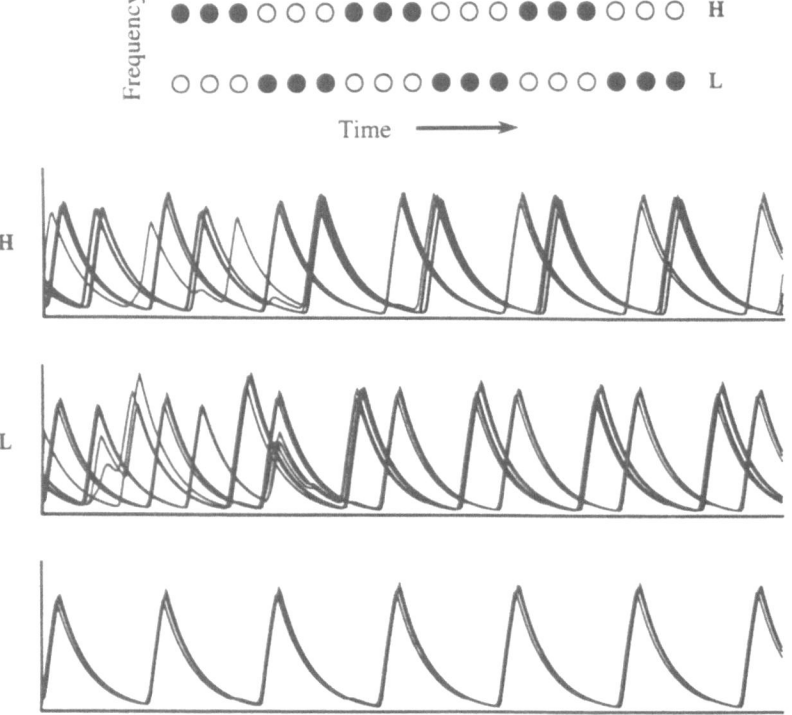

Figure 9. Responses of two corresponding frequency channels to medium-rate presentation of alternating H and L tones (3 oscillators per tone). Top panel: the stimulus pattern in the two frequency channels. Middle panels: The combined activities of all the oscillators in the high and the low frequency channels respectively. The traces show one basic delay interval after a cycle of six tones was presented. Lower panel: The combined activities of the three oscillators that were activated by a single tone. The parameter values are the same as in Fig. 6.

of Fig. 10 shows the activities of the four oscillators representing a single tone. In this case, the four oscillators were synchronized. That is, a single tone did not further break into different segments. Thus, at this rate of presentation, one tone formed a segment with a neighboring tone of a different frequency (see the top panel of Fig. 10), and the six tones in the input sequence were segregated into three segments.

From all the simulations, it can be concluded that tones can be grouped together based on their similarities in frequency, but that segmentation critically depends on the rate of presentation (in these simulations, the duration of each tone). Stream segregation is best for high rates of presentation or short duration, absent for low rates or long duration, and is intermediate for medium rates of tone presentation. This dependency on presentation rate is consistent with the psychological data on auditory stream segregation (Introduction). The simulation results demonstrate that the segmentation network (Fig. 2) is able to implement the computational properties required by the oscillatory correlation theory.

How do such computational behaviors arise in the simulations? They can be explained in terms of general principles of competition and cooperation. Adjacent oscillators in the network (Fig. 2) always synchronize because they are strongly coupled with each other (see Fig. 5). For fast presentation rates, successive tones of the same frequency are separated only by two oscillators (the top panel of Fig. 8). Thus, with strong coupling, all tones of the same frequency are grouped into the same segment. Slowing down the rate of presentation increases the distance between successive tones of the same frequency, and thus reduces their coupling. Recall that due to dynamic normalization each oscillator has the same overall amount of incoming dynamic links (Eq. 2), so the reduced coupling between successive tones of the same frequency will increase the relative coupling between tones across the different frequency channels. This explains why, at slower presentation rates, tones that are adjacent in time are more likely to be grouped together. In the case of Fig. 10, only tones that are adjacent in time are grouped into the same segments. The global inhibitor serves to desynchronize oscillator

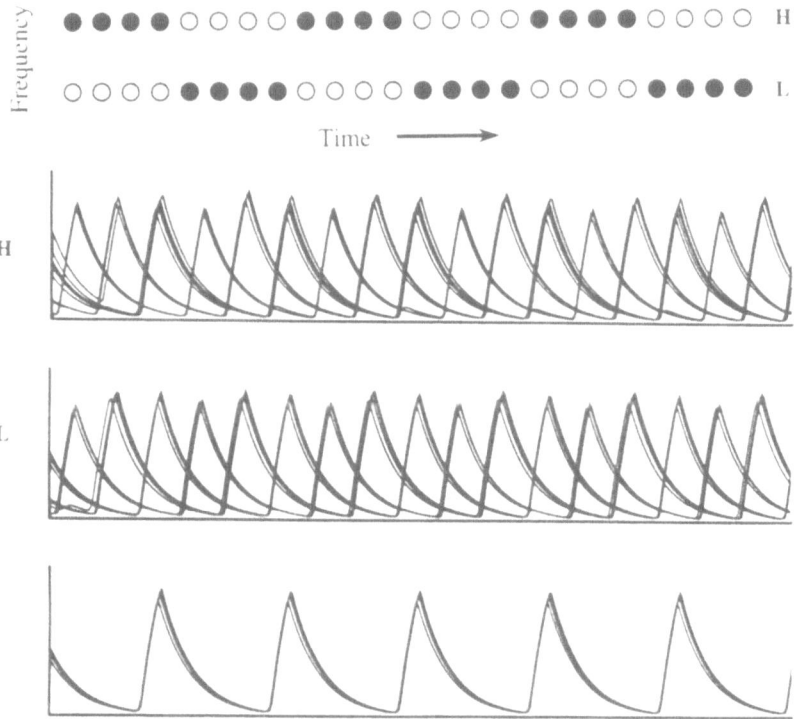

Figure 10. Responses of two frequency channels to slow presentation of alternating H and L tones (4 oscillators per tone). Top panel: the stimulus pattern in the two frequency channels. Middle panels: The combined activities of all oscillators in the high and the low frequency channels respectively. The traces show one basic delay interval after a cycle of six tones was presented. Lower panel: The combined activities of the four oscillators that were activated by a single tone. The parameter values are the same as in Fig. 6.

groups that are not strongly coupled, and thus segregate the stimuli on the entire network into different segments. Without it, the stimulated oscillators in the network would be all synchronized regardless of which tones they represent, since these oscillators would form a locally coupled population by the Gaussian connectivity pattern (Fig. 5), which we know will reach global synchrony (see Wang 1993a; Wang, in press).

CONCLUSIONS AND FUTURE DIRECTIONS

Although only the preliminary simulation results of a simplified model of temporal segmentation were presented here, the model is not limited by the stimuli used. For example, the three high (low) tones do not have to trigger the same frequency channel. As long as they trigger nearby frequency channels, temporal segmentation based on frequency proximity will occur. This is because synchronization of oscillators depends on the strengths of the connections between them, and oscillators with similar frequency coordinates have relatively large mutual connection strengths in the model. Two oscillators need not have exactly the same frequency in order to have strong coupling. By the same token, each tone need not be a pure tone. A tone with frequency modulation will work just as well. These observations have been confirmed by further simulation results. In essence, the permanent connectivity pattern of a Gaussian distribution (Fig. 5) strongly biases towards the grouping of sounds that have continuous frequency transitions, which is consistent with the analysis of speech perception (Handel, 1989; Bregman, 1990).

The global inhibitor plays the role of breaking groups of oscillators with weak mutual coupling into different segments, and introduces rich oscillatory behavior within the network. As previously noted, without the global inhibitor, oscillators with nonzero connections eventually will reach phase synchrony. The global inhibitor adds competition into the oscillator network so that only relatively strong coupling leads to entrainment. Thus, the segmentation network incorporates both competition and cooperation (see Arbib, 1989, for a general discussion of competition and cooperation in the brain). One interesting outcome of such competition and cooperation is that the network does not segment tones based on their absolute distances in the frequency domain, which would have been the case without global competition. Instead, the distances that are needed to pass the segmentation threshold depend on the presentation rates. Thus there is an interaction between frequency proximity and presentation rate for temporal segmentation. Such "trading" between frequency and time seems to be present in human stream segregation (Jones, 1976).

Although structurally similar, the global inhibitor of this model serves an entirely different role than the global comparator in the comparator model of Kammen et al. (1989). The comparator in their model is used to synchronize activated oscillators on an oscillator network, while our global inhibitor serves to desynchronize oscillators so that they form separate synchronized groups. Group synchronization through a mechanism similar to a comparator has also been suggested by Llinás and his colleagues (Ribary et al., 1991; Llinás et al., 1993). Synchronization is achieved in our model by lateral connections on the network, which is consistent with the suggestions of Engel et al. (1991a, 1991b). Both the comparator and the global inhibitor are assumed to be located in the thalamus which provides the structural requirements.

As noted previously, our model predicts that the global inhibitor oscillates with a frequency double that of the oscillators on the network when stream segregation occurs. This observation suggests an hypothesis that could be tested experimentally. Assume that we deal with only two streams and that the oscillator network is located in the auditory cortex. Our model predicts that auditory cortex would oscillate at 40 Hz but that the thalamus would oscillate at 80 Hz. The comparator model would predict that the thalamus and auditory cortex oscillate at the same frequency. To determine which prediction is correct, one would need to make sure that stream segregation does occur in the experiment. Otherwise, both models would predict the same frequency of oscillation in the auditory cortex and the thalamus. Admittedly, the experiment would be difficult to conduct because stream segregation so far is only a psychological phenomenon. However, both models are also assumed to be applicable to visual segmentation, so animal experiments in the visual domain should be able to verify the contrasting models.

To achieve a more realistic model of auditory pattern segmentation, the basic architecture of Fig. 2 must be extended in order to incorporate other qualities of auditory stimuli, such as amplitude, rhythm, harmonics, timbre, etc. However, the basic principle underlying this set of simulations, namely, that connection strengths provide the basis for encoding similarities among auditory stimuli, should be applicable in general. Temporal segmentation as discussed so far concerns only the relationships between input stimuli. Psychological data show that temporal segmentation is also influenced by prior knowledge of a subject (Bregman, 1990). We previously proposed a model of segmentation based on the patterns stored in associative memory (Wang et al., 1990), where a mixture of multiple memory patterns can be segmented by a network of oscillators which encodes a number of patterns. In the future, these two types of segmentation must be integrated into a coherent model. Such a model should be able, for example, to simulate the data on melody segmentation observed by Dowling (1973) and Dowling et al. (1987).

SUMMARY

Temporal segmentation is critical for complex temporal pattern processing and auditory perception. This chapter reviews pertinent psychological data on auditory scene analysis, and a number of computational models that address the issues of temporal segmentation. The chapter presents a novel neurocomputational theory, namely the oscillatory correlation theory, of how temporal segmentation might be achieved in the brain. On the basis of the theory, we present a generic neural network framework for temporal pattern segmentation. The architecture consists of a laterally coupled two-dimensional neural oscillator network with a global inhibitor. One dimension represents time and the other represents frequency. We show that this architecture can group auditory features into a segment by phase synchrony and segregate different segments by desynchronization, both in real time. The network simulates the psychological observation that temporal segmentation critically depends on the rate of presentation. Although the exploration of this approach to studying temporal segmentation is only at the beginning stage, it seems promising in terms of both neural plausibility and computational effectiveness.

ACKNOWLEDGMENTS

The author is very grateful to Dr. Mari Jones for stimulating discussions, and to her and the editors of the book, particularly Dr. Ellen Covey, for their critical reading of earlier drafts and very constructive suggestions. The preparation of this paper was supported in part by ONR grant N00014-93-1-0335 and NSF grant IRI-9211419.

REFERENCES

Arbib, M.A., 1989, "The Metaphorical Brain 2: Neural Networks and Beyond", Wiley Interscience, New York, NY.

Beauvois, M.W., and Meddis, R., 1991, A computer model of auditory stream segregation, *Quart. J. Exp. Psychol.* 43 A:517.

Bregman, A.S., 1990, "Auditory Scene Analysis", MIT Press, Cambridge MA.

Bregman, A.S., 1978, The formation of auditory streams, *in*: "Attention and Performance VII", J. Requin, ed., Lawrence Erlbaum Associates, Hillsdale NJ.

Bregman, A.S., and Campbell, J., 1971, Primary auditory stream segregation and perception of order in rapid sequences of tones, *J. Exp. Psychol.* 89:244.

Bregman, A.S., Abramson, J., Doehring, P., and Darwing, C.J., 1985, Spectral integration based on common amplitude modulation, *Percept. Psychophys.* 37:483.

Buhmann, J, 1989, Oscillations and low firing rates in associative memory neural networks, *Phys. Rev. A* 40:4145.

Cherry, E.C, 1953, Some experiments on the recognition of speech, with one and with two ears, *J. Acoust. Soc. Am.* 25:975.

Cole, R.A., and Scott, B., 1973, Perception of temporal order in speech: the role of vowel transitions, *Can. J. Psychol.* 27:441.

Crick, F., 1984, Function of the thalamic reticular complex: The searchlight hypothesis, *Proc. Natl. Acad. Sci. USA* 81:4586.

Crick, F., 1994, "The Astonishing Hypothesis", Scribner, New York, NY.

Dannenbring, G.L., and Bregman, A.S., 1976, Effect of silence between tones on auditory stream segregation, *J. Acoust. Soc. Am.* 59:987.

Dear, S.P., Simmons, J.A., and Fritz, J., 1993, A possible neuronal basis for representation of acoustic scenes in auditory cortex of the big brown bat, *Nature* 364:620.

Dowling, W.J., 1973, The perception of interleaved melodies, *Cognit. Psychol.* 5:322.

Dowling, W.J., Lund, K.M-T., and Herrbold, S., 1987, Aiming attention in pitch and time in the perception of interleaved melodies, *Percept. Psychophys.* 41:642.

Eckhorn, R., Bauer, R., Jordan, W., Brosch, M., Kruse, W., Munk, M., and Reitboeck, H.J., 1988, Coherent oscillations: A mechanism of feature linking in the visual cortex? *Biol. Cybern.* 60:121.

Ellias, S.A., and Grossberg, S., 1975, Pattern formation, contrast control, and oscillations in the short term memory of shunting on-center off-surround networks, *Biol. Cybern.* 20:69.

Engel, A.K., König, P., Kreiter, A.K., and Singer, W., 1991, Synchronization of oscillatory neuronal responses between striate and extrastriate visual cortical areas of the cat, *Proc. Natl. Acad. Sci. USA* 88:6048.

Engel, A.K., König, P., Kreiter, A.K., and Singer, W., 1991, Interhemispheric synchronization of oscillatory neuronal responses in cat visual cortex, *Science* 252:1177.

Freeman, W.J., 1991, Nonlinear dynamics in olfactory information processing, *in*: "Olfaction", J.L. Davis, and H. Eichenbaum, eds., MIT Press, Cambridge MA.

Galambos, R., Makeig, S., and Talmachoff, P.J., 1981, A 40-Hz auditory potential recorded from the human scalp, *Proc. Natl. Acad. Sci. USA* 78:2643.

Gottlieb, Y., Vaadia, E., and Abeles, M., 1989, Single unit activity in the auditory cortex of a monkey performing a short term memory task, *Exp. Brain Res.* 74:139.

Gray, C.M., König, P., Engel, A.K., and Singer, W., 1989, Oscillatory responses in cat visual cortex exhibit inter-columnar synchronization which reflects global stimulus properties, *Nature* 338:334.

Guckenheimer, J., and Holmes, P., 1983, "Nonlinear Oscillations, Dynamical Systems and Bifurcations of Vector Fields", Springer-Verlag, New York, NY.

Handel, S., 1989, "Listening: An Introduction to the Perception of Auditory Events", MIT Press: Cambridge, MA.

Hocherman, S., and Gilat, E., 1981, Dependence of auditory cortex evoked unit activity on interstimulus interval in the cat, *J. Neurophysiol.* 45:987.

Hopfield, J.J., and Tank, D.W., 1989, Neural architecture and biophysics for sequence recognition, *in*: "Neural Models of Plasticity", J.H. Byrne, and W.O. Berry, eds., Academic Press, San Diego CA.

Jones, M.R., 1976, Time, our lost dimension: toward a new theory of perception, attention, and memory, *Psychol. Rev.* 83:323.

Jones, M.R., Jagacinski, R.J., Yee, W., Floyd, R.L., and Klapp, S.T., 1994, Tests of attentional flexibility in listening to polyrhythmic patterns, *J. Exp. Psychol.: Human Percept. Perform.*, in press.

Jones, M.R., Kidd, G., and Wetzel, R., 1981, Evidence for rhythmic attention, *J. Exp. Psychol.: Human Percept. Perform.* 7:1059.

Jones, M.R., and Yee, W., 1993, Attending to auditory events: the role of temporal organization, *in* "Thinking in Sound", S. McAdams and E. Bigand, ed., Clarendon Press, Oxford, UK..

Kammen, D.M., Holmes, P.J, and Koch, C., 1989, Origin of oscillations in visual cortex: Feedback versus local coupling, *in*: "Models of Brain Functions", R.M.J. Cotterill, ed., Cambridge University Press, Cambridge UK.

Kandel, E.R., Schwartz, J.H, and Jessell, T.M., 1991, "Principles of Neural Science (3rd Ed.)", Elsevier, New York, NY.

Llinás, R., and Ribary, U., 1993, Coherent 40-Hz oscillation characterizes dream state in humans, *Proc. Natl. Acad. Sci. USA* 90:2078.

Madler, C., and Pöppel, E., 1987, Auditory evoked potentials indicate the loss of neuronal oscillations during general anesthesia, *Naturwissenschaften* 74:42.

Mäkelä, J.P., and Hari, R., 1987, Evidence for cortical origin of the 40 Hz auditory evoked response in man, *Electroencephalogr. Clin. Neurophysiol.* 66:539.

McKenna, T.M., Weinberger, N.M., and Diamond, D.M., 1989, Responses of single auditory cortical neurons to tone sequences, *Brain Res.* 481:142.

Miller, G.A., and Heise, G.A., 1950, The trill threshold, *J. Acoust. Soc. Am.* 22:637.

Milner, P.M., 1974, A model for visual shape recognition, *Psychol. Rev.* 81:521.

Morgan, D.P., and Scofield, C.L., 1991, "Neural Networks and Speech Processing", Kluwer Academic, Norwell MA.

Murthy, V.N., and Fetz, E.E., 1992, Coherent 25- to 35-Hz oscillations in the sensorimotor cortex of awake behaving monkeys, *Proc. Natl. Acad. Sci. USA* 89:5670.

Parsons, T.W., 1976, Separation of speech from interfering speech by means of harmonic selection, *J. Acoust. Soc. Am.* 60:911.

Popper, A.N., and Fay, R.R., eds., 1992, "The Mammalian Auditory Pathway: Neurophysiology", Springer-Verlag, New York, NY.

Press, W.H., Teukolsky, S.A., Vetterling, W.T., and Flannery, B.P., "Numerical Recipes in C: The Art of Scientific Computing", 2nd Ed., Cambridge University Press, Cambridge, UK.

Rabiner, L.R., and Juang, B.H., 1986, An introduction to hidden Markov models, *IEEE Acoust., Speech, Signal Process. Magazine* 3:4.

Rasch, R.A., 1978, The perception of simultaneous notes such as in polyphonic music, *Acustica* 40:22.

Ribary, U., Ioannides, A.A., Singh, K.D., Hasson, R., Bolton, J.P.R., Lado, F., Mogilner, A., and Llinás, R., 1991, Magnetic field tomography of coherent thalamocortical 40-Hz oscillations in humans, *Proc. Natl. Acad. Sci. USA* 88:11037.

Roberts, B., and Bregman, A.S., 1991, Effects of the pattern of spectral spacing on the perceptual fusion of harmonics, *J. Acoust. Soc. Am.* 90:3050.

Somers, D., and Kopell, N., 1993, Rapid synchronization through fast threshold modulation, *Biol. Cybern.* 68:393.

Sompolinsky, H., Golomb, D., and Kleinfeld, D., 1990, Global processing of visual stimuli in a neural network of coupled oscillators, *Proc. Natl. Acad. Sci. USA* 87:7200.

Sporns, O., Gally, J.A., Reeke Jr., G.N., and Edelman, G.M., 1989, Reentrant signaling among simulated neuronal groups leads to coherency in their oscillatory activity, *Proc. Natl. Acad. Sci. USA* 86:7265.

Tank, D.W., and Hopfield, J.J., 1987, Neural computation by concentrating information in time, *Proc. Natl. Acad. Sci. USA* 84:1896.

van Noorden, L.P.A.S., 1975, "Temporal Coherence in the Perception of Tone Sequences", Ph.D dissertation, The Institute of Perception Research, Eindhoven, The Netherlands.

von der Malsburg, C., 1981, The correlation theory of brain function, *Internal Report* 81-2, Max-Planck-Institut for Biophysical Chemistry, Göttingen, Germany.

von der Malsburg, C., and Schneider, W., 1986, A neural cocktail-party processor, *Biol. Cybern.* 54:29.

Waibel, A., Hanazawa, T., Hinton, G.E., Shikano, K., and Lang, K.J., 1989, Phoneme recognition using time-delay neural networks, *IEEE Trans. Acoust. Speech. Signal Process.* 31:328.

Wang, D.L., 1993a, Modeling global synchrony in the visual cortex by locally coupled neural oscillators, *Proc. 15th Ann. Conf. Cog. Sci.*, 1058.

Wang, D.L., 1993b, Pattern recognition: Neural networks in perspective, *IEEE Expert* 8:52.

Wang, D.L., 1994, Auditory stream segregation based on oscillatory correlation, *Proc. IEEE 1994 Workshop on Neural Networks for Signal Processing*, 624 .

Wang, D.L., 1995, Emergent synchrony in locally coupled neural oscillators, *IEEE Trans. Neural Networks*, in press.

Wang, D.L., 1995, Temporal pattern processing in neural networks, *in*: "Handbook of Brain Theory and Neural Networks", M.A. Arbib, ed., MIT Press, Cambridge MA.

Wang, D.L., Buhmann, J., and von der Malsburg, C., 1990, Pattern segmentation in associative memory, *Neural Computat.* 2: 95. Reprinted *in*: "Olfaction", J.L. Davis, and H. Eichenbaum, eds., MIT Press, Cambridge MA.

Warren R.M., Obusek C.J., Farmer R.M., and Warren, R.P., 1969, Auditory sequence: Confusion of patterns other than speech or music, *Science* 164:586.

Weintraub, M., 1986, A computational model for separating two simultaneous talkers, *in*: *IEEE ICASSP*, Tokyo, 81.

Wilson H.R., and J.D. Cowan, 1972, Excitatory and inhibitory interactions in localized populations of model neurons, *Biophys. J.* 12:1.

Winer, J.A., 1992, The functional architecture of the medial geniculate body and the primary auditory cortex, *in*: "The Mammalian Auditory Pathway: Neuroanatomy", D.B. Webster, A.N. Popper, and R.R. Fay, eds., Springer-Verlag, New York, NY.

TOWARD SIMULATED AUDITION IN OPEN ENVIRONMENTS

Robert F. Port, Sven E. Anderson, and J. Devin McAuley

Department of Computer Science
Program in Cognitive Science
Indiana University
Bloomington, IN 47405

INTRODUCTION

This paper considers the problem of hearing in the broadest possible terms. What kind of information is available in sound about events in the environment? What kind of cognitive mechanisms could extract this information? To address such global problems, we think it is essential to construct simulations of auditory processing, but to evaluate progress, real-world audition is difficult to deal with. To test the simulations, it is useful to develop artificial auditory environments that are simplified, yet retain certain critical properties of natural auditory environments, such as the property of *openness*. In this paper, we shall schematize the general properties of auditory environments and some major classes of information about time that are relevant for perception. Our theme is that to understand auditory cognition, we not only need to understand auditory processing mechanisms, we also need a clear idea of the problems that must be solved by an auditory system. If the task can be clearly defined, some hypotheses can be formulated about the global dynamic properties of neural systems that are potentially capable of performing these tasks. These systems may then serve as a starting point for developing models of neural mechanisms.

AUDITORY ENVIRONMENTS AND AUDITORY PATTERNS

In order to gain a picture of how audition works, two classes of information are needed. First, we need to understand how the ear and peripheral auditory system process sound. Second, we need to have an idea of the kinds of events that generate distinctive sound patterns, that is, the acoustic events that occur in the auditory environment, and a description of what their properties are. With these two tasks accomplished, we can begin to develop a plausible model of the auditory system.

The first task is to model low-level processing related to the transduction from acoustic space into neural space. Research in auditory psychophysics over the past hundred years has led to the development of some practical mathematical models that simulate basic auditory capabilities (e.g., Shamma, 1985a, 1985b; Hermansky, 1990; Patterson, 1990). We understand the basic properties of critical bands, of masking, and of binaural phenomena such as localization and comodulation masking release (see Moore, 1993, Yost and Watson, 1987). Psychophysical research has typically employed very simple stimuli, such as steady-state tones or white noise to reveal the processing capabilities of the auditory system. Processing audio signals in a way that simulates the ear is now straightforward and can be done in real time on a laboratory workstation. However, an understanding of the sensory transduction mechanism is only a beginning. Charles Watson tells the story that one of the pioneer psychophysicists of this century, Lloyd Jeffress, used to ask his students "What are the three most important aspects of sound?", to which his students would often respond, "Frequency, amplitude, and phase." Jeffress would say, "No! Where it is, whether it can eat you, and whether you can eat it".

Practically speaking, how can the broad issues of the ecological relevance of auditory cognition be addressed? After all, we do not yet have a very clear idea of the acoustic ecology that is relevant to our species, or to any species. The information that is potentially available to real nervous systems about biologically relevant environmental events is not yet characterized. As pointed out years ago by Gibson (1968), it is methodologically dangerous to try to build models for supposed perceptual processes when you don't understand the information structure that is actually available in the stimulus. One can easily find oneself trying to solve the wrong problem. Nevertheless, several researchers have begun to construct insightful taxonomies of ecological acoustics (Handel, 1989; Gaver, 1993; Yost, 1991; Yost, 1992). Knowledge of the physics of sound imposes many constraints on how it can be processed. In addition, there are now many well-understood constraints from neuroscience. Auditory mechanisms are limited by their implementation in nervous tissue. We believe that if we take all these constraints into account and design functional mechanisms for simplified perceptual tasks that nevertheless capture critical features of the ecological problem, we should be able to find solutions that will help us understand the function of real auditory cognition systems.

Developments in mathematical model construction within the class of dynamical systems – that is, models resembling so-called "neural networks" (e.g., Grossberg, 1986; Govindarajan, et al., 1994; Marshall, 1990; Anderson, 1994) – have made it possible to attempt direct simulation of auditory pattern recognition. However, the usefulness of computational modeling of audition depends directly on the appropriateness of the tasks the model is asked to perform. As we will show below, models of pattern recognition that start off by attempting to group together those acoustic micro-events that regularly occur together have great advantages over those that attempt to simply buffer everything that happens.

What is a Sound Pattern?

What is the primary task of an auditory system? Our real-world acoustic environment contains many events that occur as sequences over time. Aside from Jeffress' very basic questions, one task is to use sound to identify types of events, that is, to recognize something that has occurred before. It is the task of identifying the most pattern-like events, the ones that relate the present to our previous experience with the world. What basic mechanisms would make this possible? One popular guess by psychologists and lay people alike has been to suppose that we first just record audio inputs in some kind of buffer, a short-term auditory memory. For the last 50 years, we have had technology that permits us to make static, spatial records of real acoustic events. When we look at two-dimensional hard copies of these waveforms, or at three-dimensional sound spectrograms, we see "auditory patterns". But we must be careful. The theoretical determination of what constitutes an actual "pattern" for an animal is rather tricky. In fact, we believe it is impossible to define the notion of a pattern without assuming a particular theory about how one can be recognized.

To see why the definition of a pattern is such a problem, imagine that we record two seconds of sound from an elementary school cafeteria during lunch, and make a sound spectrogram (a plot of frequency and intensity as a function of time) of these 2 seconds. The resulting picture will obviously be a complex spatial pattern with many kinds of structure visible in it. The critical question is whether the original two seconds of sound in the cafeteria constitute a "pattern" in a useful sense. It was a record of a historical event, but was it a pattern? Surely we should say that before it was recorded it was not a "pattern". It was merely the accidental co-occurrence of several acoustic events that *are* genuine patterns: clattering dishes, squeaking chairs, children talking, running feet, air conditioners, etc. Each of these alone is a genuine pattern since they recur in much the same form on many occasions. Various individual patterns on this recording obviously have many different sets of invariances. Thus, the sound of a child skipping across the room would be the same pattern even if performed at a different rate of locomotion. Similarly, the sound of a particular English word, e.g., "baseball", would be invariant not only across a range of speaking rates, but also across differences in voice between a child and a teacher. These differences in invariance properties imply that the definition of specific patterns is problematic. (See Port, 1986, for further discussion of the issue of linguistic invariance).

Of course, the piece of paper containing the sound spectrogram of the complex event has a "pattern" on it in some sense. This object has many invariances. For example, the pattern will remain on the paper indefinitely. We can copy it, or memorize the image well enough to assert that we have seen this particular sound spectrogram before, etc. But in contrast with the spectrogram, it seems clear that the original superimposed sound waves in a particular room at a particular time should *not* be called a pattern. There was a historical event when all those familiar sounds (and perhaps some unfamiliar ones) occurred in a particular temporal collocation. Thus, not everything about an event sequence can be called a "pattern". Although we cannot nail down *a priori* exactly what constitutes a pattern, we do insist that a pattern must have some degree of invariance associated with it. We must not confuse the invariance of a piece of paper with the invariances of a historical event, and, of course, we must not assume that because we find it technologically convenient to make a spatial pattern of sounds, that systems for auditory cognition must do so as well.

In our view, only when specific events are statistically correlated in some way can there be a "pattern". There must be some kind of temporal predictability or causal relationship with another event (e.g., simultaneity, sequentiality, etc.) for the collocation of acoustic features to become a pattern. Some kinds of correlational and causal relations between acoustic events are easy even for primitive learning systems to recognize while others are far more difficult. In fact, some correlations and causally related structures are so difficult that even modern humans, with their access to various cognitive technologies such as mathematics, and mechanical technologies such as audio recorders, have not recognized them yet. This is why the definition of "pattern" cannot be separated completely from the cognitive system that recognizes them.

When auditory perception is approached from this point of view, the phenomenon of auditory stream segregation (Bregman, 1992) is just what should be expected for an animal that lives in an environment in which independent patterns may overlap to varying degrees. From an ethological point of view, a major task for an auditory system is to detect and use as much information about the actual correlational and causal relationships in sound as possible. It must recognize the invariances of these patterns. At the same time, it must avoid being distracted by the configuration of accidentally superimposed acoustic patterns and events that are, as far as the animal is concerned, completely random. Thus, the patterns that an animal can recognize depend both on the cognitive capabilities of the animal and on the particular contingencies of the environment in which the animal lives.

Open Environments.

The auditory system of a mature animal is always well-customized for its particular acoustic environment. A new pet cat arriving in one's house may at first startle when the refrigerator turns on or the hot water pipes creak. It soon learns to ignore these sounds but will attend closely to subtle sounds that happen to be correlated with feeding, such as opening of the refrigerator door, or the clink of its food dish.

A major goal of auditory research should be to understand how an auditory system can function in an "open environment". By an *open environment* we mean, first, a context in which there are many events and subevents that recur with various probabilities such that only some relationships are temporally linked to each other as "patterns". Second, the distribution of these patterns and pattern components changes slowly over time. New ones will appear and familiar ones will sometimes change their internal structure or disappear.

Events in an open acoustic environment will recur with a variety of statistical distributions making some of them easier to learn than others. Any natural or artificial perceptual system will be able to make use of only some correlations. In addition, the reinforcement properties of the sounds will vary greatly as well. Darwinian principles will assure that selection has major effects on the tendency of any particular pattern to be learned or adapted to.

One approach to research on hearing is to simulate an auditory system that can support behavior through continuous adaptation in an open acoustic environment. Most of the time, when a spectro-temporal event occurs in such an environment, it is either some known pattern or an accidental superposition or sequential configuration of known patterns. This problem is ubiquitous among animals in any environment. As an example, imagine a "forest environment" that contains three species of vocalizing monkeys, 10 species of singing birds, 10 sound-generating insects plus occasional rain, thunder and steady wind. One or more of these sound sources may produce a sound at any time, and each has a certain probability of occurrence. Still, some new or rare animal might make noise in this region of forest at any time and should be detected at first as something novel, then be gradually learned and adapted to as one of the patterns that occur from time to time. This is important whether the new animal happens to be potential prey, in which case it should be followed, whether it is a potential predator, in which case it must be hidden from, or whether it is one of the set of ignorable animals.

For a perceptual system to deal with an open environment, it requires the ability to recognize familiar patterns despite adventitious collocation with other patterns, plus capability for continuous slow adaptation to changes in the environment (Grossberg, 1987). Adaptive perceptual processing in such an environment requires recognizing the statistical regularities in stimulation without any explicit teaching.

Our interest is in how temporal aspects of open acoustic environments impose constraints on practical models for auditory cognition. Thus far, we have restricted our investigation to complex non-speech sound patterns that range in duration from roughly 1/10 sec up to several seconds, plus sequencing of these patterns in a periodic structure. We avoid the complexities of binaural hearing and focus on monaural information. We assume an ear-like spectral analysis system that analyzes the temporal microstructure of the acoustic waveform into spatially parallel spectral channels. One useful approach to research on auditory cognition may be this kind of "biologically-inspired engineering" – building simulated auditory systems that customize themselves for simplified acoustic environments that are nevertheless open. Such a program may help us see what methods might effectively solve the problem of information gathering in any fully open environment.

Artificial Acoustic Environments.

When considering natural or artificial perceptual systems, it is useful to consider the entire experience of the system, whether in training or under test, as its "stimulus environment". For example, a neural network that is presented with various acoustic"training stimuli" and then with "test stimuli" could be said to have been installed in a particular acoustic environment – albeit a peculiar one. The network adapts to stimuli for a while and then is evaluated on further stimuli. Any animal is similarly adapted to a specific acoustic environment or set of environments. In the training and testing of a neural network model for speech recognition or other pattern recognition tasks, the typical acoustic environment is far from open in our sense. Laboratory constructed training and testing environments consist of nothing more than a set of targets and distractors that are presented one at a time with a very high signal/noise ratio. If a target were to overlap with another target, or if there were considerable background noise, most artificial systems would fail to recognize their target patterns. More naturalistic presentation would allow for overlap of targets and for unpredictable time of occurrence of the targets (cf. the psychophysical "method of free response" which incorporates some of these uncertainties (Watson and Nichols, 1976)).

The primary function of an auditory system is to recognize patterns, that is, information about significant events in the environment. If the environment is open, then the statistical properties of "noise patterns" are every bit as important for successful function as the statistical properties of the patterns that are thought of as targets.

General properties of an open acoustic environment are due to the nature of sound itself (cf., Yost, 1992). First, sound waves normally interact as a three-dimensional linear, additive system. The sounds generated by independent events occurring at different locations are simply superimposed on one another. Separating independent sound sources is thus a ubiquitous problem to be solved (cf. Govindarajan, et al., 1994). According to Bregman (1990), separation of distinct "auditory objects" is based on two processes. *Schema-based segregation* of auditory streams is based on familiarity with specific patterns, while *primary auditory stream segregation* is based on spectrotemporal neighborhoods of primitive acoustic features. In our view, there can be no clear distinction between these two cases. Both processes for pattern extraction are based on statistical regularities in stimulation.

A second major feature of our environment is that events can be periodic over a wide range of time scales – from microseconds to years. Further, an event that is repeated may be as simple as a sinusoidal pressure change or as rich as an arbitrarily complex pattern in frequency and time. In addition, the rate of periodicity may itself vary over time, e.g., faster and slower footsteps. It would be very useful for an auditory system to be able to exploit such regularities by "locking in" to periodically repeating patterns of whatever shape – including periodic squeaks, chirps, miaows, etc. – at a broad range of periods. Our environment is cluttered with rate-varying sound patterns that are periodic in complex ways. Competing periodicities of events that are spectrotemporally complex also need to be differentiated.

Given these properties of the acoustic environments in which animals live, how might research on actual cognitive auditory systems be conducted? Obviously, any real environment – whether it is the hallway outside my lab, my back yard, or the Amazon jungle – will be much too complex to work with initially. A practical method for research would be to use a sound synthesis programming language, such as *C-sound* (Vercoe, 1986) to produce a specific inventory of complex sounds that occur independently with specified statistical constraints. Such an almost infinitely variable environment could exhibit as much complexity as we think our simulated auditory systems can handle. Such an environment generator could provide a flexible test bed for evaluation of self-organizing auditory perceptual systems – systems that attempt to make sense of a rich acoustic environment. Research on such simulated acoustic environments is likely to be very informative about the natural function of the auditory systems of terrestrial and marine animals.

Thus far, our goal has been to clarify the problems in audition faced by an animal living in an open environment, and the consequences of these problems for models of hearing. The perceptual system needs to know about patterns much more than about raw auditory events. It seems that familiarity with whatever patterns occur frequently would be useful, but that only a subset of patterns should "attract attention" by demanding a behavioral response. Examples of important patterns would be those that elicit prey-related and predator-related behaviors. We propose that a practical method for research on a fully functioning auditory system is to construct artificial, open auditory environments and then to test various systems for auditory cognition.

One specific interest in this paper is the neglected problem of how timing information in auditory patterns can be extracted and used by auditory pattern recognizers. Thus the next section will address the issue of what kind of temporal information can be perceived by real auditory systems and exploited by simulated ones.

TEMPORAL ASPECTS OF AUDITORY COGNITION

One of the most fundamental but overlooked problems of general perception is the treatment of patterns that extend over time. The problem arises in all modalities, but is most striking in audition, since steady state sounds are rare. How is information about the temporal extent of an event extracted and used? The traditional engineering approach to recognizing temporal patterns has been to first use an "assignment clock" to label raw stimulus events as having occurred at a particular point in time (see Port, Cummins and McAuley, 1995, for further discussion). The time labeling is typically assumed to take place in advance of any pattern recognition – it merely serves the purpose of representing the temporal distribution of sound in spatial terms. After generation of a spatial representation, pattern analysis begins. This is the strategy traditionally used in speech recognizers, like Hearsay-II (Lesser et al., 1975) and dynamic time-warping models (Sankoff and Kruskal, 1983). Essentially the same idea underlies the psychological theories of "short term auditory store" (Massaro, 1972; Baddeley, 1992) and "precategorical acoustic store" (Crowder and Morton, 1969). Unfortunately, this strategy is biologically infeasible, since, as far as is known, animals lack any "assignment clock" (Port et al., 1995) and lack any spatial representation of time for raw stimulation. More importantly, as we saw above, it is informationally inappropriate since animals have little use for all the temporal detail of a tape-recorder memory. Mere events are not necessarily very informative. Animals do need to detect information in the form of patterns, and they need to identify patterns at the earliest possible point in time, not after the pattern is completed.

If there is no assignment clock independent of the stimulus pattern, how can the temporal structure of a complex pattern be extracted? The answer is that there is no general answer, only a collection of different answers. In this section, we specify some basic kinds of temporal information that can arise in an open environment but which do not require an assignment clock.

The description of temporal information on any particular time scale seems to depend on an inventory of "features" extracted at a shorter time scale that have some kind of spatial description. A spatial specification is critical because neural systems must work without physically moving any object from one place to another. They can only move excitation and inhibition along hardwired, slow physical channels. It seems that to define a temporal pattern, one must logically have spatially specified elements first. Then changes in these elements over time can be used to define patterns in time. Auditory systems, for example, extract some spatial description of spectral shape that can be used as a basis for defining temporal cues. Thus, some self-organizing or adaptive process gives rise to a set of spatially distributed elements that correspond to spectral shapes, e.g., specific cells or populations of cells whose activity specifies their spectral identity. One example of a simple spatiotemporal feature is a "spectral peak" – a spectral region of relatively greater energy. Another possible kind of spectral primitive is something resembling the "spectral profile" of Green (1976). Profiles are level-invariant spectral shapes defined across a broad frequency band that take subjects many hours of experience to learn well enough to achieve asymptotic performance. Methods for implementing some of these spectral shape descriptions using neurally plausible mechanisms are well known. For example, we know that lateral inhibition across a set of nodes representing energy levels in a spectrum can quickly change a spectrum with a few slight prominences into a display with sharp peaks that stand out with respect to neighboring frequencies. This process can be seen in adaptive resonance models (Grossberg, 1987) and many others. There may also be methods for learning profile-like spectral descriptors.

Given that there are short-duration spatiotemporal primitives, temporal information on a longer time scale can be obtained from fast-acting lateral connections across the spatial map. We will describe a general method for controlling the dynamics of such lateral connections to achieve recognition of various kinds of spectral patterns that exist over time. We will also describe methods for rhythm recognition that are not dependent on spatial maps at all. Because they imply rather different kinds of dynamical mechanisms, we classify auditory information into three types (see Figure 1): *acoustic motion*, *rhythm*, and *sequences*. Most sounds that display changes over time seem to fall into one or more of these classes. Each of these three types of information implies a particular kind of dynamic mechanism for system adaptation. In this section, we will sketch some examples of each of the three types of information. In the following section, we will suggest some general characteristics of the dynamical mechanisms for extraction of the information.

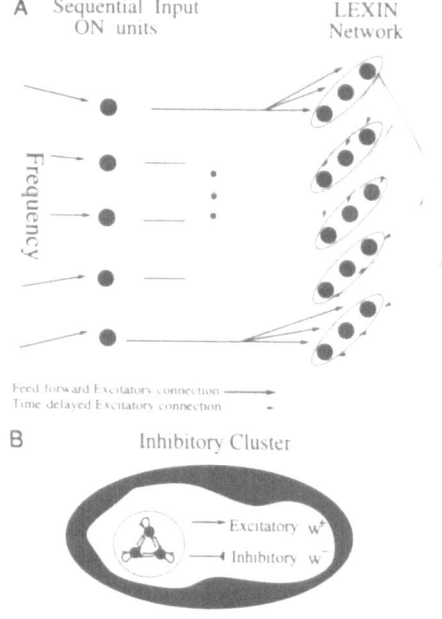

Figure 1. Three basic types of temporal information important for biological systems. The simplest is detection of auditory transients, or acoustic motion. The second requires entrainment to a periodic pattern so that events can be assigned a particular phase angle as a measure of time. The third is location of an item in a particular position in a serial sequence.

Acoustic Motion.

Given a set of spectral elements with spatial identities, such as relative energy peaks, the lateral connections between the peak identifying units can be used to construct a mechanism to identify change – and rate of change – of the peak location across the spectrum. This system could detect a change in the frequency of a salient spectral event over a brief period of time, or a change in the intensity of a frequency component over a brief period of time. The mechanisms for detecting spectral change are not especially difficult to design (Grossberg, 1976; Smythe, 1988; Grossberg and Rudd, 1992), and neural circuits capable of picking up this kind of information have been discovered in the sensory systems of many animals (Delgutte and Kiang, 1984).

Detection of a moving spectral peak, or other spectral feature, over time requires two steps. First, the spectral feature must be extracted. Second, either inhibition or excitation of similar feature detectors in neighboring frequency regions must occur after an appropriate time delay (Grossberg and Rudd, Anderson, 1994). For example, using cells with variable time delays, feature detectors for upward and downward moving peaks can be constructed.

Just as features of static spectral shape distributed over time give rise to dynamic auditory features such as a "tweet" or a "chewp", the "tweet" and "chewp" features potentially provide the building blocks of higher-level and longer-duration spectro-temporal patterns.

Rhythm.

Recognition of a rhythm is more complex. A minimal rhythm can be defined as a series of regularly repeated similar events. When such a *primary series* of regular prominent events occurs, other, nonprimary events can be localized in time by measurement of their phase angle with respect to the primary series. Notice that Western musical notation for time is, in part, a notational implementation for the phase angles of note onsets and offsets with respect to the musical measure. For example, in two measures of 4/4 time, a series of three half notes begin at 0, π and 2π radians and the third half note ends at 3π ($= \pi$ of the second measure). Like musical rhythms, phase angle is a time measure that is invariant under a change in rate. In general, rhythmic sounds occur whenever sufficiently large physical objects oscillate mechanically. Rhythmic events arise in the environment due to footfalls, swaying branches, swinging animal limbs, bouncing objects, etc. Reciprocal engines and other kinds of machinery also tend to produce rhythmically structured patterns that are invariant across changes in rate. Within the nervous system, it seems that cells can oscillate at any of a broad range of rates. When a complex event occurs, such as the periodic production of footfalls in quadruped locomotion, the relative durations of certain internal events can be measured by how much phase angle they occupy with respect to an underlying beat duration – where a beat is a psychological rate that is easily tapped to with a finger or a hand.

While listening to a horse, one can tell a trot from a gallop, even when they occur at the same period, by the phase-angle pattern of the footfalls. Thus, it seems that musical notation might be a practical notation for representing various animal gaits. Instead of soprano, alto, tenor, and bass, we might have left-front, right-front, left-rear, and right-rear, with the entire gait cycle defining a measure. A waltz differs from a foxtrot or a tango in the number of beats per measure and the phase angle at which various event types occur within the primary metrical cycle.

The simplest mechanism that can "recognize" periodic events would be a mechanical resonator - a device that "rings" or responds strongly at a particular frequency. Thus, one way to recognize simple periodic events is to construct a set of pretuned mechanical oscillators, like the undamped strings of a harp or piano. Each string responds strongly when mechanically moved at its natural frequency. If you utter a vowel loudly near a piano with the dampers lifted, certain strings will respond strongly. Change the vowel, and different strings become active. There are several basic limitations to this method, however. First, mechanical oscillators integrate over many cycles, so they cannot exploit the internal phase angles of events within the metrical period. Second, each resonator would have to be independent of the others. Third, a simple acoustic resonator can only "recognize" a single kind of input: oscillations that come at its own resonant frequency. Thus, if the frequency of the excitation changes, a new resonator becomes sensitive. To recognize higher-level identities across a range of rates, a more sophisticated mechanism is required.

A system built of oscillators that can change their resonant frequency would be useful for many purposes – especially if such oscillators lie upstream from detectors for more primitive events such as acoustic motion. By an *adaptive oscillator*, we mean a unit or set of units that adjusts its natural resonant frequency to match periodic components of input signals – that is, a unit that will entrain itself to any frequency over some range. Such a device can become tuned to a pattern at one rate and then remain tuned to the periodicities in the pattern as the rate changes. In this way, since the same unit remains active when rate changes, the identity of the pattern across the change in rate is explicitly retained.

A mechanism for constructing such an adaptive oscillator will be described below. To describe a simple problem that cannot be handled by a bank of resonators or filters, let us imagine the task of detecting the presence of a repeated series of rising glissandos, something that sounds like "tweet, tweet, tweet...". Let us assume that there is a need to distinguish this series from a competing sequence of falling glissandos over the same frequency range, something that sound like "chewp, chewp, chewp...". In order to track the "tweet" sequence and not be confused by the spectrally and temporally overlapping "chewp" sequence, one would first need to detect a single tweet and use the output of the "tweet-detector" to excite an adaptive oscillator which then quickly predicts when subsequent tweets will occur. The output of this oscillator, integrated over time, indicates that a series of "tweets" is under way. The "chewps" will similarly be identified by a "chewp"-detecting unit and will be ignored by the "tweet" subsystem. This mechanism solves a very difficult perceptual task that could not be handled by a model that simply integrated spectral information over time.

Periodic rhythmic patterns are only part of the story. More abstract temporal patterns are invariant not just across uniform changes of rate, but across highly nonlinear changes in duration. Sometimes we need to recognize patterns in time that are defined only by their serial order, that is, sequences.

Sequences.

When most people think of "temporal pattern recognition", they may think immediately of sequence recognition, that is, the serial order of events without concern about their relative spacing in real time. Linguists, for example, have always viewed the problem of speech and language perception as strictly a sequence problem. Words and phonemes are assumed to come in sequence such that the relative spacing of the words in time does not affect the meaning of a sentence. Absolute time properties and rhythmic properties are assumed to have only marginal cognitive relevance. We suspect that linguists underestimate the importance of rhythm and other kinds of timing information in human language (Port and Dalby, 1982; Port et al., 1987; Port and Cummins, 1992), but we do not deny that there are some aspects of language that are adequately described as sequences.

Ordered events define the most abstract kind of time, since it is the form of temporal measurement to which the most powerful distorting transformations can be applied without significant effects (Port, 1986). For example, if one is only interested in the order of the events in ABC, then, by hypothesis, the duration of the event called A and the temporal spacing between A and B is ignored. Real sequence recognition is particularly difficult for a dynamical or network-like system, in contrast to computational systems, because sequence recognition requires very powerful processing mechanisms that are relatively insensitive to temporal detail – something that a dynamical system like a nervous system will naturally find difficult to do.

Logically, differentiating one sequence from another requires saving up either the raw input signal or some description of it for a sufficient duration to differentiate the pattern from its competitors. Thus the set of target patterns themselves determine the amount of memory required. How can position in a sequence be found? If the input sequence arrives with the pattern elements already distinguished, then the length of the pattern is simply the number of events: AABB is longer than AB. Of course, many natural patterns that are "the same" may appear at a range of rates. This is true of human speech, familiar footsteps, and many other periodic sounds including music. Recognition of such patterns presents serious difficulties for sequence description since the amount and form of memory required depends on the rate of presentation of the pattern. If memory is to be measured in terms of intrinsic pattern elements rather than in absolute time, then the system must be able to recognize each pattern element as it arrives – not a simple matter. Thus, sequence recognition in real time poses many fundamental problems. When absolute durations are not useful, then an *a priori* clock that measures seconds is also not useful. If there is no repeated series of similar events, the signal cannot serve as its own clock.

The three types of temporal pattern information just described are conceptually quite distinct, however, the mechanisms for their analysis may not be naturally separable. We suspect that most of the temporal information used by auditory systems falls into one of these three types. Particular sound patterns, including speech, birdsong, other animal sounds, natural ecological events, etc, can be defined by combining several of these classes of information. These three main kinds of temporal pattern information suggest there may be three general types of recognition architecture, all of which depend on dynamics for their proper function. Some illustrative mechanisms that could plausibly be implemented in neural machinery will be shown in the next two sections.

MOTION AND SEQUENCE: THE "LEXIN" MODEL

What kind of system could recognize particular patterns of auditory motion? What is necessary in order for a system that begins with spatial layout of frequency components to recognize a spectrotemporal sequence? In this section we will review some general mechanisms for pattern recognition and show how they work. There are now many systems that are capable of detecting the motion of a stimulus pattern primitive such as a dot for vision, or a spectral peak in audition, across a spatial map of points on the retina or frequencies in a tonotopic auditory map. It is known that the brain contains visual maps (e.g., Barlow and Levick, 1965) and auditory frequency maps (e.g., Whitfield and Evans, 1965). Computational models of systems that detect direction and rate of change have been available for some time (e.g., Wilson, 1989, Smythe, 1987, Grossberg and Rudd, 1992).

We have developed a model that exhibits many useful properties for recognition of motion patterns. It exhibits some of the features that should be expected in a fully functional model of auditory motion and auditory sequences. For example, it can learn to recognize glissando-like patterns, given that they occur frequently in the system's auditory environment. Although the equations are continuous, the stimuli are defined in discrete time, and the model was implemented in discrete time.

One essential property of the model is that it self-organizes to recognize the frequent spatiotemporal patterns that are presented to it. This is important because it should be assumed that real auditory environments are in constant flux. A perceptual system must constantly adapt to long term features of the environment. Our method for adaptation follows principles developed by Grossberg, Carpenter and others (e.g., Grossberg, 1987; Carpenter and Grossberg, 1987).

The Lexin Model

Recent progress in modeling the detection of motion in vision has demonstrated how a network that operates in an environment of moving, spatially coherent objects, can "learn" to encode direction and speed of motion (Marshall, 1990; Marshall, 1995; Grossberg and Rudd, 1992). The model we will describe is inspired by Marshall's *Exin* networks (Marshall, 1990) and is derived from variants by Sven Anderson (Anderson, 1992, Anderson, 1994). The model is illustrated schematically in Figure 2. On the left in Figure 2A, there is a tonotopic array – the output of a frequency analyzer (the ear) – that receives external stimulation. For each of the frequency units there is a cluster of units that are excited by that frequency unit. The units within each cluster inhibit each other in proportion to their activation. Cluster dynamics is competitive; that is, if one unit in a cluster has higher activation level than the others, it will quickly suppress other units via inhibition. So far, we have discussed only external input that tonotopically projects to all clusters and is the same for all units within a cluster. However, units also receive excitation from the units of neighboring clusters, as shown on the right in Figure 2A. These lateral excitatory connections to all the units in neighboring tonotopic clusters, up to some limited distance, are critical because if one unit in a cluster that receives external stimulation simultaneously receives the most lateral excitation, then it will be able to inhibit its partners more that it is inhibited by them – and will thus win the competition. The pattern of weights on the lateral connections determines which sequential patterns will be recognized by the units. Because lateral excitation exhibits some amount of time delay, temporal patterns of frequencies that rise and fall at particular rates can be distinguished if the weights on the lateral connections are chosen or set appropriately.

Initially, the lateral excitatory connections between units of the motion detection layer are connected with small connection strengths in a distribution that favors small inter-unit distances. During exposure to sequential patterns, these connections organize themselves to represent the spatio-temporal correlations between lateral and afferent excitation, that is, correlations between inputs at the frequency present during the current time step with frequencies that have occurred in the recent past. Within a cluster, the unit that receives the most afferent *and* lateral excitation will suppress all the other units in its cluster. It will, as a result, "learn" more strongly than the other units in its cluster. Competition among the units in a cluster ensures that they will eventually encode different patterns that involve specific correlations between the current input and some other recent input. The activation level of

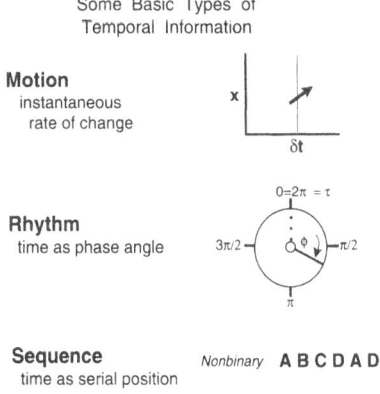

Figure 2. **A** Layer of units responsible for different types of sequence recognition. Actual networks usually consist of a larger number of clusters and units per cluster. **B.** Inhibitory clusters are enclosed by an ellipse and illustrated in the exploded view. All excitatory connections from a single unit of one cluster are shown.

each unit in the inhibitory cluster obeys a differential equation consisting of excitatory and inhibitory terms. Equation parameters are chosen to ensure that below a certain threshold all units in a cluster are active, but that above that threshold the units will compete in a way that allows only one member of the cluster to "win". The selection of the most strongly activated unit in a cluster must be rather rapid relative to the duration of the pattern to be recognized.

For unit v_j in cluster C_l,

$$\tau \dot{x}_j = -\gamma x_j + f[Bx_j{}^3(t) + \sum_i w_{ij}^+ x_i(t-k) - \sum_{i \in C_l} w_{ij}^- x_i{}^2(t) + \Theta_j + I_j(t)], \qquad (1)$$

where $I_j(t)$ is external input to v_j, Θ_j is a bias, and f is the linear threshold function,

$$f(\xi) = \begin{cases} 0 & \xi \le 0 \\ \xi & 0 < \xi < 1 \\ 1 & \xi \ge 1. \end{cases} \qquad (2)$$

The parameters τ and γ are the time constant and the decay rate of the unit, respectively. The parameter B scales unit self-excitation, and is balanced against other parameters to achieve a threshold above which competition exists. These parameters are fixed across all units. The weights on the lateral excitatory connections are w_{ij}^+ from unit i (not equal to j) to j. The weights w_{ij}^- are the inhibitory connection strengths from within each cluster. Inhibitory connections were all set to a single value. Notice that the w^- term uses $x_i(t)$ while w^+ uses $x_i(t-k)$, where k was set to 18 time steps in the first two simulations below. This means that intra-cluster competition is based on current excitation combined with lateral excitation from 18 time steps in the past. When there is strong input activation, the "winning" unit of a cluster quickly saturates and suppresses the other units, whereas at low input activation values, all units in a cluster will remain active for considerably longer and transmit that excitation laterally to other clusters. It can be seen that the excitatory and inhibitory connections play rather different roles. Lateral excitation serves to carry information about earlier events in a pattern, while inhibition forces the system to make some sort of choice.

The behavior desired of the cluster units implies a set of constraints on parameter choices that will guarantee (1) decay of unit activation x_j to zero in the absence of external stimulation, (2) competition within a cluster when stimulation exceeds a threshold, and (3) low level activation that quickly decays when stimulation is below a minimum threshold. Other choices of model parameters determine the rate at which competition is to occur.

One of the most important features of this model is that it adapts to its environment over many trials. The units in each cluster eventually specialize for statistically common spectro-temporal events. This adaptation is achieved by self-organized "learning" of excitatory connection weights between clusters. The equation for adaptation of excitatory connection strengths is based on competitive learning as developed by Carpenter and Grossberg (1987) and modified by Marshall (1990). The network "learns" incrementally on each simulation step. During the course of "learning", the connection strengths encode the spatio-temporal correlations that occur when delayed lateral excitation is strongly correlated with afferent activation from input to the motion detection layer. Because shorter connections across the frequency scale are initially stronger, units are more likely to encode local transition information, although this tendency may be overcome after sufficient adaptation.

Connection strengths come to reflect the results of unit competition and also enhance its effect. Thus, whenever a unit dominates other units in its cluster during presentation of a particular pattern, competitive learning guarantees that it will encode this pattern slightly better on future trials. This behavior can be tested by studying how the system adapts to sets of simple input patterns – as illustrated in the following simulations.

Selectivity for Direction of Motion: Simulation 1

The self-organizing properties of the model can be demonstrated most clearly when it is exposed to an environment containing simple stimuli that sweep across the "receptive fields" of all the frequency units. The patterns used for these simulations are defined in discrete time. Our network contained 10 frequency-sensitive input units that were connected to 10 clusters of 3 units each. The stimuli rose or fell monotonically at 3 different rates, corresponding to 8, 9 or 10 time steps for each frequency step. Figure 3 illustrates four of the patterns used. Partial frequency series were also presented to mitigate edge effects. Individual frequency components were produced by activating an input unit for 5 time steps followed by a variable duration null ("quiet") interval.

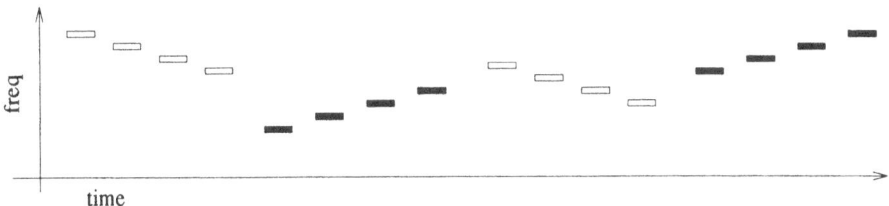

Figure 3. Examples of stimuli used to induce motion sensitivity. Each rising or falling series of frequency steps ("sweeps") has a characteristic slope. The network was presented with only one "sweep" series at a time.

Prior to "learning", no units in the network exhibited a suprathreshold response to any pattern. After about 200 "sweep" stimuli were randomly presented, all of the clusters in the network had "selected" one unit that responded to rising patterns and one that responded to falling patterns. The third unit typically remained uncommitted. We can study the results of this "learning" by looking at the response of each unit to a long rising pattern followed by a long falling pattern, as shown in Figure 4. The response of the first six clusters to upward and downward patterns is shown. To the right of the activation traces, units have been labeled with arrows to indicate their direction sensitivity, either rising or falling. Close examination reveals that during presentation of the first component of the rising pattern, all of the units in Cluster 1 were weakly activated. When the second frequency was presented, one unit in Cluster 2 showed greater activity and the other two were somewhat suppressed. When the third frequency occurred, the middle unit of Cluster 3 was able to quickly shut off the other units in Cluster 3.

The explanation for this behavior lies in the connection weights from units in neighboring frequency clusters. These reveal how the units were able to exhibit complementary sensitivities to direction of motion. Figure 5 displays the connection weights between individual units in the clusters. Connection strengths are proportional to the areas enclosed

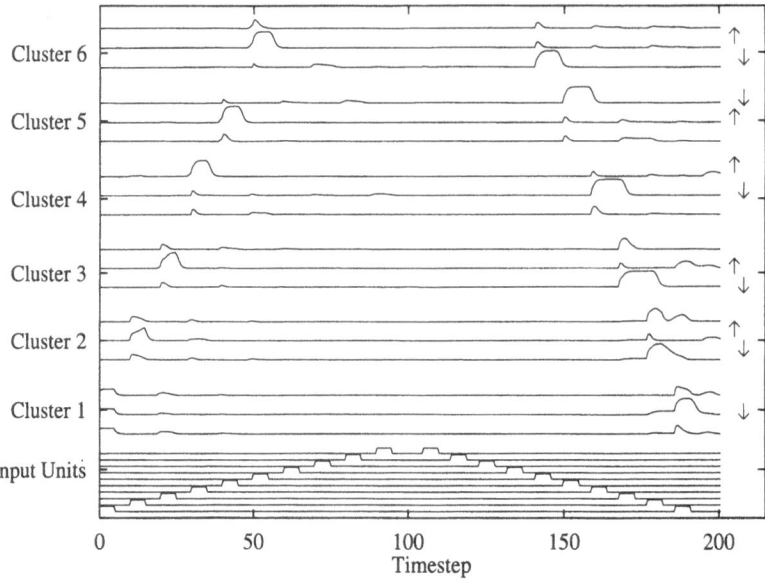

Figure 4. Time-series plot of unit activations for the first 6 (of 10) clusters during presentation of a rising and falling stimulus having slope 1/10 and -1/9, going up and down. Activation levels in the 10 input units appear along the bottom of the figure ordered from low to high frequency. The clusters also are numbered from low to high frequency. Arrows to the right of the traces indicate whether units respond selectively to rising or falling spectral motion.

by the boxes. Excitatory connections are shown as open boxes and inhibitory connections as filled boxes. The row and column numbers correspond to the source and target units.

Tracing the pattern of activation for a rising input pattern reveals how the response of the network becomes increasingly "confident" as a pattern unfolds, and reveals how the recognition layer is able to link activations to form a unique representational "chain". When Units 1-3 (Cluster 1, the cluster that responds to the lowest input frequency) receive afferent stimulation, all 3 units become moderately activated. Because the threshold for competition is not exceeded, none of the units "win" (lower left corner of Figure 4). From the weights diagram (Figure 5), we see that the activation of "source" Units 1-3 causes Unit 5 in Cluster 2 to become more active than Units 4 and 6 – this is represented by the larger open squares under destination Unit 5 of Cluster 2. Because Unit 5 receives some lateral excitation from Units 1-3 in addition to its afferent input, the competitive dynamics of Cluster 2 causes Unit 5 to dominate Units 4-6 even more. Unit 5 is then able to excite Unit 8 in Cluster 3 because 8 is more strongly excited by Units 4-6 than either Unit 7 or 9 are.

The structure of the chaining effect is also apparent in the pattern of connection strengths in the column corresponding to Unit 12. This unit receives its strongest connectivity from Unit 8, thus forming a chain of units that respond optimally to rising stimuli. A similar argument holds for the responses of Units 14 and 17, each of which responds best to rising stimuli. The competitive learning rule allowed this network to structure itself in just 200 exposures so that it could "identify" and differentiate upward and downward patterns over a modest range of presentation rates. In this way, the network is capable of making certain generalizations about its auditory environment without explicit teaching.

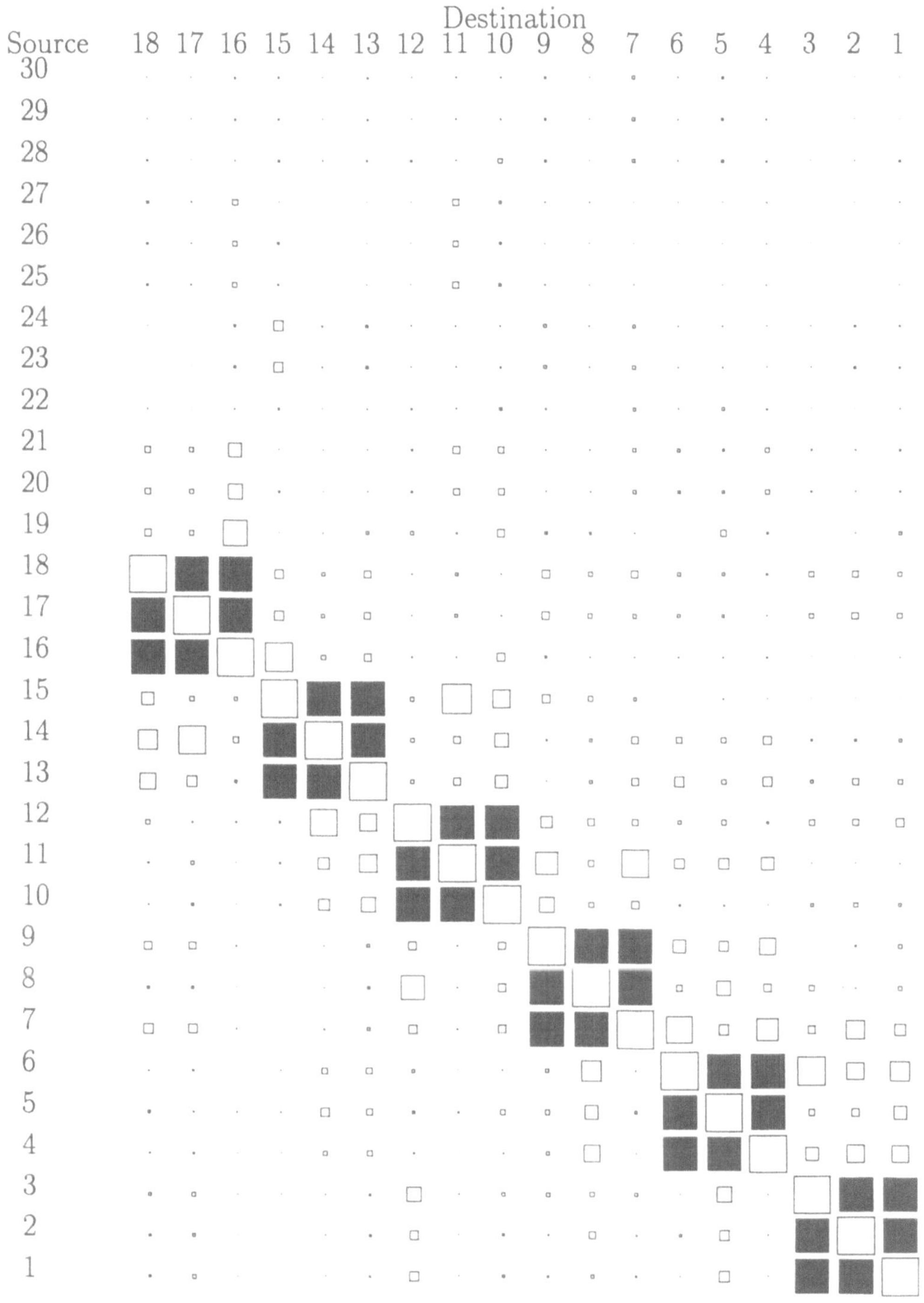
Destination
Source
18 17 16 15 14 13 12 11 10 9 8 7 6 5 4 3 2 1
30 29 28 27 26 25 24 23 22 21 20 19 18 17 16 15 14 13 12 11 10 9 8 7 6 5 4 3 2 1

Figure 5. Lateral connection strengths. Only the connections to the first 6 clusters are shown, with each three units corresponding to a cluster. Excitatory connections are indicated by open boxes; inhibitory connections are indicated by filled boxes. The area of each box is proportional to the "learned" magnitude of the connection strength w_{ij^+}. The row and column indices specify the source and destination units. Thus, each column contains a graphic display of all lateral connections to the unit labeled by that column index.

Complex Sequence Detectors: Simulation 2

Can this selectivity now be exploited to identify patterns at the next level of complexity? Can the same learning mechanisms be used to tune specific units for more complex sequential patterns – including patterns that involve frequency jumps and patterns that terminate in the same frequency but differ at some earlier point in time? For the next simulation, we prepared an environment that contained only the frequency patterns that can be represented as the sequences 2-1-5, 3-1-5 and 4-1-5.

The chaining solution to motion detection described above is a viable means for complex sequence recognition that does not employ extensive short-term memory for raw, unprocessed information. A network of 5 input units (for the 5 frequency channels) and 5 clusters of 3 units each was presented with the 3 different complex sequences. After approximately 600 randomly ordered sequence presentations, the response of the network to the three sequences settled to that shown in Figure 6. The three input patterns appear strung together along the bottom as 2-1-5-3-1-5-4-1-5. Even though all of the patterns ended in 5, note that a different unit in Cluster 5 responds maximally to each of the three sequences.

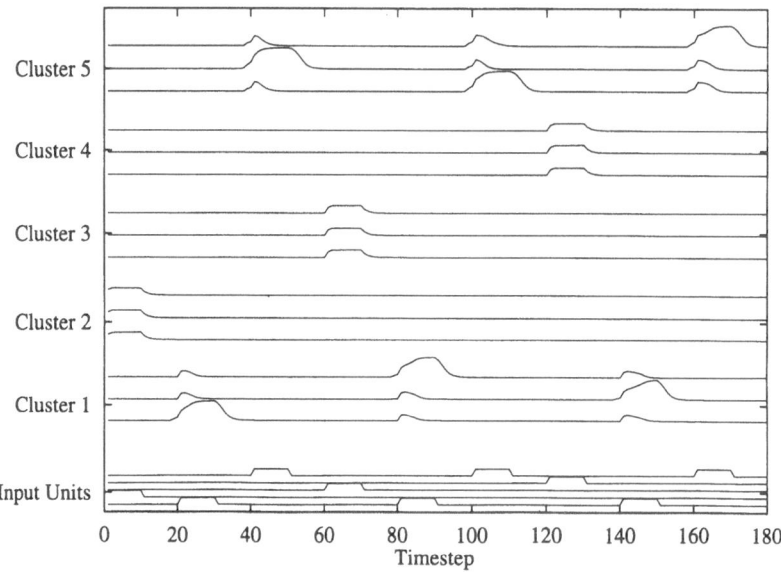

Figure 6. Responses of sequence identification clusters to three sequences: 2-1-5, 3-1-5, and 4-1-5, presented as a continuous string. The first input of each pattern yielded only a subthreshold response, while the third element in each pattern produced the strongest and longest duration response.

By induction, we may speculate that networks of this type can correctly identify complex sequences of arbitrary degree, as long as each cluster is large enough to have a unit dedicated to recognition of each "basic pattern" that is statistically common in the input. The chaining solution to sequence recognition requires no more memory than its longest time-delay. Complex patterns are differentiated at the point in time where they first differ – thereafter,the representations are nearly orthogonal since only one unit among all units in a cluster is active at any time.

Sequence-Selective Units: Simulation 3

An apparent shortcoming of the chaining solution is its dependence on the fixed duration delays along the lateral connections of the cluster array, or sequence recognition layer. If a single component is omitted for some reason such as noise from another source, this change might be sufficient to cause the response to a pattern to be completely destroyed. Worse still, although strong competition is required for the early stages of adaptation, too much competition can result in the false-alarm recognition of a whole sequence from as few as two of its components. For example, the representation of 1-2-3-4 may be indistinguishable from an input of 3-4. There are several ways to increase memory for context, but the simplest is to employ a range of time delays rather than a fixed delay for all connections. Since the competitive adaptation rule we used enforces sharing of total connection strength among all connections to a unit, a similar process can be directly applied to units linked by connections having several different delays. This simple change implements the learning of spatio-temporal correlations at multiple time scales across multiple units.

The enhanced contextual sensitivity of a network that employs as few as two time delays is demonstrated in Simulation 3. A network of 7 frequency (input) units and 7 clusters of 4 units, all of which had complete sets of excitatory connections at delays of both and time steps, was exposed to the four patterns drawn in Figure 7. These patterns are somewhat more realistic than the ones used in the previous simulation because they are of longer duration.

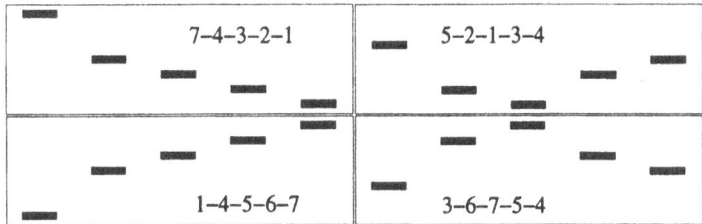

Figure 7. Four patterns used to examine unit selectivity. Like some auditory patterns, they are approximately spectrally continuous since frequencies change to near neighbors rather than jumping around. Each pattern is labeled with its numeric description.

After about 100 sequence presentations, most of the clusters had developed a unique representational node for sub-sequences of the long 5-step patterns – quite similar to those observed in Simulations 1 and 2. This can be illustrated by looking at the behavior of a single node, unit 19, which happened to be the recognition node for the pattern 5-2-1-3-4. In order to assess the selectivity of this unit, various sequence fragments that resemble the trained sequence to varying extents were presented.

Each of the 7 frames of Figure 8 is centered over the presentation of a sequence used to activate Unit 19 after training on the 4 patterns. The first frame on the left shows the time course of activation of Unit 19 on presentation of the entire (trained) sequence 5-2-1-3-4. This unit reaches its peak activation for this pattern. Frame 2 illustrates response to simple afferent stimulation by frequency 4 only. Activation remains near 0.3, below the threshold for intracluster competition. The partial sequence shown in the third frame, 3-4, elicits slightly more activation which is then gradually enhanced through intracluster competition. As more context is added (Frames 4 and 5), the unit response nearly equals that for Frame 1, although it is of shorter duration. In the final two frames, single components in the original sequence were replaced by null input. If both the third and fourth components of the pattern were omitted (not shown here), Unit 19 response reached a level of only 0.54, intermediate between its response to 4 and 3-4. This behavior seems close to what a perceptual system ought to do. It exhibits considerable robustness in the face of variability and incompleteness in otherwise familiar temporal patterns.

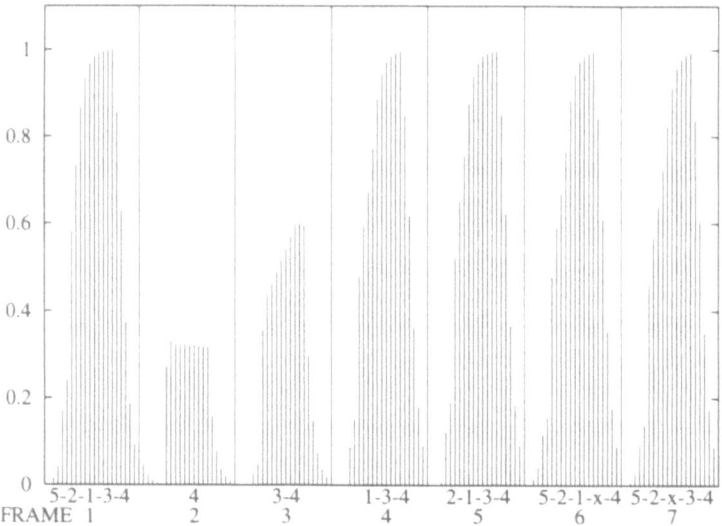

Figure 8. Response of Unit 19 to various sequences ending on the frequency of the afferent input to Unit 19. Each impulse corresponds to one simulation time step. Unit activation level is given on the ordinate. Below each frame, the numerical description of the sequence presented (i.e., the numerical order of active afferent units) is provided. The letter X in a sequence indicates that no input was presented at that time step. The numeral below each sequence descriptor indicates the frame number. For example, frame 2 shows that in the absence of prior input, Unit 19 responds to frequency 4 alone at a level of about 0.3.

Discussion of the Sequence Recognition Model

The most important feature of the Lexin model is its ability to self-organize and thereby adapt to whatever regularities occur in different environments. Recognition of spatial patterns and temporal patterns both require some kind of internal representation, implemented in the current model as the activation of one particular unit. Additionally, self-organization for the analysis of sequential input requires the system to determine when a series of elements constitutes a sequence. That is, it requires actually parsing input sequences. The model we have presented employs competition to achieve this parsing and assignment of an internal representation to patterns that frequently recur. The hard-wired time delays are not ideal, and they seriously limit the temporal resolving power of internal representations. Whether or not a sub-sequence is represented through assignment of a node depends on its frequency of occurrence and on the number of units available in each cluster. As the number of sub-sequences approaches the number of units per cluster, less frequently occurring sub-sequences will fail to acquire a unique internal representation. A question for future research is whether such a system can continuously learn so that it can remain optimized for an environment which itself changes slowly.

The effects of self-organization can be further enhanced if the output of one cluster array provides input to another cluster array having delays on a different and longer time scale. For example, the motion-detection network described in Simulation 1 operates over a very limited time scale, leading to the linearly independent representation of upward and downward frequency "sweeps". A network designed to recognize trill-like sounds would more easily develop "trill-selective" units if it were given the output of the motion-detection layer than if it merely had raw sequences of simple tones. If the nervous system employs matched spectral and temporal scaling, sequential patterns and feature detectors may not be analyzed as unitary objects but rather as constructs arising from several intermediate stages of representation involving longer and longer time scales. Another issue for future research is the extent to which the model can be extended to stimuli having much greater complexity.

Finally, the simulations demonstrate unit selectivity for sequential patterns. We showed that unit response is strongest for complete patterns and becomes weaker as context is removed. In the case where expected stimulation is missing, selectivity still results from non-optimal afferent input enhanced by "learned" intracluster competitive dynamics. The tonotopic layout of the initial input representation is maintained at all higher layers because units respond weakly to afferent activation within their range of frequency tuning. This means that a unit selective for a sequence that ends on a particular component will also respond weakly to sequences that lack that final component, because lateral excitation is sufficient to cause suprathreshold activation. In real neural networks, more complex types of learning are probably responsible for unit selectivity.

The model demonstrates quite robust performance on sound patterns having moderate complexity. In other simulations, this system performed well despite poor signal/noise ratios. This framework may accurately model some aspects of human auditory perception. Models of this class may also prove effective for practical engineering applications in rich, acoustically cluttered environments.

RHYTHM WITH ADAPTIVE OSCILLATORS

Thus far, we have considered two of the three basic types of temporal information in audition: the motion of primitive features such as spectral peaks, and sequences of simple events. We showed that units in the first-level cluster of the Lexin model develop selectivity to frequency changes that occur frequently, and that a second layer can recognize longer sequences of pattern primitives . However, an open environment will exhibit many other kinds of significant events, including patterns that repeat periodically. These patterns could be repeated frequency sweeps (e.g., "tweet, tweet, tweet"), noise bursts (e.g., footsteps), and many others. There are numerous ways in which such physical events can be generated – by animals of all kinds as well as by inanimate objects (see Gaver, 1993 for discussion). Many periodic events retain their identity despite changes in the rate of the periodicity. If variation in rate is a pervasive property, then the most useful measure of time relationships would be phase angle rather than absolute time or serial position. However, in order to calculate phase angle, some reference duration is needed for comparison to the input stimulus, that is, the input pattern should define its own measurement unit. This way, the time of occurrence of the next event can be predicted without measuring the duration in absolute units. It seems possible that humans may employ such a method of time measurement for a wide range of tasks.

The General Model

Our focus in this section is on the problem of how an oscillator could be entrained by an external periodic input pattern. Although it may eventually be possible to integrate such oscillators into a general model of the auditory system, for the moment our aim is to produce the simplest possible "front end" or "preprocessor", one that supplies periodic pulses with variable amplitude. Nevertheless, we might also imagine a more complex preprocesser – one made up of components resembling those in the Lexin model, for example, that would produce a pulse-like periodic signal each time some complex sequence occurred in the stimulation. One simple way to obtain a response to periodic input is to build a "bank of oscillators", like the undamped strings of a piano, such that one oscillator will respond to each frequency across the range. Such a system, however, cannot have the required flexibility. Therefore, in this section, we describe a general class of *adaptive oscillators* which track underlying periodicities in rhythmic patterns in spite of variability in the timing of those patterns (McAuley, 1995). In recent years, a number of different adaptive oscillator models have been explored, largely independently, by several groups of investigators (Large and Kolen, 1994; Large, 1994; McAuley, 1993; McAuley, 1994a,b). The model described here was partially inspired by a detailed neural model of how pacemaker cells in certain invertebrates respond to periodic electrical stimulation (Torras, 1985).

An adaptive oscillator is a device that fires spontaneously at some particular rate, but will modify this rate when stimulated at a different frequency. In its simplest form, the adaptive oscillator will do two things in response to an input pulse. First, it will instantly reset itself to phase 0 when the input arrives, and second, it will use the phase of the input pulse relative to the beginning of its cycle to change its own natural period to be a little closer to the "perceived" periodicity of the input. This behavior assures that the adaptive oscillator will eventually entrain to a periodic component of the input pattern.

Adaptive oscillators belonging to the class that we will describe (McAuley, 1995), share six primary properties. (1) Each oscillator has a characteristic resting period to which it gradually returns in the absence of input. (2) Each oscillator has a periodic activation function. (3) Each oscillator is phase-coupled with the input using a phase-resetting procedure such that when the combined value of input and activation (total activation) exceeds a threshold, the oscillator resets its phase to zero, beginning its cycle again. (4) Each oscillator retains a memory of the phase at which previous phase-resets occurred. This memory provides a measure of how well entrained it is. (5) Each oscillator's output is used as feedback to modify its period, which also serves to better align the oscillator with future inputs. (6) The output of each oscillator can be used to modify the shape of the activation function, so that as the oscillator becomes better entrained by a rhythmic pattern, the time window for expected future inputs becomes narrower.

For the purposes of describing the general mechanism, let the activation function of the adaptive oscillator be

$$a(t) = (1 + \cos(\frac{2\pi t}{\Omega(t)}))/2 \qquad (3)$$

where $\Omega(t)$ is the oscillator's period, initially equal to a resting value: $\Omega(0) = \Omega$.

As outlined above, each oscillator adapts by adjusting the phase and period of its activation function in response to perturbations from discrete inputs. Inputs in the range [0,1] represent event onsets with variable intensity values. This assumption is consistent with the view that rhythmic organization is primarily determined by the temporal pattern of event onsets (Handel, 1993). Thus, rhythmic patterns can be represented as a sequence of pulses. Figure 9A shows the input representation for a test pattern of four equally spaced (isochronous) tones, with a 300 ms interval between tone onsets, i.e., a 300 ms inter-onset interval (IOI). Formally this pattern is represented as

$$i(t) = \begin{cases} I & \text{if } t = nT \\ 0.0 & \text{otherwise} \end{cases} \qquad (4)$$

where T is the 300 ms IOI, and I is the intensity of the nth input pulse.

The adaptive oscillator utilizes a discrete form of phase coupling called phase-resetting. In the phase-resetting model, each weighted input, or coupling strength $w_i i(t)$, is added to the activation of the oscillator, providing a measure of total activation. The total activation of the system is then compared to a threshold of 1.0. If the total activation exceeds this threshold, the oscillator resets its phase to zero, beginning its cycle again. Thus, phase-resetting is defined by the following piecewise function

$$\phi(t) = \begin{cases} 0 & \text{if } a(t) + w_i i(t) > 1.0 \\ \phi(t) & \text{otherwise.} \end{cases} \tag{5}$$

Figure 9 shows the behavior of a model that only resets phase. An oscillator with a 500 ms period resets its phase in response to an isochronous pattern with a 300 ms IOI. Since the input pattern sequence is isochronous all of the phase resets occur at the same phase.

The system needs some measure of how well it is synchronized to its inputs. The degree of synchronization with a rhythmic pattern can be measured by maintaining a memory of the phase at which the input events forced the oscillator to reset. The symbol $\phi^r(n)$ is used to indicate the phase of the nth input pulse $i^r(n)$ forcing the total activation above threshold. The superscript r indicates that the phase and the associated input correspond to a phase-reset. The degree of synchronization is measured by the output $o(n)$, defined on $[0,1]$ according to

$$o(n) = (1 - w_o)o(n - 1) + w_o(1 - 2|\phi^r(n)|). \tag{6}$$

The parameter w_o establishes a weighting between the current reset phase $\phi^r(n)$ (scaled between 0 and 1 by $1 - 2|\phi^r(n)|$) and the memory of the previous reset phase $o(n - 1)$. Thus, the output is computed like a recursive smoothing filter. The output varies between 0 for poor synchronization and 1 for perfect synchronization. The synchrony measure is plotted in Figure 9C; it remains low in this case.

In a refinement of the basic mechanism, it turns out that the output of the adaptive oscillator can be used as feedback to modulate the shape of the activation function (cf., Large, 1994; McAuley, 1995). The width of the "receptive field" for the phase angle of input that will cause continued adaptation to the input pattern can be varied so that it is wider when entrainment is poor, becoming narrower as entrainment and predictability improve. How does the actual entrainment process take place?

Process of Entrainment

Two properties characterize period coupling in an adaptive oscillator. (1) The adaptive oscillator uses its output $o(n)$ (its measure of synchronization) as a "teaching" signal to determine how much to adjust its period. (2) The sign of the phase $\phi^r(n)$ (the reset phase of the nth input event) is used to determine the direction of the period change (increase or decrease). Both properties are expressed by the period coupling term

$$\mathcal{P} = \phi^r(n)(1 - o(n)). \tag{7}$$

This choice of P guarantees that the oscillator does not adjust its period when either $o(n) = 1$ (where it attains perfect synchrony) or $\phi'(n) = 0$ (where the current input coincides with the beginning of the oscillator's cycle). For negative reset phases, the input is assumed to be "early" with respect to the beginning of the oscillator's cycle, and the oscillator's intrinsic

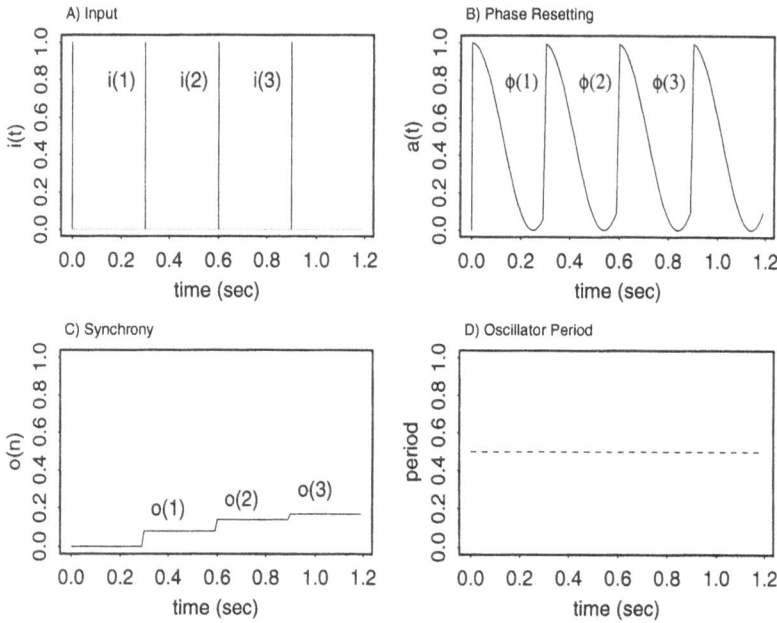

Figure 9. Phase-resetting of an oscillator with a 500 ms period in response to a rhythmic pattern with a fixed 300 ms inter-onset interval, but no adaptation of period.

period is shortened, speeding it up. Conversely, for positive reset phases, the input is assumed to be "late" with respect to the beginning of the oscillator's cycle, and the oscillator's intrinsic period is lengthened, slowing it down. The amount of period adjustment is inversely related to the output. If the output is large, the oscillator adjusts its period by only a small amount. If the output is small, the oscillator makes a much larger change in its period, enabling it to search the space of possible entrainment frequencies.

A complete description of period coupling in the adaptive oscillator is given by

$$\frac{\delta\Omega}{\delta t} = \alpha\frac{\Omega}{2}\mathcal{P}[\phi^r(n), o(n)]\mathcal{M}[i^r(n)] - \beta(\Omega - \overline{\Omega})(1 - \mathcal{M}[i^r(n)]). \tag{8}$$

In this differential equation, the period coupling term is scaled by $\Omega/2$ to ensure that the period changes by at most one half of the oscillator's cycle, by an input impulse response function M to spread the change in the oscillator's period over the entire cycle, and by the entrainment rate α. A decay rate β is included in Equation 8, so that in the absence of input, a penalty is incurred for large differences between the adapted and intrinsic periods, so that the oscillator will gradually return to its resting rate.

Figure 10 shows how an adaptive oscillator with both phase resetting and period coupling, and an intrinsic period of 500 ms responds to an isochronous input pattern with a 300 ms IOI. Panel 10A shows the isochronous input pattern. Panel 10B shows the response of the adaptive oscillator while it is entraining to this pattern. Both the oscillator response and the input pulses are shown. Period coupling forces the beginning of the oscillator's cycle on successive input pulses $i^r(n)$ to move closer to synchrony with the input. The precise change in the oscillator's period ($\Delta\Omega$) for each time step (ΔT), in the implementation of the adaptive oscillator, is determined using a discrete approximation to Equation 8. As the oscillator entrains itself to the input pattern, its period quickly approaches 300 ms (the IOI of the input pattern), as shown in Panel 10D. Panel 10C shows how the output changes rapidly from 0 toward 1.0, reflecting successful synchronization. Approximately one cycle after the last input,

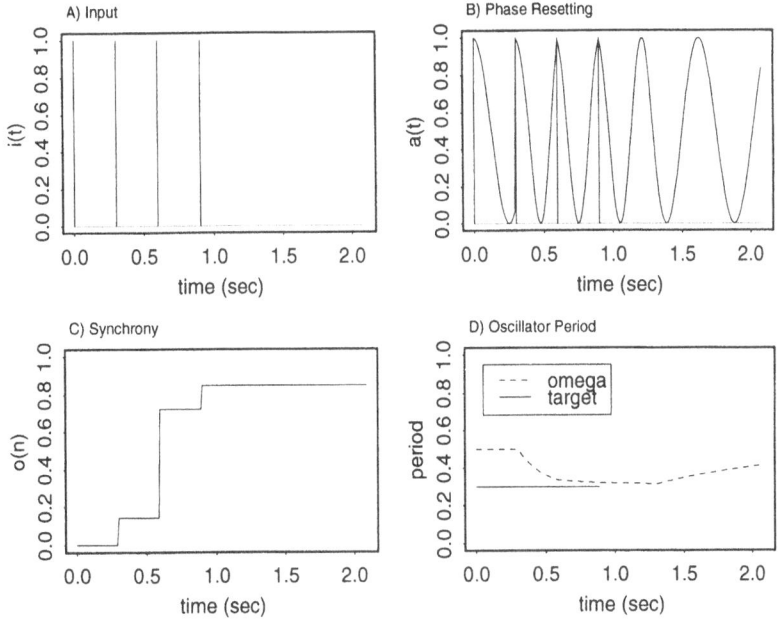

Figure 10. Responses of a phase-resetting and period-adapting oscillator with a 500 ms intrinsic period to isochronous input pulses (300 ms inter-onset interval). **A.** Input signal. **B.** Phase-resetting with period adaptation. **C.** Output signal. **D.** Period entrainment and decay.

the period begins to return to its resting value of 500 ms. This example shows the behavior of the system under the simplest conditions. But what if the period of the input pattern were somewhat variable or noisy? From the response of the oscillator in Panel 10D, it should be easy to see that modest amounts of variability will have almost no influence on the system. In fact, it tracks input sequences with moderate timing variability just as well as those that are perfectly isochronous. The instantaneous rate value simply fluctuates around the region of the mean rate of input. But what if the input patterns were much more complex, with a broad range of intervals between input pulses? The system might conceivably bounce around aimlessly. In the next section we show how adaptive oscillators behave when exposed to rhythmic patterns of varying complexity.

Entrainment to Complex Rhythms: Simulation 4

Povel and Essens (1985) tested the relative difficulty experienced by human listeners in repeating a set of systematically constructed rhythmic tone patterns. Each pattern was 16 time units long, where a time unit was 200 ms. Some of the patterns were easy for listeners to pick up and imitate while others were quite difficult as measured by how accurately listeners could reproduce the pattern. Povel and Essens performed these experiments to test a particular theory about how such patterns could be learned. An alternative explanation was suggested by McAuley (1995), who evaluated the ability of the adaptive oscillator to be entrained by the same set of temporal patterns.

Figure 11. The stimulus set used by Povel and Essens (1985).

Each pattern in this data set consists of a repeating sequence of intervals 16 beats long, using a 200 ms beat period. For each pattern, the repeated sequence of inter-onset intervals is a permutation of the set of intervals lasting 16 beats. Presenting the intervals in terms of the number of beats, the list is 4 single-beat time units (that is, four 200-ms intervals), two 2-beat (400 ms) intervals, one 3-beat (600 ms) interval, and one 4-beat (800 ms) interval. A set of 35 distinct permutations of these intervals, illustrated in Figure 11, were used as inputs to the adaptive oscillator model .

The patterns in the figure are ordered according to their approximate rhythmic complexity from easy to difficult (cf., Povel and Essens, 1985). Povel and Essens' subjects listened to as many cycles as they wished through each 16-beat pattern of short beeps and then tried to reproduce it with a tone-producing key. As the authors pointed out, when trying to memorize one of these patterns, a listener finds that a sense of periodic beats is evoked. When asked to imitate one of these patterns, most subjects engage in some kind of "cognitive oscillation" – a periodicity that seems to provide a framework for remembering where the beeps should occur. In fact, most subjects will tap out beats with their finger or a foot at roughly every 400 or 800 ms (every second or fourth beat), consistent with a musical meter. What we would like to suggest is that some process that can be modeled as an adaptive oscillator underlies subjects' performance on rhythm learning, and even rhythmic listening. If this idea is correct, the set of patterns that an adaptive oscillator has trouble with may be the same ones that subjects have difficulty with.

For this simulation (McAuley, 1995), a single adaptive oscillator was exposed to 8 cycles of each of the 35 rhythmic patterns. The intrinsic period of the oscillator was 500 ms, halfway between the maximum and minimum intervals contained in these input patterns. Before the presentation of each pattern, the period of the oscillator was initialized to its resting value. It was found that when the input stimuli were perfectly timed (that is, all intervals were exact multiples of 200 ms), the model would sometimes oscillate between target frequencies, that is, it would repeatedly speed up and then slow down. However, when five percent noise was added to the input patterns, the model behaved more stably and more like human listeners. Thus, we will report the behavior of the system with a small amount of temporal noise added to the input. Although the noise should probably be added internally to the model, it has exactly the same stabilizing effect when added to the input periods.

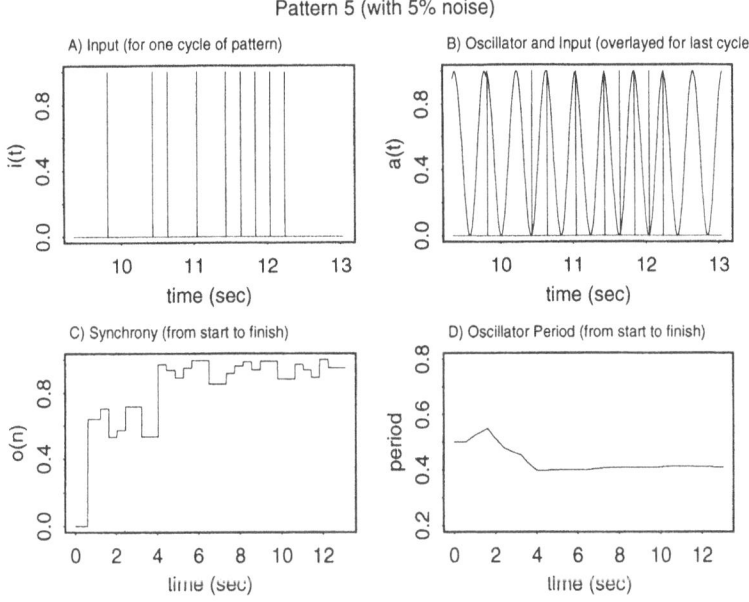

Figure 12. Performance of the adaptive oscillator on Pattern 5 in the 5% noise condition.

The main question addressed in the simulation was whether the adaptive oscillator could achieve and maintain stable entrainment for complex patterns of varying rhythmic complexity. Further issues were whether the model could entrain at the same period that human subjects do, and whether the model would find the same alignment or specification of "downbeats" that human listeners do. A complete study of these issues for this data set is reported in McAuley (1995), so here we report only a few examples to show that without any particular adjustments other than the addition of noise, the model's behavior is similar to that exhibited by human subjects.

Povel and Essens (1985) found that when subjects reproduced the rhythmic patterns, some were much more difficult than others. Their performance measure was the sum of deviations in ms between the time of beep onsets in the presented pattern and the times of the beats in the reproduced patterns, across 4 pattern repetitions. The resulting scores ranged from a minimum of about 140 ms to a maximum of 250 ms. To compare just two specific patterns, Pattern 5, the fourth easiest pattern, had a mean deviation score of about 158 ms; Pattern 12, ranked 15th in difficulty, had a mean deviation of about 230 ms. McAuley (1995) conducted a simulation of one aspect of the task performed by human subjects: to entrain an oscillator to the pattern.

The behavior of the model in response to Pattern 5 is shown in Figure 12. In this figure, one cycle of the test pattern is displayed in Panel 12A, where the pattern begins just before the 10 sec point. Panel 12B displays the response of the oscillator during the last cycle of the four cycle test pattern, thus ignoring the initial transient. The first pulse of the pattern, which may be slightly out of position because of noise, occurs again just before the 10 sec point on the time scale. What does the observed alignment correspond to? If we use a 1 to represent a pulse, a 0 to represent a silent beat, and commas to mark measure boundaries (where each measure is two cycles of the oscillator), then the oscillator found the following structure: 1001, 1010, 1111, 1000. This is about the only alignment we can easily find for this sequence. Panel 12C displays the output of the oscillator, expressed as its measure of synchrony, through all four cycles of the test pattern. Panel 12D shows the corresponding changes in oscillator period over this same interval. Since one cycle of the pattern takes 3.2 seconds, Panel 12C shows that after completion of about a cycle and a half through the pattern, entrainment reaches its maximum. Since there is noise in the input, the output will continue to bounce around somewhat. Panel 12D shows that the period moved fairly steadily from 500 ms to 400 ms. This result shows that the adaptive oscillator finds an appropriate alignment of beats, one that likely agrees with your own "tapping" intuitions.

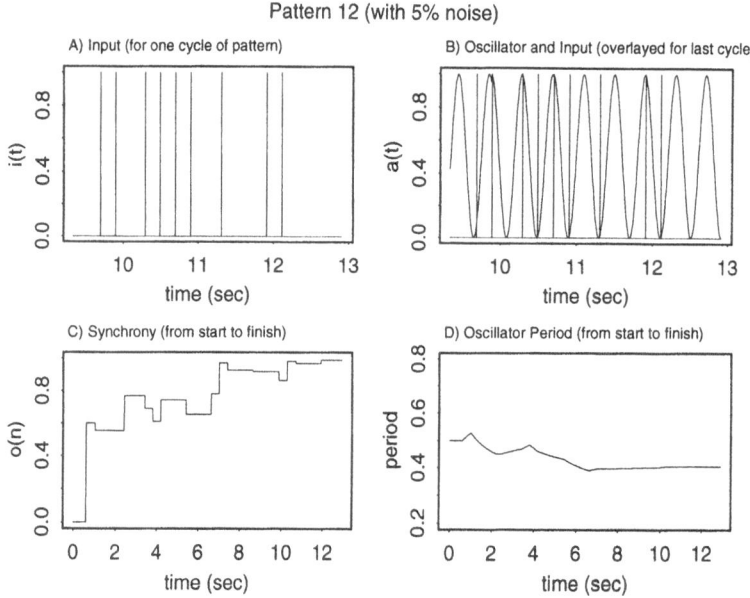

Figure 13. Performance of the adaptive oscillator on Pattern 12 in the 5% noise condition.

According to Povel and Essens (1985), Pattern 5 was relatively easy but Pattern 12 was more difficult. Figure 13 shows the performance of the model on this pattern. Panel 13C shows that it took almost three cycles through the pattern before the output approached a maximum. The period of the oscillator also shows several rises and falls as the model "searches" for a suitable rate and phase alignment. The solution found by the model in Panel 13B begins again just before the 10 sec point. This alignment has an "upbeat" and looks like this: 1, 1011, 1101, 0011, 000. This is certainly one possible alignment of beats. In other runs, the model also sometimes found the alignment that we prefer intuitively, one that locates the downbeat one step to the left: 1101, 1110, 1001, 1000. In any case, our point here is only that the adaptive oscillator, with no special additional assumptions or parameter tweaking, exhibits behavior that resembles human response to these moderately complex patterns. Furthermore, the model finds some patterns more difficult than others, in roughly the same way human listeners do, and it finds beat alignments that are intuitively reasonable.

We have presented only a brief summary of these data, but McAuley (1995) shows that the model found human-like solutions of frequency and phase alignment for all 35 of the patterns used by Povel and Essens. These results suggest that the model captures some essential features of human processing of rhythmic patterns.

Conclusions about Adaptive Oscillators for Rhythm.

Is there any more direct evidence that human subjects employ adaptive oscillators when listening to periodic sound patterns? Several additional lines of evidence are discussed in McAuley (1995). The model is obviously highly specific, so it allows many very precise qualitative predictions for experiments. One appropriate task is the discrimination of a change in the rate of production of two series of clicks. For example, Drake and Botte (1993) had subjects listen to several cycles of equally spaced tones and, after a pause, listen to a second set. The subjects were asked to discriminate a change in the tempo of the second series relative to the first. The base tempo of the first series varied over a wide range. It was found that, across subjects, the overall period that permitted the most accurate performance lay between 300 and 800 ms – that is, in the region of the 500 ms resting rate that was used for the Povel and Essens task. Drake and Botte found that performance improved uniformly as the number of cycles through the pattern increased. In a simulation of the discrimination task, the adaptive oscillator was trained using a two-alternative forced-choice task with an adaptive-tracking procedure (Levitt, 1971) to discriminate tempo changes. For the model, optimal tempo discrimination was centered at its preferred oscillation rate, with worse performance at faster and slower tempos. Increasing the number of pattern cycles improved performance more for fast tempos than for slow tempos, providing a qualitative match to the data reported by Drake and Botte (McAuley, 1995).

If some process that can be modeled as an adaptive oscillator, or set of oscillators, plays a role in the performance of rhythmic tasks by human listeners, then a number of specific predictions can be developed. For example, as the interstimulus interval increases, the activity of an adaptive oscillator should decay back to its preferred rate. We could even predict that performance would be optimum when the duration of interstimulus interval is at or near a multiple of the period of the standard sequence. If some other interstimulus interval is used, the system must reset to phase 0 at the onset of the second interval. This might be expected to reduce performance somewhat. The adaptive oscillator models under development should be very useful in accounting for human performance on rhythmic tasks.

Not only are adaptive oscillators potentially useful for describing the perception and performance of music, but they may also be useful for recognizing and producing complex hierarchically structured patterns such as human speech, independently of rate (see Port, Cummins and Gasser, 1995).

We have shown how an adaptive oscillator effectively follows rather low frequency periodic events – ones that are typically thought of as 'envelope properties' of sound rather than spectral properties. However, we should point out that it is possible that this same model could operate at much higher frequencies to perform analysis and prediction at frequencies that overlap the range of mechanical spectrum analysis, thus helping to differentiate signals in the same frequency channel that have different phase relationships. Thus, adaptive oscillators might help solve the monaural cocktail party problem.

CONCLUSIONS

We have discussed a number of issues that are relevant to the development of a model auditory system that adapts to a relatively open auditory environment. First, we considered the practical question of what kind of information is likely to be useful given the statistical properties of open environments. We proposed that in order to do a satisfactory job at auditory perception in typical acoustic environments, a system would have to include, in addition to mechanisms for identification of static features, methods for recognizing acoustic motions, acoustic sequences and rhythmic structures. We have tried to show how it should now be possible to design a research program that would seek to model audition in very general biological terms. Rather than just modeling the recognition of small subsets of the patterns humans are capable of differentiating under highly idealized conditions, we would like to attempt to simulate the general skills of animals living in a complex open environment. In tackling such a problem, we consider it essential to address the problem of patterns that are distributed in time.

This line of thinking has led us to consider the ways in which temporal patterns can be defined for animals. We conclude that there are three basic types of temporal information: acoustic motions over short time intervals, rhythmic patterns, and sequential patterns. Complex structures like music, speech and birdsong can be defined in terms of these basic types of information. None of these classes of information depend on absolute measurement of time in seconds. Biological systems have no suitable clocks for making such measurements, and they lack any means to read and utilize absolute time measurements even if they were made. Instead, it is likely that animals' auditory systems employ timing measurements that are achieved by synchronizing an internal oscillator to an externally presented periodic signal. Thus, when there is no periodic structure to lock in to, performance is likely to be seriously degraded.

We reviewed two models developed in our laboratory to deal with certain aspects of all three types of temporal information. The Lexin model incorporates mechanisms based on lateral connections between spatially arrayed frequency-sensitive units, and a self-organization process to recognize common sequential patterns in its auditory environment. It discovers statistically prominent frequency states and frequency motions, and assigns representational nodes to those complex events. Adaptive oscillators adjust their spontaneous firing rate to match an input period and can be employed to characterize repetitive patterns in the input.

Further development of models like these, especially more comprehensive ones that integrate the individual models in some appropriate fashion, should permit much more sophisticated simulations of the interpretation of complex sound patterns in an open acoustic environment. Of course, there are still many daunting challenges to be addressed in dealing with truly open environments. For example, we have suggested nothing about how one of our systems could deal with multiple, overlapping auditory patterns. We know that human audition is capable of remarkable selectivity in filtering out competing patterns, but a single Lexin module can apparently only recognize a single familiar pattern at a time. Thus, to avoid confusion from simultaneous competing patterns, we must imagine parallel sequence recognition models that are able to use their own predictive capabilities to suppress one another's inputs so that each replication of the model will track only one familiar pattern at a time. Govindarajan et al., (1994) have recently proposed a model for auditory stream segregation with properties similar to those of the Lexin model, but with stream selectivity as well. This work is only a beginning, but we hope that it is the beginning of sophisticated modeling of the many rich problems in auditory processing that lie before us.

ACKNOWLEDGMENTS

The authors thank Gary R. Kidd, Charles S. Watson, Fred Cummins and Bryan Gygi for helpful comments on portions of this paper. Fred Cummins helped with formatting and in many other ways. This work was supported by the Office of Naval Research Grant No. N0001491-J1261 to Robert Port, the Indiana University Graduate School, the Armed Forces Communications and Electronics Association to Sven Anderson, and by the National Institute of Mental Health Fellowship Grant No. MH10667-01 to J. Devin McAuley. Sven Anderson's current address is the Department of Anatomy, University of Chicago, 1025 E. 57th St, Chicago, IL 60637.

REFERENCES

Anderson, S., 1992, Self-organization of auditory motion detectors, in: "Proceedings of the Fourteenth Annual Conference of the Cognitive Science Society", Lawrence Erlbaum Associates., Hillsdale, NJ.

Anderson, S., 1994, "A Computational Model of Auditory Pattern Recognition", PhD thesis, Technical Report No. 112, Cognitive Science Program, Indiana University, Bloomington, IN.

Baddeley, A., 1992, Working memory. *Science*, 255:556.

Barlow, H. B., and Levick, W. R., 1965, The mechanism of directionally selective units in a rabbit's retina. *J. Physiol.*, 173:477.

Bregman, A. S., 1990. "Auditory Scene Analysis: The Perceptual Organization of Sound", Bradford Books, MIT Press, Cambridge, MA.

Carpenter, G., and Grossberg, S., 1987, A massively parallel architecture for a self-organizing neural pattern recognition machine. *Computer Vision, Graphics and Image Processing*, 37:54.

Crowder, R., and Morton, J., 1969, Precategorical acoustic storage. *Percept. Psychophys.*, 5:365.

Delgutte, B., and Kiang, N. ,1984, Speech coding in the auditory nerve: I. Vowel-like sounds. *J. Acoust. Soc. Am.*, 75:866.

Drake, C., and Botte, M.-C., 1993, Tempo sensitivity in auditory sequences: Evidence for a multiple-look model. *Percept. Psychophys.*, 54:277.

Gaver, W. W., 1993, What in the world do we hear? An ecological approach to auditory perception. *Ecol. Psychol.*, 5:1.

Gibson, J. J., 1968, "The Senses Considered as Perceptual Systems", Harcourt Brace, New York, NY.

Govindarajan, K. K., Grossberg, S., Wyse, L. L., and Cohen, M. A., 1994, A neural network model of auditory scene analysis and source segregation, Technical Report CAS/CNS-TR-94-039, Center for Adaptive Systems, Boston University, Boston, MA.

Green, D., 1976, "An Introduction to Hearing", Lawrence Erlbaum Associates, Hillsdale, NJ.

Grossberg, S., 1976, Adaptive pattern classification and universal recoding. *Biol. Cybern.*, 23:121.

Grossberg, S., 1986, The adaptive self-organization of serial order in behavior: Speech language, and motor control, in: "Pattern Recognition by Humans and Machines: Speech Perception", E. Schwab, and H. Nusbaum, eds., Academic Press, Orlando, FL.

Grossberg, S., 1987, Competitive learning: From interactive activation to adaptive resonance. *Cognitive Sci.*, 11:23.

Grossberg, S., and Rudd, M. E. (1992). Cortical dynamics of visual motion perception: short-range and long-range apparent motion. *Psychol. Rev.*, 99:78.

Handel, S., 1989, "Listening: An Introduction to the Perception of Auditory Events", Bradford Books/MIT Press, Cambridge, MA.

Handel, S., 1993, The effect of tempo and tone duration on rhythm discrimination. *Percept. Psychophys.*, 54:370.

Hermansky, H., 1990, Perceptual linear predictive (PLP) analysis of speech. *J. Acoust. Soc. Am.*, 87:1738.

Large, E. W., 1994, "Dynamic Representation of Musical Structure", Ph.D. Thesis, The Ohio State University, Columbus, OH.

Large, E. W., and Kolen, J. F., 1994, Resonance and the perception of musical meter. *Connection Sci.*, 6:177.

Lesser, V. R., Fennel, R. D., Erman, L. D., and Reddy, D. R., 1975, Organization of the Hearsay-II speech understanding system. International Conference on Acoustics, Speech, and Signal Processing, 23:11.

Levitt, H., 1971, Transformed up-down methods in psychoacoustics. *J. Acoust. Soc. Am.*, 49:467.

Marshall, J., 1990, Self-organizing neural networks for perception of visual motion. *Neural Networks*, 3:45.

Marshall, J., 1995, Adaptive perceptual pattern recognition by self-organizing neural networks: Context, uncertainty, multiplicity, and scale. *Neural Networks*, in press.

Massaro, D. W., 1972, Perceptual images, processing time, and perceptual units in auditory perception. *Psychol. Rev.*, 79:124.

McAuley, J. D., 1993, Learning to perceive and produce rhythmic patterns in an artificial neural network., Technical Report 371, Computer Science Department, Indiana University, Bloomington, IN.

McAuley, J. D., 1994a, Finding metrical structure in time. *in:* "Proceedings of the 1993 Connectionist Models Summer School", M.C. Mozer, P. Smolensky, D.S. Touretzky, J.L. Elman, and A.S. Weigend, eds, Lawrence Erlbaum Associates, Hillsdale, NJ.

McAuley, J. D., 1994b, Time as phase: A dynamic model of time perception. *in:* "Proceedings of the Sixteenth Annual Meeting of the Cognitive Science Society", Lawrence Erlbaum Associates, Hillsdale, NJ.

McAuley, J. D., 1995, "On the Perception of Time as Phase: Toward an Adaptive Oscillator Model of Rhythm". Ph.D. Thesis, Cognitive Science Technical Report, TR-137, Indiana University, Bloomington, IN.

Moore, B. C. J., 1989, "An Introduction to Psychology of Hearing", third edition, Harcourt Brace Jovanovich, New York, NY.

Patterson, R. and Holdsworth, J., 1990, "An Introduction to Auditory Sensation Processing". MRC Applied Psychology Unit, Cambridge, UK.

Port, R., 1986, Invariance in phonetics, *in:* "Invariance and Variability in Speech Processes", J. Perkell, and D Klatt., eds, Lawrence Erlbaum Associates, Hillsdale, NJ.

Port, R., and Cummins, F., 1992, The English voicing contrast as velocity perturbation, *in:* "Proceedings of the 1992 International Conference on Spoken Language Processing", J. Ohala, T. Nearey, B Derwing, M. Hodge, and G. Wiebe, eds, University of Alberta, Edmunton.

Port, R., Cummins, F., and Gasser, M., 1995, A dynamical approach to rhythm in language: Toward a phonology of time. *in:* "Proceedings of the Chicago Linguistic Society", Chicago Linguistic Society, Chicago, IL (in press).

Port, R., Cummins, F., and McAuley, J. D., 1995, Naive time, temporal patterns and human audition, *in:* "Mind as Motion: Explorations in the Dynamics of Cognition", R. Port, and T. van Gelder, eds, MIT Press, Cambridge, MA.

Port, R. and Dalby, J., 1982, C/V ratio as a cue for voicing in English. *J. Acoust. Soc. Am.*, 69:262.

Port, R. F., 1990, Representation and recognition of temporal patterns. *Connection Sci.*, 2:151..

Port, R. F., Dalby, J., and O'Dell, M., 1987, Evidence for mora timing in Japanese. *J. Acoust. Soc. Am.*, 81:1574.

Povel, D.-J. and Essens, P., 1985, Perception of temporal patterns. *Music Perception*, 2:411.

Sankoff, D. and Kruskal, J. B., eds, 1983. "Time Warps, String Edits and Macromolecules: The Theory and Practice of Sequence Comparison", Addison-Wesley, Reading, MA.

Shamma, S. A.,1985a, Speech processing in the auditory system I. The representation of speech sounds in the responses of the auditory nerve. *J. Acoust. Soc. Am.*, 78:1612.

Shamma, S. A., 1985b, Speech processing in the auditory system II. Lateral inhibition and the central processing of speech evoked activity in the auditory nerve. *J. Acoust. Soc. Am.*, 78:1622.

Smythe, E., 1987, The detection of formant transitions in a connectionist network, *in:* "Proceedings of the First IEEE International Conference on Neural Networks", San Diego, CA.

Smythe, E. J., 1988, Temporal computation in connectionist models. Technical Report 251, Indiana University, Computer Science Department, Indiana University, Bloomington, IN.

Torras, C., 1985, "Temporal-Pattern Learning in Neural Models", Springer Verlag, Berlin.

Vercoe, B. L., 1986, C-sound. Technical report, Experimental Music Studio, Media Laboratory, Massachusetts Institute of Technology, Cambridge, MA.

Watson, C. S. and Nichols, T. S., 1976, Detectability of auditory signals presented without defined observation intervals. *J. Acoust. Soci. Am.*, 59:655.

Whitfield, I. C. and Evans, E. F., 1965, Responses of auditory cortical neurons to stimuli of changing frequency. *J. Neurophysiol.*, 28:655.

Wilson, M. A., Bhalla, U. S., Uhley, J. D., and Bower, J. M., 1989, GENESIS: a system for simulating neural networks, *in:* "Advances in Neural Information Processing Systems I", D.S. Touretzky, ed, Morgan Kaufmann, San Mateo, CA.

Yost, W., 1991, Auditory image perception and analysis: the basis for hearing. *Hearing Res.*, 56:244.

Yost, W., 1992, Auditory perception and sound source determination. *Psychol. Sci.*, 1:179.

Yost, W. A. and Watson, C. S., eds , 1987, "Auditory Processing of Complex Sounds", Lawrence Erlbaum Associates, Hillsdale, NJ.

TEMPORAL PATTERNING IN A SMALL
RHYTHMIC NEURAL NETWORK

Scott L. Hooper

Neurobiology Group
Department of Biological Sciences
Irvine Hall
Ohio University
Athens, OH 45701
USA

INTRODUCTION

Rhythmic motor patterns such as walking, flying, and chewing form a large part of the behavioral repertoire of most animals. It is now clear that the fundamental characteristics of such motor patterns (their rhythmicity and the sequence of movements that make up the pattern) result from the activity of relatively small neural networks that produce appropriately ordered motor outputs in the absence of timed input from either the periphery (e.g., proprioceptive inputs) or the rest of the nervous system (e.g., descending pathways) (Delcomyn, 1980). Over the last twenty years the mechanisms underlying the ability of these central pattern generator (CPG) networks to inherently produce ordered rhythmic motor outputs has been intensively studied, and a fairly deep understanding of this process has been achieved for several invertebrate model systems.

Neural network *rhythmicity* can arise in two different ways. The first is from the presence in the network of endogenously oscillatory neurons that alternately depolarize and repolarize and hence drive the rhythm. The second is from network based rhythmicity in which no single neuron of the network is inherently an oscillator; instead, the rhythmicity in such a system arises from the interactions among the network's neurons. Neuronal *firing order* within a multi-phase pattern is determined both by the pattern of synaptic connections within the network, and by the time constants of the active cellular properties of the network's neurons. Examples of such properties are post-inhibitory rebound, plateau potentials (see below), and synaptic desensitization. Finally, many rhythmic networks have highly interconnected, non serial synaptic connectivity patterns. In such *distributed* networks it is generally impossible to attribute specific network functional characteristics to the activity of any single neuron, and each neuron contributes to many aspects of the network's output. Thus, the activity in the network results from the collective interactions among the many neurons that make up the network.

No rhythmic motor pattern is produced in a perfectly stereotypical, unchanging manner; instead, the pattern's overall cycle frequency and the phase relationships among the movements that make it up continually change to match motor pattern output to behavioral requirements. *A priori*, there are two different ways in which this control could be achieved. The first is dictatorial control, in which descending or proprioceptive input would have access to the final common pathway to the effector muscles (typically the appropriate motor neurons), and would alter the motor pattern by directly controlling the activity of this pathway, in essence bypassing the CPG network entirely. This mechanism requires either that more central networks continually calculate the appropriate muscle movements or that proprioceptive input pathways be precisely correct; it completely defeats the presumed computational advantage represented by having a dedicated central pattern generating network in the first place. The second is delegative control, in which the controlling input would trigger a given type of change (e.g., speed up) by altering only a few neuronal or synaptic properties of the CPG network. In this scenario, the CPG network architecture is such that in response to these changes, the network now inherently produces (i.e., calculates) the new motor pattern required. This mechanism is extremely advantageous computationally in that the input merely serves to alert the CPG of the required change; the reconfigured CPG then performs all the tedious calculations required to put it into action. Put another way, a CPG would no longer just produce a motor pattern; it would instead produce, depending on the input it receives, all of a set of related motor patterns (e.g., walking forward, backward, uphill, and downhill; and all at whatever speed was required).

Studies on a variety of invertebrate preparations suggest that single neural networks do in fact produce multiple outputs as a result of the activation of modulatory inputs arising from either the center or the periphery, and that they do so in the delegative fashion outlined above. However, the active cellular properties and distributed nature of these networks make the mechanisms underlying this process difficult to understand intuitively. The fact that activity in these networks depends on synaptic contacts and active neuronal responses that each have their own specific time constants means that postsynaptic neurons respond actively (and in their own good time) to the inputs they receive. The fact that these networks are distributed means that changes in the activity of one neuron in the network will induce changes in the activity of the rest of the network, which will in turn feed back on the original neuron and further alter its activity; this process continues until a new network equilibrium is reached. It follows that neural networks do not function in a gearlike fashion in which alterations in the activity of a single neuron explain in any obvious way the observed changes in the activity of the network as a whole.

However, just as small invertebrate model systems were invaluable in elucidating the fundamental mechanisms underlying ordered rhythmic motor pattern generation, they are also providing insight into the mechanisms underlying regulation of the frequency and phase of these motor patterns. The rest of this chapter will be devoted to one of the best understood of such systems, the pyloric network of the stomatogastric system of decapod crustacea such as lobsters and crabs (Harris-Warrick et al., 1992a; Hartline, 1979, 1987; Miller, 1987; Miller and Selverston, 1982a,b; Selverston et al., 1976; Selverston and Miller, 1980; Selverston and Moulins, 1987). I shall first introduce the network, then consider control of the frequency and duty cycle of the network's pacemaker neuron, then phase control of the networks follower neurons, and finally phase regulation in the network as a whole as cycle frequency changes.

THE PYLORIC NETWORK

The pyloric network of the stomatogastric nervous system of decapod crustacea such as lobsters and crabs produces rhythmic contractions of the striated muscles of the pyloric chamber of the stomach. This motor pattern separates chewed food into three streams, one for absorption, one for further chewing, and one for excretion (Johnson and Hooper, 1992). Even *in vitro*, isolated from sensory and central nervous system inputs (Selverston et al., 1976), the network continues to generate rhythmic motor patterns similar to those in the intact animal (Rezer and Moulins, 1983). Figure 1 (left panel) shows a schematic of the stomatogastric ganglion (STG), which contains all the pyloric network neurons, and the nerves in which the pyloric axons run. The middle panel shows simultaneous intracellular (top three traces) and extracellular (bottom two traces) recordings of the pyloric pattern. The pyloric

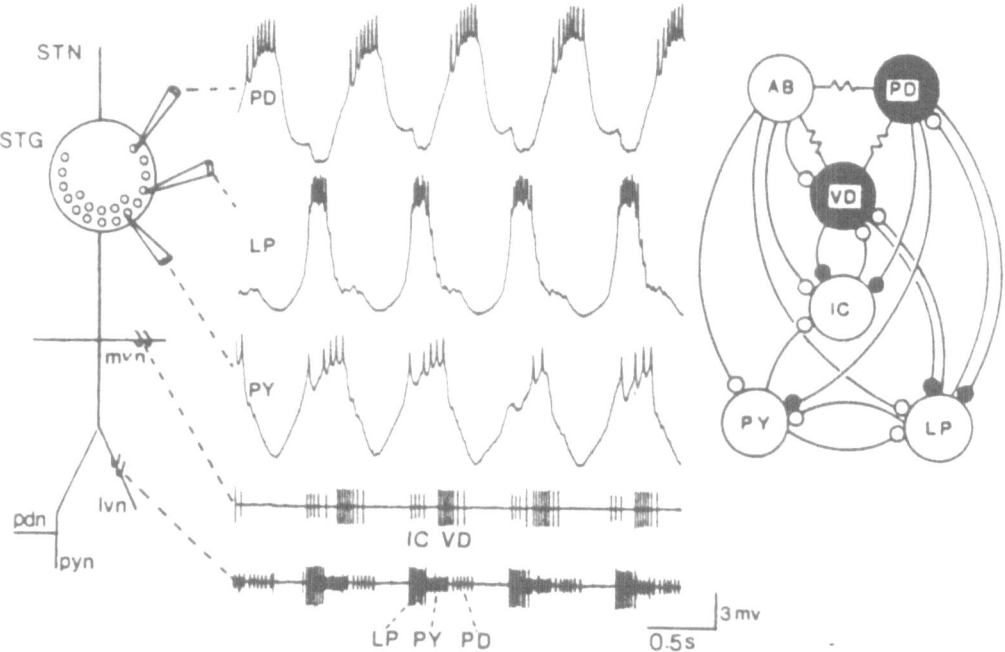

Figure 1. The pyloric network. The stomatogastric nervous system lies on the surface of the lobster stomach, and consists of four ganglia (the paired commissural ganglia, the esophageal ganglion, and the stomatogastric ganglion), the nerves that connect the ganglia, and the nerves that innervate the stomach and its muscles. The left panel is a schematic of the stomatogastric ganglion (STG) and the nerves in which the pyloric neurons send their axons. The cell bodies of all pyloric network neurons are located in the STG. The middle panel shows simultaneous intracellular (top three traces) and extracellular (bottom two traces) recordings of the pyloric neural output. The right panel shows the pyloric network synaptic connectivity. Resistor symbols are electrical coupling, filled circles are cholinergic synapses, open circles glutamatergic synapses. All chemical synapses in the network are inhibitory. See text for full description. Nerve abbreviations: STN, stomatogastric; mvn, medial ventricular; lvn, lateral ventricular; pyn, pyloric; pdn, pyloric dilator. Figure taken (with permission) from Hooper and Marder, 1987.

pattern is a repeating sequence of alternating bursts of action potentials from the pyloric motor neurons. The pattern has approximately three parts. The pyloric dilator (PD) neurons define the beginning of the pattern; they and the anterior burster (AB) interneuron (not shown) fire together. The lateral pyloric (LP) and inferior cardiac (IC) neurons fire next, and finally the ventricular dilator (VD) and pyloric (PY) neurons fire.

The synaptic connectivity of the pyloric network is shown in Fig. 1, right panel. The pyloric network consists of the pacemaker AB interneuron and 5 classes of motor neurons. The pattern of synaptic connectivity among the pyloric neurons (Eisen and Marder, 1982; Selverston et al., 1976) and many of their cellular properties are known (Gola and Selverston, 1981; Raper, 1979; Russell and Graubard, 1987; Russell and Hartline, 1978). All the chemical synapses in the network are inhibitory; the AB, LP, PY, and IC neurons use glutamate as a transmitter, and produce fast, short lasting inhibition of their follower neurons; the PD and VD neurons use acetylcholine as a transmitter, and produce slow, long lasting inhibitions (Marder and Eisen, 1984a). The network is highly distributed: a completely connected 6 neuron network would have 30 synapses; the pyloric has 19.

The network's rhythmicity depends on regenerative membrane currents in its neurons. The concerted action of these currents leads to these neurons expressing at least three types of active properties. First, they can all be endogenous oscillators (Bal et al., 1988; Miller and Selverston, 1982a). Second, they can all have plateau properties. A plateauing neuron has two quasistable membrane potentials (a hyperpolarized rest potential and a depolarized plateau potential). The neurons make transitions between the two states in response to synaptic input, postinhibitory rebound, or externally applied current injection. These transitions are regenerative, i.e., depolarization from rest above a threshold voltage activates voltage -dependent depolarizing conductances that drive the neurons to the fully depolarized plateau, and relatively small hyperpolarizations from the plateau induce an active repolarization to the rest state (Russell and Hartline, 1978, 1982). Third, they all can exhibit postinhibitory rebound (Selverston et al., 1976). These properties all have their own time constants, and result in these neurons having very non-linear input to output relationships that differ in sign, amplitude, and time; thus, in the sense used in the introduction, the pyloric network is a distributed network composed of neurons and synapses with different inherent time constants.

The combined effects of these active properties and the network's interconnectivity pattern largely explain the observed pyloric pattern. The AB neuron is generally the fastest oscillator; the PD neurons fire with it because the AB and PD neurons are electrically coupled. The AB and PD neurons inhibit all the other pyloric neurons, which therefore fire out of phase with the pacemaker group. The rest of the network's neurons (the VD, IC, PY, and LP neurons) fire due to plateau potentials. These potentials are triggered by postinhibitory rebound induced by inhibition received from the neurons that fire before them in the pyloric pattern (for the LP and IC neurons, inhibition from the pacemaker group; for the VD and PY neurons, a combination of the inhibition they receive from the pacemaker group and the later inhibition they receive from the LP and IC neurons – see Fig. 1, middle panel for the firing order of the various pyloric neurons).

A full explanation of the firing order of the pyloric networks neurons, however, requires a detailed examination of the differing time constants and strengths of the network's various synapses, and of the differing active properties of the networks neurons. For example, consider why the VD and PY neurons fire late in the pattern. The VD neuron might be expected to fire immediately after the pacemaker group's activity because it receives only the AB neuron's short acting glutamatergic inhibition. However, in fact the LP and IC neurons fire next because their combined inhibition is sufficient to prevent the VD neuron from firing first. The PY neurons, alternatively, fire late in the cycle because these neurons have cellular properties that delay their postinhibitory rebounds (Hartline, 1979). When they do rebound, they inhibit the LP and IC neurons, and this frees the VD neuron to fire as well. Finally, the PY and VD neurons are turned off by the next AB/PD neuron depolarization, and the cycle repeats.

Both the expression of active properties in these neurons and the strength of their synaptic connections is under modulatory control. Figure 1 and the explanation given above represent a *specific* state of the pyloric network, that is, when the network is receiving input from other stomatogastric nervous system ganglia through the stomatogastric nerve (STN). When STN conduction is blocked, the expression of active properties in all the pyloric neurons is decreased; pyloric cycling slows down and sometimes stops. Depending on the species, the

PY and IC neurons may lose the ability to produce plateau potentials and therefore become silent (Selverston et al., 1976). These results imply that the STN contains input fibers to the STG that alter pyloric network activity. Immunohistochemistry and, in some cases, high performance liquid chromatography, have been used to identify the nature of the inputs. These studies have identified literally scores of chemical and peptidergic neurotransmitter substances, including dopamine, octopamine, serotonin, histamine, acetylcholine, proctolin, and FMRF-like and CCK-like peptides, that could serve as putative neuromodulators and are present in the STN and STG (Harris-Warrick et al., 1992b).

Bath application of these substances to STGs that are isolated from STN input alters the synaptic strengths and/or active properties of the pyloric neurons in specific ways (Dickinson et al., 1990; Dickinson and Nagy, 1983; Flamm and Harris-Warrick, 1986a,b; Hooper and Marder, 1987; Marder and Eisen, 1984b). Since the phase relationships and burst durations of the pyloric pattern are due to these active properties and synaptic strengths, each different modulator drives the network to a new functional configuration that produces a different motor pattern. It is important to note that we do not yet know the activity patterns of these input fibers in the intact animal, and so these experiments measure the possible range of patterns the network can produce, not necessarily the actual patterns produced *in vivo*. Reassuringly, however, work by Nusbaum and Marder (1989) shows that stimulation of the neurons that give rise to the proctolinergic input to the STG induces changes in pyloric network activity that is similar to that achieved by bath application of proctolin; this result shows that, at least in some cases, bath application of neuromodulators and activity of neuromodulatory inputs have similar effects on pyloric output.

An example of these multiple output patterns is shown in Figure 2, which shows the alterations in pyloric output induced when pilocarpine (a muscarinic agonist) and the neuropeptides proctolin and FMRFamide are bath applied to the STG of the crab, *Cancer irroratus*. Each modulatory substance induces specific, reproducible changes in pyloric output. Pilocarpine increases pyloric frequency without causing any large change in pyloric phase relationships. Proctolin increases pyloric frequency slightly, but its major effect is to dramatically increase the relative burst duration of LP neuron firing. Finally, FMRFamide application also slightly increases pyloric frequency, but the major effect is a dramatic decrease (in this case complete cessation) of LP neuron firing. As noted above, these changes in neuronal activity are a collective response of the network to changes that the modulators induce in only a few of the network's neurons (see below), and thus one cannot assume that changes in any given neuron's activity are due to that neuron having receptors to, and thus being directly altered by, the modulator in question. For example, is the increased cycle frequency and PD neuron firing during application of pilocarpine due to a direct excitatory effect of pilocarpine on the PD neurons, or rather only an indirect effect mediated through the AB-PD neuron electrical coupling of a pilocarpine mediated excitation of the AB neuron?

Luckily, it is possible in this system to selectively delete by photoinactivation any desired neuron from the network. This is done by injecting the neuron with Lucifer Yellow and then illuminating the STG with blue light (Miller and Selverston, 1979). In this manner the role any given neuron plays in the whole network response to application of the various modulators can be assessed. Furthermore, by combining photoinactivation of specific neurons with bath application of picrotoxin to the STG to block glutamatergic synapses (Bidaut, 1980), any desired neuron can be isolated from network derived synaptic input, and thus the individual direct response of that neuron to a given treatment may be determined. These techniques have been successfully used to elucidate the cellular and synaptic bases underlying several of this networks different output patterns. I shall first consider the mechanisms for frequency control in the network.

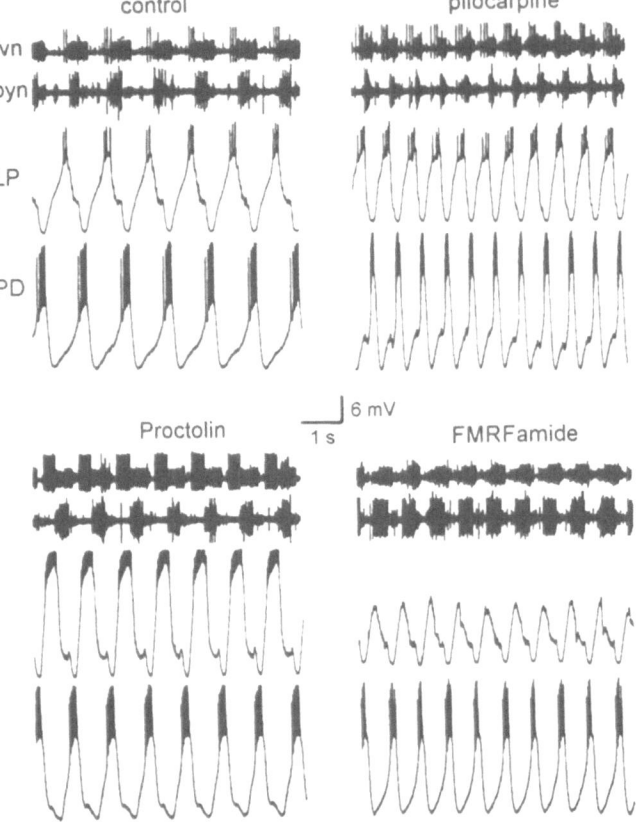

control   pilocarpine

lvn
pyn

LP

PD

6 mV
Proctolin   1 s   FMRFamide

Figure 2. The pyloric network is capable of producing multiple output patterns. Pilocarpine (10^{-4} M) increases pyloric frequency but has little effect on neuronal phasing. Proctolin (10^{-6} M) increases LP neuron activity whereas FMRFamide (10^{-6} M) decreases LP neuron activity. E_{max}: (control and FMRFamide) PD, -65 mV; LP, -74 mV; (pilocarpine): PD, -57 mV, LP, -69 mV; (proctolin): PD, -63 mV; LP, -70 mV. Figure taken (with permission) from Marder and Hooper, 1985.

CONTROL OF PACEMAKER FREQUENCY AND BURST LENGTH

Although all the pyloric neurons can be endogenous oscillators and the network architecture itself is appropriate for network based rhythmicity, generally the AB neuron serves as the network pacemaker. However, that is not to say that feedback from the rest of the network, in particular the electrically coupled PD neurons, does not profoundly affect AB neuron oscillation characteristics. One of the first unambiguous demonstrations of the effects of this feedback was found during the investigation of the cellular bases of proctolin's effects on the pyloric network in the lobster, *Panulirus interruptus* (Hooper and Marder, 1987). Proctolin application increases the frequency of slowly cycling (~1 Hz) pyloric networks to about 1 Hz, but has little effect on the frequency of pyloric networks that in control conditions are cycling at frequencies of 1 Hz or higher. This effect on frequency suggested that either or both of the electrically coupled AB and PD neurons were directly affected by proctolin, and

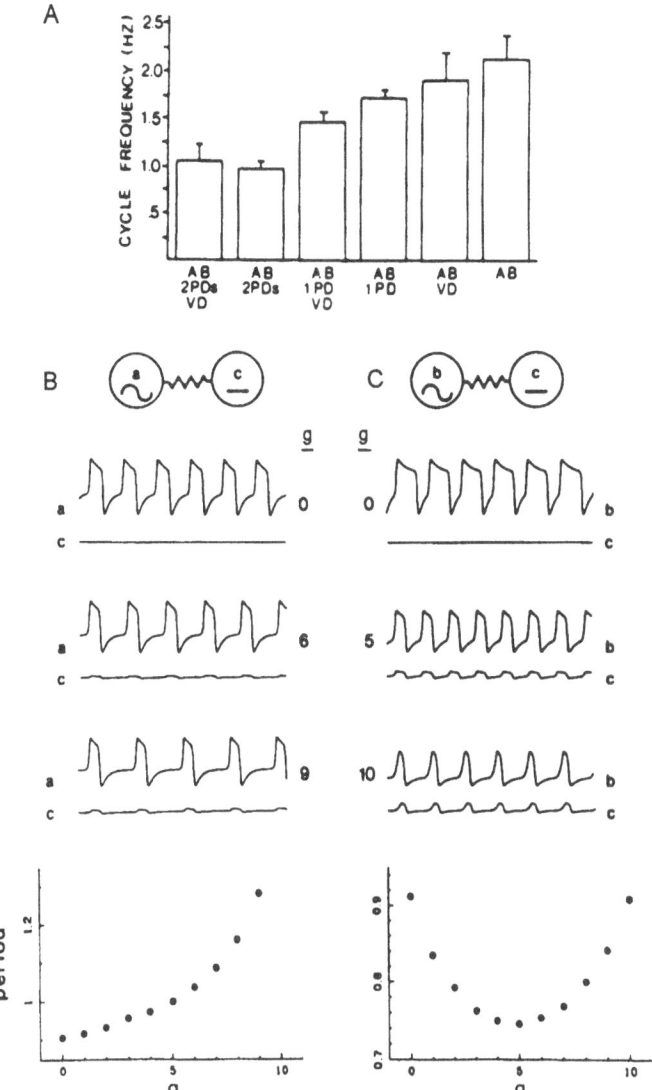

Figure 3. Input from the rest of the pyloric network helps determine pyloric pacemaker frequency. **A.** The AB neuron cycle frequency in the presence of proctolin in networks containing the neurons noted under the bars. Decreasing the number of neurons electrically coupled to the AB neuron resulted in greater cycle frequencies. **B, C.** Electrical coupling of a non-oscillatory neuron to an oscillator can either decrease or increase the oscillator's frequency. In **(B)** the oscillator waveform (trace a) is dominated by the depolarizing phase of the oscillation and electrical coupling to a passive neuron (trace c) results in increasing period as the coupling strength increases (bottom graph). In **(C)** the oscillator waveform (trace b) is originally dominated by the hyperpolarizing phase of the oscillation, and electrical coupling to a passive neuron (trace c) decreases the oscillation period (compare the top two traces to the middle two, see graph). Once the oscillators burst length is sufficiently reduced so that the oscillation waveform is dominated by its depolarizing phase, further increase in coupling strength increases oscillation period (bottom two traces, graph). Figure taken, with permission, from Marder et al., 1992.

experiments in which either the PD or AB neurons were isolated from the network (by photoinactivating electrically coupled neurons and applying PTX to block glutamatergic synapses) revealed that in fact only the AB neuron directly responded to proctolin. However, proctolin induced the isolated AB neuron to cycle at almost 2 Hz, approximately twice as fast as in the intact network. Sequential deletion of the electrically coupled PD and VD neurons from the network showed that this difference between the intact and isolated case was due to a drag on the AB neuron by a current sink from the electrically coupled PD neurons. (See Fig. 3A, note that the cycle frequency of the AB neuron increases as the network size is decreased). Thus, in the presence of proctolin, the PD neurons serve as frequency governors, allowing proctolin to increase the frequency only in slowly cycling preparations.

This observation led Kepler et al. (1990) to wonder whether electrical coupling of non-oscillatory neurons always would slow down an endogenous oscillator neuron (Fig. 3B). To test this hypothesis Kepler et al. built several mathematical models of two neuron networks composed of an oscillator and a non oscillatory neuron, and tested the effect on the oscillator frequency of changing the coupling strength between the two neurons. The rather unexpected result was that, depending on the voltage waveform of the oscillator neuron, electrical coupling to a non-oscillatory neuron can either decrease or *increase* the oscillator's cycle frequency. Further investigation of the model showed that the effect of electrical coupling depended on whether the waveform of the endogenous oscillator in the absence of coupling had a longer inward, and hence depolarizing current (Fig. 3B), or a longer outward, and hence hyperpolarizing, current phase (Fig. 3C).

This result can be understood as follows. The effect of electrical coupling is to introduce a persistent outward current into the endogenous oscillator neuron. This current prolongs the depolarization phase, and shortens the hyperpolarization phase, of the neuron's oscillations by approximately equal percentages. In neurons in which the depolarizing phase of the oscillation is longer than the hyperpolarizing phase, the additional outward current therefore prolongs the depolarization time more than it shortens the hyperpolarization time, and hence the oscillator's cycle frequency decreases. (This can be seen in Fig. 3B, which compares the activity of the electrically coupled neuron pair as coupling strength is increased). Alternatively, in neurons in which the hyperpolarizing phase of the oscillation dominates, the additional outward current due to the electrical coupling shortens the hyperpolarizing phase more than it prolongs the depolarizing phase. The neuron's burst length thus decreases more than the interburst interval increases, and hence the cycle frequency increases (This can be seen in the first four traces in Fig. 3C, which compare the activity of the neuron pair under 0 coupling strength (top two traces) with the activity under moderately strong coupling (middle two traces)). Note that under moderately strong coupling (middle traces), the burst length is so shortened that the interburst interval (depolarization phase) now dominates the oscillation waveform. Further increases in coupling strength now decrease oscillation frequency (compare the bottom traces to the middle traces) because the outward current prolongs the depolarization time more than it shortens the hyperpolarization time, as in Fig. 3B. The dependence of cycle period on coupling strength for the two types of oscillators is summarized in the bottom two graphs.

Kepler et al. performed similar analyses using a variety of different model neurons; the dependence of frequency change on oscillator waveform was robust, and did not depend on the precise characteristics of the model neurons used. Finally, these results also have important implications for situations in which oscillator waveform, as opposed to electrical coupling strength, is changed. Before this work one might have thought that changes in the ratio of hyperpolarizing to depolarizing phase in an oscillator waveform would not change the oscillator frequency. However, the results of Kepler et al. show that in networks in which the oscillator is electrically coupled to other neurons, such changes in oscillator waveform will alter the effect of the electrically coupled neurons on the oscillator, and hence also change oscillator frequency.

Another example of how network interactions can alter oscillator frequency and burst characteristics was shown by experiments Abbott et al. (1990) carried out to determine how pacemaker burst length depends on frequency. When the AB and PD neuron group is isolated from the rest of the network and current is injected into the AB neuron to alter oscillation frequency (Fig. 4A), it is apparent that both the burst duration and the interburst interval of the AB and PD neurons change proportionally as frequency changes. The AB-PD neuronal ensemble maintains a constant ratio of burst duration to cycle period, and is thus a constant duty cycle oscillator (filled circles, Fig. 4C). Alternatively, when the same experiment is

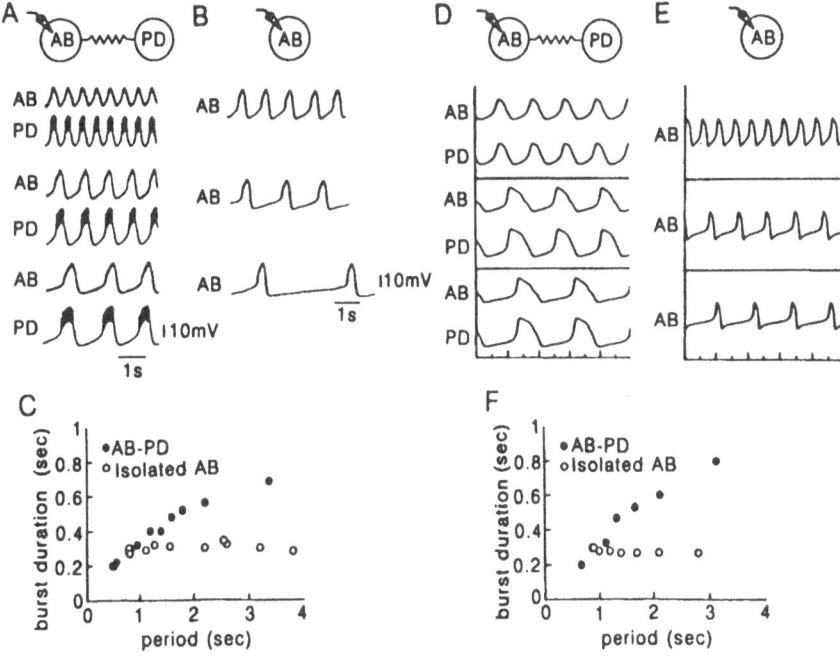

Figure 4. The PD neuron transforms the AB neuron from an oscillator with constant burst duration to an oscillator with constant duty cycle. A and B: Electrophysiological data showing the response of the AB neuron to injected current when it is electrically coupled to the PD neurons (A) and when it is isolated (B). Note that when the PD neurons are present, the burst duration of the AB neuron changes with changes in period and thus maintains a constant duty cycle. When it is isolated, its burst duration remains constant as frequency is changed (C). D and E: Computer simulations of the AB-PD neuron group. The AB model neuron was a simple oscillator, whereas the PD model neuron contained a current with time constants that were long relative to the AB neuron oscillation frequency. Note that the computer simulations reproduce well the activity seen in the real neurons. Figure taken, with permission, from Marder et al., 1993.

performed on the isolated AB neuron (Fig. 4B), AB neuron burst duration remains constant, and only the interburst interval changes (open circles, Fig. 4C).

How the PD neuron transforms the AB neuron from a constant burst duration to a constant duty cycle oscillator has not yet been determined experimentally, but modeling studies have suggested one possible mechanism. In this work the AB neuron was modeled as a simple relaxation oscillator; as with the real AB neuron, the model AB neuron was a constant burst duration oscillator (Fig. 4E; open circles, Fig. 4F). The PD model neuron was slightly more complicated in that it contained a depolarizing current that increases and then slowly inactivates with depolarization, and that decreases and from which inactivation is slowly removed (i.e., it slowly activates) with hyperpolarization. Importantly, the time constants of inactivation and activation are much greater than the burst frequency of the AB neuron (this is a simplified explanation; for a complete description see Abbott et al., 1990). When the two model neurons are electrically coupled, AB neuron depolarization depolarizes the PD neuron through their electrical coupling. This depolarization increases the PD neuron current described above, and this increase drives the PD neuron burst. During the PD neuron burst this current inactivates, helping end the PD neuron burst; during the PD neuron interburst intervals the slow current activates, and thus can again help depolarize the PD neuron when triggered by the next AB neuron burst. When the system is at equilibrium, the amounts of activation and of inactivation over a cycle are equal, and hence the average value of activation of the PD neuron current is unchanging.

If hyperpolarizing current is injected into the AB neuron to slow down the oscillation frequency, initially only the interburst interval increases and the AB-PD neuron burst length remains constant. However, now the PD neuron current has longer to activate (during the lengthened interburst intervals) than it does to inactivate (during the unchanged burst durations). As a result, the average value of activation of the PD neuron current increases. This increases the amount of depolarizing current that crosses the PD neuron membrane during the PD neuron bursts, and hence the amplitude of the PD neuron oscillations increases. This results in depolarizing current flowing across the electrical coupling from the PD to the AB neuron (because the PD neuron burst amplitude is now greater than the AB neuron burst amplitude), which causes the AB neuron burst length to increase. This process continues over several cycles until the ratio of the burst duration and interburst interval is again such that the PD neuron current inactivates and activates equally over a single oscillation cycle, and thereby attains a constant average value. As a result of this process, the electrically coupled model neuron ensemble is, like the real AB-PD neuron group, a constant duty cycle oscillator (Fig. 4D; filled circles, Fig. 4F.)

Taken together, the results presented in this section show that oscillator frequency and burst characteristics can be profoundly altered by influences from the rest of a network; these influences can function as frequency governors, determine the frequency response profile of an oscillator to external inputs, and transform constant burst duration oscillators to constant duty cycle oscillators. Consideration of the pyloric network synaptic connectivity, however, suggests that these results probably represent only a small part of the true range of network alterations of pacemaker activity present in this" simple" network. In particular, the integrative effects on pacemaker activity of the dual chemical and electrical synapses in the AB-VD neuron pair, and the feedback loops such as the AB-LP-PD-AB neuron loop are as yet completely unknown. Clearly, however, the simple formulation of a pacemaker driven network in which network frequency is set by the oscillator neuron is inadequate to explain pyloric rhythmicity control. Instead, in this network and presumably other networks in which network feedback to the oscillator is present, network oscillator function cannot be understood to even a first approximation without taking into consideration the distributed nature of the network's synaptic connectivity.

FOLLOWER NEURON PHASE CONTROL

Changes in pacemaker activity will obviously change follower neuron activity. However, as would be expected from a complex network such as the pyloric, the dependence of follower phase on pacemaker activity is not necessarily obvious. Work by Eisen and Marder (1984) on PY neuron phase control provided the first case in which pyloric follower phase regulation was understood at the cellular level (Fig. 5). In this study, dopamine was applied and an input nerve to the STG (the inferior ventricular nerve (IVN)) was stimulated. The result was that dopamine advanced PY neuron activity, and IVN stimulation retarded it. Experiments on isolated PY neurons revealed that neither treatment had direct effects on the PY neurons, so the induced changes in PY neuron activity presumably arose as an indirect consequence of changes induced in the activity of other neurons in the network. The authors showed that the PY neuron shifts arose because of changes in the relative amount of inhibition the PY neurons received from the electrically coupled AB and PD neurons. The AB neuron

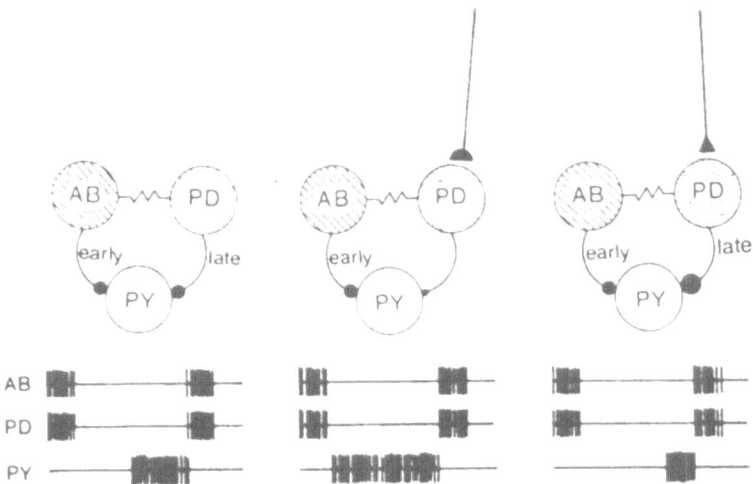

Figure 5. PY neuron phase can be altered by inputs to its presynaptic neurons. The AB neuron synaptic input is fast and short acting; the PD neuron synaptic input is slow and long lasting. Alteration of PD neuron activity (inhibition in the middle column, excitation in the right column) changes the ratio of AB and PD neuron inhibition, and hence PY neuron firing phase. Figure taken, with permission, from Marder, 1984.

is glutamatergic and the PD neurons are cholinergic; in this system transmission via glutamatergic synapses is rapid in onset and of short duration whereas cholinergic transmission is slow to onset and long lasting. The key observation was that dopamine decreased PD neuron transmitter release, and IVN stimulation increased it. Thus in the presence of dopamine, the magnitude of the long lasting inhibition from the PD neuron is reduced, so the PY neurons begin to fire earlier after the pacemaker burst (middle column, Fig. 5). Alternatively, after IVN stimulation the magnitude of the long lasting inhibition from the PD neuron is increased, thus causing the beginning of PY neuron firing to be delayed (right column, Fig. 5).

Another mechanism for controlling follower phase is suggested by the connections of another pyloric follower neuron, the VD neuron. As is seen in Figure 6, the VD neuron is not only electrically coupled to the AB neuron, but also receives an inhibitory chemical synapse from it. In addition, the VD neuron is electrically coupled to the PD neurons. Computer simulations by Mulloney et al. (1981) indicated that, depending on whether the electrical or chemical synapses are dominant, the VD neuron could be made to fire either in phase or out of phase with the AB-PD neuron bursts (compare left and right columns). The physiological relevance of these simulations was later supported by studies by Nagy and Dickinson (1983) which showed that stimulation of a modulatory input neuron, the APM neuron, induced similar shifts in VD neuron firing in the real network. Because neuron isolations were not performed in this work, the exact cellular mechanisms (e.g., decreased AB neuron transmitter release, changes in VD neuron sensitivity) of this shift are not known. However, the combination of computer simulation and electrophysiology clearly shows that combined inhibition and electrical coupling provide a powerful mechanism for creating variable phase neuronal firing.

A final example of mechanisms that control firing time comes from investigations by Hooper and Moulins (1989, 1990) on another input to the pyloric network, the lateral posterolateral nerve (lpln). Stimulation of this input induces a persistent loss of plateau potentials in the VD neuron and as a result the VD neuron no longer fires with the pyloric pattern. Concomitant with this change is a dramatic advance in IC neuron firing, as is shown in Figure 7 (raw data, top panel; phase normalized, bottom panel). Lpln stimulation had no effect on isolated IC neurons, so this phase shift is again the result of indirect network mediated effects. A series of experiments in which the VD neuron was deleted or

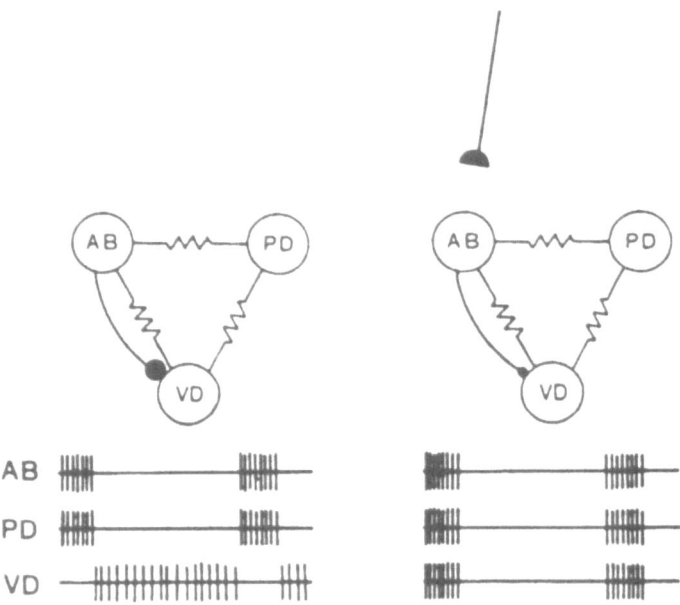

Figure 6. VD neuron phase can be altered by changing the relative importance of chemical and electrical synaptic input. The VD neuron receives both inhibitory chemical and electrical synaptic input from the AB neuron. Normally the chemical input dominates, and the VD neuron fires out of phase with the AB neuron. Inputs that increase the relative functional strength of the electrical synapse can result in the VD neuron firing in phase with the AB neuron (right column). Figure taken, with permission, from Marder, 1984.

hyperpolarized showed that the IC neuron phase advance was specifically due to the lack of VD neuron firing. However, as seems to be the usual case when dealing with distributed networks, this change is completely non-intuitive; the VD neuron normally fires after the IC neuron, and thus would presumably help *end* its bursts, yet when the VD neuron is silent it is the phase of the *beginning* of the IC neuron burst that changes. This anomaly presumably occurs because the inhibition from the AB/PD neuron ensemble to the IC neuron is insufficient to block IC neuron firing without the additional long lasting cholinergic inhibition that originates from the VD neuron (note the marked decrease in IC neuron hyperpolarization in the right column of the top panel of Fig. 7).

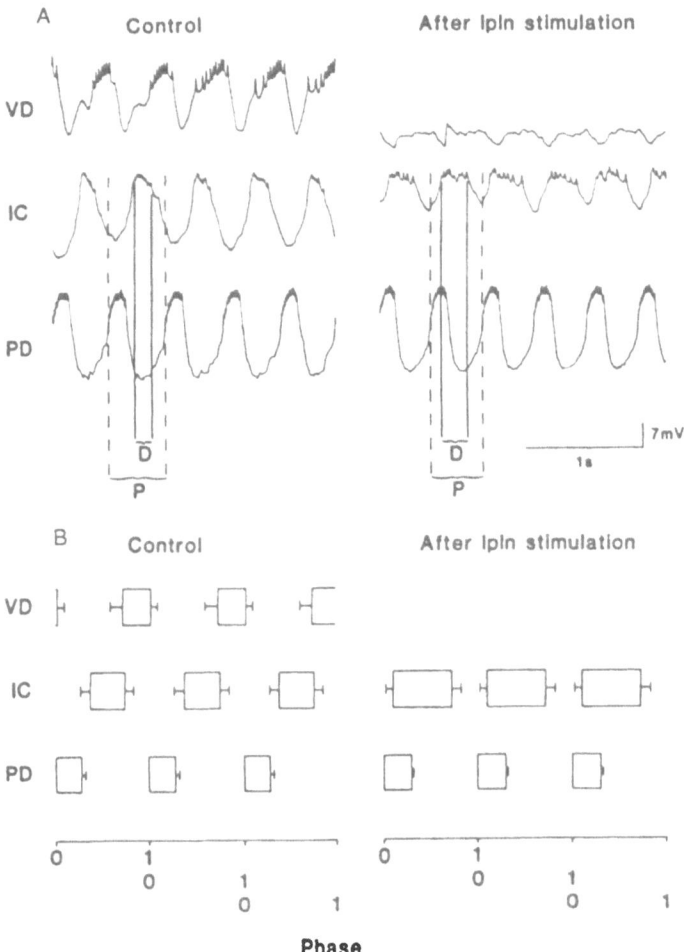

Figure 7. Sensory input induced changes in IC neuron activity are the indirect network based result of changes in VD neuron activity. **A** shows raw data; stimulation of a sensory input, the lpln, silences the VD neuron and induces the IC neuron to fire earlier in the pyloric pattern. E_{max}: (before lpln stimulation) VD, -58 mV; IC, -50 mV; PD, -60 mV. **B**. Phase normalization of the data from 7 experiments. Phase is calculated as follows: The first PD neuron spike defines the beginning of a pyloric cycle. The delay from the cycle beginning to events in the cycle (e.g., the beginning of a neuron's burst) is then measured, and the phase of the event is the delay divided by the period. Figure taken, with permission, from Hooper et al., 1990.

These authors extended these results by investigating the bases of the change in intraburst firing profile of the IC neuron that occurs after lpln stimulation (Fig. 8). Figure 8A shows the instantaneous firing frequency of the IC neuron as a function of phase when the VD neuron is a part of the pyloric pattern (the inset is representative raw data). It is apparent that the intraburst firing pattern of the IC neuron has a single peaked frequency profile. Figure 8B shows the same data after lpln stimulation when the VD neuron is inactivated. Now the IC neuron has a two peaked frequency profile. Again this effect is solely an indirect, network

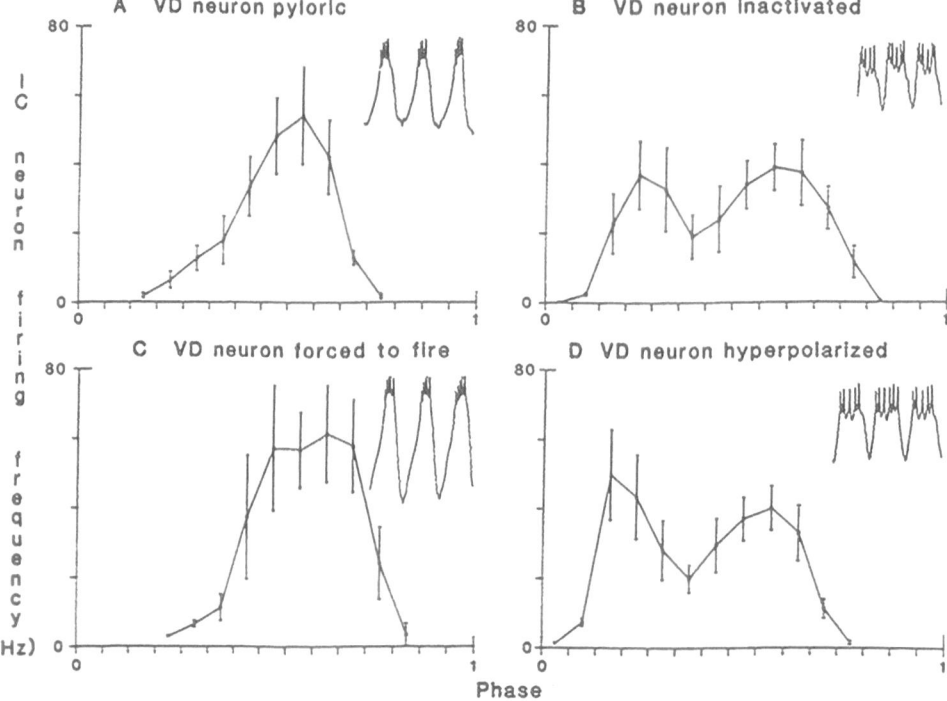

Figure 8. IC neuron intraburst firing pattern is also determined by the VD neuron. The instantaneous IC neuron firing frequency as a function of phase was determined in four different conditions: when the VD neuron was active in the pyloric pattern (**A**), when the VD neuron was silent (inactivated) after lpln stimulation (**B**), when an inactivated VD neuron was forced by current injection to fire approximately normally in the pyloric pattern (**C**), and when an active VD neuron was silenced by the injection of hyperpolarizing current (**D**). Insets show representative raw data. Note that in the absence of the VD neuron the IC neuron intraburst firing profile shifts from having a single maximum to having two maxima per pyloric cycle. All of the lpln induced changes in IC neuron activity arise as the indirect consequence of changes in VD neuron activity. Figure taken, with permission, from Hooper and Moulins, 1990.

based result of the lack of VD neuron firing, since forcing a silent VD neuron to fire restores IC neuron firing to control values (compare Fig. 8C to Fig. 8A), and silencing a firing VD neuron results in alterations in IC neuron firing similar to those induced by lpln stimulation (compare Fig. 8D to Fig. 8B). Thus, not only neuronal phase in a pattern, but also intraburst firing parameters, can be profoundly altered by indirect, network based mechanisms.

PHASE VS. CYCLE FREQUENCY

Almost all motor patterns are expressed at many different frequencies, and it is of course necessary that at all frequencies the neuronal firing order and relative timing be such that a functionally relevant motor pattern is produced. Nervous systems appear to have taken two different approaches to this problem depending on the nature of the motor pattern. Motor patterns that interact with solid substrates (e.g., walking) or that otherwise have very clear power vs. return strokes (flight) preferentially alter the duration of the stance, or power, phase as frequency changes. Motor patterns that do not have distinct power strokes or are characterized by all phases moving against considerable resistance (air stepping, underwater swimming) very often instead proportionally alter the durations of all phases of the motor pattern. Regardless of the approach, however, regulation of phase as frequency changes is a non trivial problem in distributed networks composed of neurons with active properties. This problem basically arises because neurons in such networks actively transform the inputs they receive into outputs, and they do so according to the inherent time constants of their synaptic connections and their own membrane characteristics. As a result, for phase to be appropriately regulated as the timing of the inputs these neurons receive changes, the time constants governing the response of these neurons must also change appropriately, and it is not intuitively clear how this might be done.

To make this issue concrete, consider the hypothetical network outlined in Figure 9A. This network produces a rhythmic output as a consequence of the long lasting cellular properties of its neurons and their synaptic interconnectivity (the dynamical issues raised here would apply to any network containing neurons with slow cellular properties). Neuron A is an endogenous burster with a period of 1 sec and a burst duration of 333 msec. Neurons B and C are plateauing neurons with postinhibitory rebound. This means that they respond to the release of inhibition by depolarizing above resting potential to a quasi-stable suprathreshold membrane potential, where they remain until they are inhibited. Neuron C recovers from neuron A's inhibition 333 msec later than does neuron B. Thus, at the end of neuron As burst, neuron B fires immediately after neuron A stops and neuron C begins to fire 333 msec later. Neuron C shuts off neuron B, which thus has a 333 msec burst length; the next neuron A burst turns off neuron C, and so neuron C also fires for 333 msec. The middle column of Figure 9B shows a block diagram of this network's output. What happens if current is injected into neuron A without any phase and duty cycle compensatory mechanisms present in the network? If we double the period to 2 sec (right column, Fig. 9B), neurons A and B still burst for only 333 msec. However, neuron C fires until it is turned off by the next neuron A burst, and thus fires for 1.33 sec. Alternatively, if we halve the period to 0.5 sec (left column, Fig. 9B), the next neuron A burst occurs only 167 msec after the end of the first. This next neuron A burst shuts off neuron B, which fires for only 167 msec, and neuron C doesn't fire at all. Clearly, without compensatory mechanisms phase and duty cycle are not maintained as frequency changes, and if this were a motor pattern, the resulting movements at different frequencies would be very different.

Any number of mechanisms extrinsic to the network (e.g., sensory feedback) could underlie such regulation. However, many of the well described isolated rhythmic networks (pyloric (Abbott et al., 1990; Eisen, 1984; see below), leech swimming (Pearce and Friesen, 1985), lamprey swimming (Grillner et al., 1987), crab gill bailer (R. DiCaprio, personal communication)) intrinsically regulate phase and duty cycle. What *intrinsic* network compensatory mechanisms are needed for this regulation? To return to our example, assume that phase and duty cycle are strictly maintained. First, the oscillator neuron itself must compensate; when its period is doubled or halved, its burst length must also double (to 666 msec) or halve (to 167 msec). Second, since neuron C shuts off neuron B, and thus determines neuron B's burst duration, neuron C's delay to firing after neuron A must also change. For instance, during high frequency oscillation neuron A's burst duration will be only 167 msec. To maintain the pattern, neuron B must also fire only for 167 msec. Consequently, neuron C's delay after neuron A must halve to only 167 msec so that neuron C inhibits neuron B after only 167 msec.

Several lines of evidence argue that the pyloric network undergoes phase compensation as its cycle frequency changes. Chronic recordings in intact animals show that the pyloric motor pattern displays considerable variability that is associated with the behavioral state of the animal. In this work only LP and PY neuron phases were reported; in the pyloric motor pattern observed in recently fed animals these neurons phase compensate over a two-fold frequency range (Rezer and Moulins, 1983). Early work by Eisen and Marder (Eisen, 1984; Eisen and Marder, 1984) showed that phase and duty cycle are maintained over the three-fold range of frequencies spontaneously present in *in vitro* preparations. As noted earlier, Abbott et al. (1990) showed that the AB/PD neuron group maintains constant duty cycles over a four-fold frequency range when pyloric frequency is altered by current injection into the

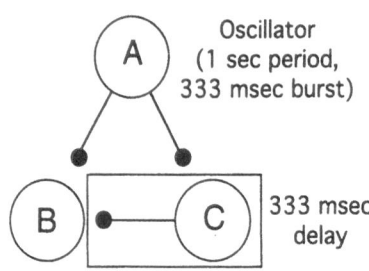

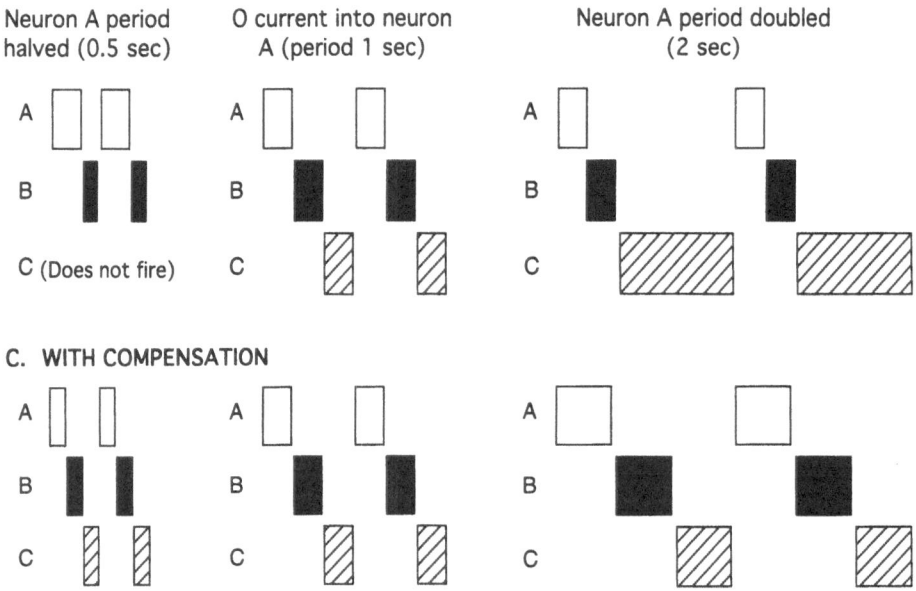

Figure 9. Phase maintenance is not automatic in networks composed of neurons with active properties. A. A hypothetical network with active properties. B. The patterns this network produces when the oscillator is depolarized (left column), at rest (middle column), or hyperpolarized (right column). Note that the patterns produced are very different. C. The patterns this network would produce if perfect phase compensation occurred. See text for details.

pacemaker AB neuron. Finally, data from my laboratory shows that when frequency is altered in this manner, all the pyloric neuron types show phase and duty cycle regulation over at least a three-fold frequency range. Figure 10 shows extracellular recordings of the activity of the PD neurons (top traces), the LP and PY neurons (middle traces; LP neuron spikes are large, PY neuron spikes are small), and the VD and IC neurons (bottom traces; large spikes are from the VD neuron, small spikes are from the IC neuron). The three panels show the pyloric pattern when depolarizing current is injected into the AB neuron (left panel, cycle frequency 1.4 Hz), when no current is injected (middle panel, cycle frequency 0.74 Hz), and when hyperpolarizing current is injected (right panel, cycle frequency 0.5 Hz). Both neuronal burst duration and delay from inhibition of several of the neurons clearly increase with increasing cycle period, and decrease with decreasing cycle period. The bottom three panels show phase normalized plots of the data above them. It is apparent that, unlike the result that would be

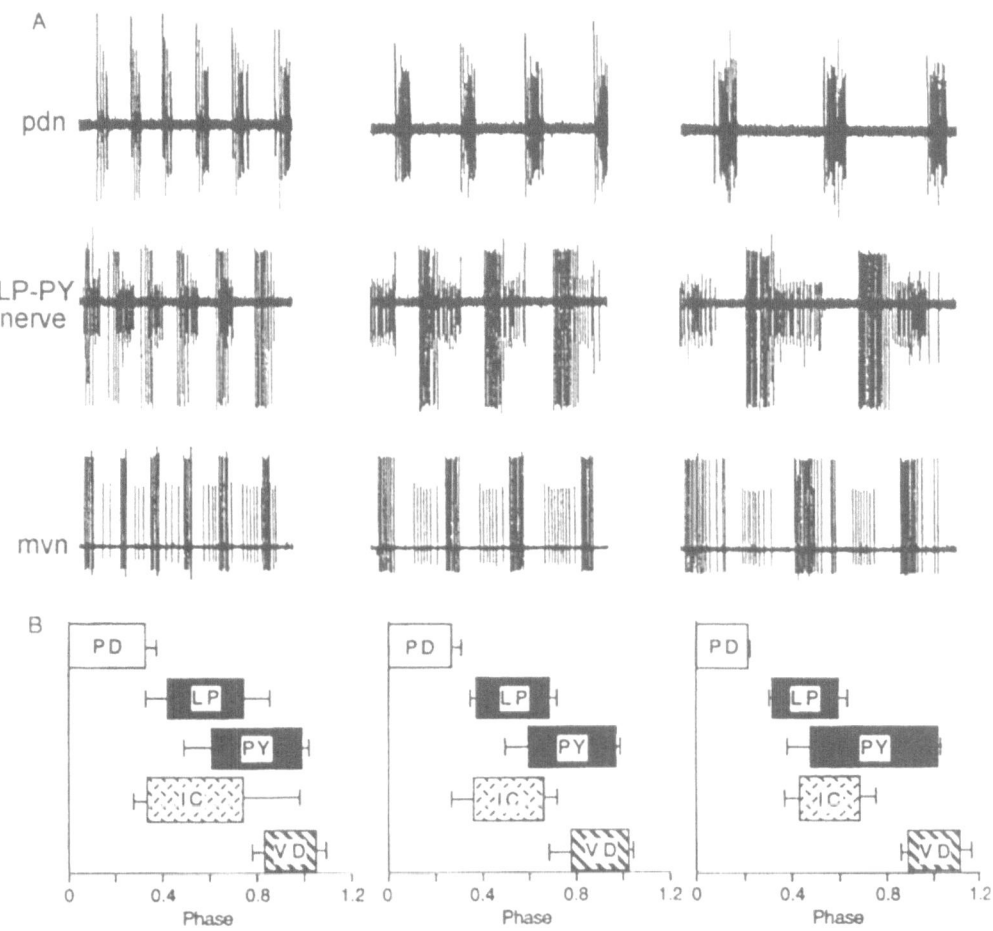

Figure 10. Phase as frequency changes in the pyloric network. **A**. Raw data showing the burst durations and delay to firing of the pyloric neurons as pyloric frequency is altered by current injection into the pacemaker AB neuron. **B**. Phase normalized plots (see Fig. 7 legend) of the data above. It is apparent that, unlike the case shown in Fig. 9B, the pyloric network continues to produce a similar pattern as pyloric frequency is altered.

expected without phase compensatory mechanisms, the pyloric pattern is largely maintained as pyloric frequency is changed.

How does the network perform this task? The short answer is that we do not yet know. However, given that we know the synaptic connectivity of the pyloric network, and a great deal about the membrane and synaptic properties of its neurons, we can at least define the problem in very specific terms. First, the pacemaker group, as noted earlier, is a nearly constant duty cycle oscillator. We do not yet know whether the modeling studies outlined above correctly represent the solution the real network employs, but it is clear from experimental evidence that constant duty cycle activity depends critically on the electrical coupling of the pacemaker AB neuron to the PD neurons. This is interesting because our modeling studies suggest that it is difficult to build a biologically plausible model oscillator that keeps phase constancy in isolation. It is thus possible that part of the reason the pyloric network has a multineuronal pacemaker group is because network feedback to the oscillator neuron is essential to produce constant duty cycle oscillations; further work in similarly well defined model systems will be necessary in order to determine whether this is a general feature of constant duty cycle systems.

Having a constant duty cycle pacemaker is only part of the solution to the problem of constructing a constant duty cycle motor pattern, however. If we next consider the neurons that fire immediately after the pacemaker (the LP and IC neurons), we note that they do not fire immediately after the pacemaker burst, but instead fire with a delay. For perfect phase compensation, this delay would also change as cycle frequency changes. In the pyloric network, however, recent work from my lab shows that this delay does not change as pyloric frequency changes; the phase compensation of these neurons solely results from the changing burst length of the pyloric pacemaker group. A consequence of this is that phase compensation for these two neurons is less than perfect. However, since the delay to firing of the LP and IC neurons after pacemaker firing is relatively short in comparison to the pacemaker burst length, this error is relatively small, as can be seen in Fig. 10B.

If we consider the neurons that fire next in the cycle (the VD and PY neurons) the situation is different, as these neurons do show phase compensation of their delays to firing, i.e., their delay to firing increases and decreases as pyloric period is increased and decreased. How this compensation is achieved is difficult to understand for two reasons. First, these neurons inhibit, and are inhibited by, the neurons that fire before them (the LP and IC neurons). Second, these neurons begin to inhibit their postsynaptic targets well before they begin to fire, since all the pyloric network neurons release transmitter as a graded function of membrane potential (Graubard, 1978). Thus as the VD and PY neurons begin to depolarize due to their own rebound properties, a competition occurs among the inhibition they receive, their own tendency to depolarize, and the inhibition they produce in the LP and IC neurons. This complex interplay of synaptic, cellular, and network properties is very non-intuitive, and as yet we cannot even begin to suggest how it occurs. However, as a result of the size and advantages of the pyloric network, at least here it can be studied experimentally by using the cell deletion and isolation experiments and computer simulations that have been used to study other aspects of pyloric network activity.

CONCLUSIONS AND FUTURE DIRECTIONS

It is often argued that by studying small model systems, insights of general relevance can be gained. The pyloric network has been the subject of intense study for almost twenty years; what generally applicable conclusions about temporal organization and processing by nervous systems have resulted? First, this work demonstrates the critical importance of knowing a neural network's architecture in terms of synaptic connectivity and cellular properties if one wishes to understand how the network operates. Repeatedly in the examples given above the activities of various neurons have changed in response to various treatments even though the neurons in question were not affected directly by the treatment. In the absence of a sufficiently detailed understanding of the network, any number of different explanations for these observations could have been postulated, and it would have been impossible to determine which was the more correct, or even whether all possible mechanisms had been suggested. The pyloric network is a very small network compared to many others that generate biologically interesting phenomena; the fact that the processes that underlie the activity of even such a small network are non-intuitive and complex underscores the importance of understanding the fine structure of a system in order to understand the behavior of the system as a whole.

Second, this work demonstrates just how complex and non-intuitive distributed systems are, particularly when composed of neurons and synapses with both short and long lasting time constants. Distributed systems by their very nature function collectively, and are thus largely resistant to reductionist analysis. The presence in such networks of neurons and synapses with specific inherent time constants ranging from very short to very long (with respect to the times over which the network processes or generates information) makes this analysis enormously more difficult because it means that the computation is not only distributed across the network, but also distributed across time. However, it may be precisely this complexity that is the basis for the control of pyloric phase, burst duration, and frequency that has been shown above.

Third, this work supports the concept of delegative control as outlined in the introduction. It is clear that many to most of the changes that occur in pyloric network activity in response to a treatment are not due to direct effects of the treatments on the neurons themselves. Instead these changes occur as a result of indirect, network mediated effects that result from changes in the activities of the relatively few neurons that are directly affected by the treatment. This implies that the pyloric network has evolved in such a way that, for each treatment, certain neurons serve as entry points to the network. The network's architecture is such that the network then automatically and collectively moves to an appropriate configuration that produces the desired output. Too few other networks are understood in sufficient detail to know how general this process is, but the possible computational advantages of such an approach are compelling.

Fourth, this work has demonstrated several specific neuronal properties and synaptic connectivities that seem to be associated with specific functional operations. First, the presence of both short and long lasting synaptic inputs to the same follower neuron is an effective way to adjust neuron firing phase. Second, combining electrical coupling and chemical inhibition can also provide a variable phase synapse. Third, oscillator frequency and duty cycle are highly affected by electrical coupling to other neurons. In particular, such coupling can both increase and decrease oscillator frequency, and transform a constant burst duration oscillator into a constant duty cycle oscillator.

Finally, what can we expect from studies of the pyloric network in the future? Perhaps the fundamental problem neurobiology faces today is understanding how form creates function; that is, how do synaptic connectivity and cellular properties lead to computation? It highlights our ignorance that even today, after twenty years study of small model neural systems, I can only name a few cases in which specific neuronal architectures give rise to specific functions and that in general, no one can predict what a neural network will do just from looking at its connectivity and neuronal properties. Presumably, however, rules do exist; a specific synaptic connectivity pattern when coupled with a set of cellular properties leads to a specific set of network functional attributes. A combination of electrophysiology and computer simulation is being used in several laboratories that study the pyloric and other model neural network systems to reveal exactly these relationships; given the recent increases in computer power and our increasing understanding of distributed systems we can expect this work to continue apace, and that our understanding of how network architecture leads to network output will correspondingly increase.

However, this will not be sufficient for us to truly understand how form leads to function unless two other as yet relatively little investigated areas are addressed. The first is an elucidation of pyloric network anatomy. As yet we do not know whether each different pyloric neuron type has a canonical structure; we do not know if the different electrical and chemical inputs on a neuron are segregated to different parts of its arbor, and we do not know where the various conductances that give rise to active neuronal properties are located in the neuron. Until these questions begin to be answered, we do not know how cable properties may affect neuronal function, and we cannot make detailed multicompartmental models of the neurons. The second area that needs to be investigated is the transform between neural network output and movement. Although we presume that the changes in neural output I have described lead to changes in motor pattern, we actually do not know even qualitatively how muscle contraction depends on neural input, nor how changes in muscle contraction affect movement production. Until these neuromuscular and biomechanical issues are addressed, we will never be able to describe behavior on the cellular level even for this small and well defined system. These issues are, however, presently being actively investigated in my and other laboratories, and the likelihood of rapid progress is good.

REFERENCES

Abbott, L.F., Hooper, S.L., Kepler, T., and Marder, E., 1990, Oscillating networks: modeling the pyloric circuit of the stomatogastric ganglion, *in*: "Proceedings of the International Joint Conference on Neural Networks", I.E.E.E., Ann Arbor, MI.

Bal, T., Nagy, F., and Moulins, M., 1988, The pyloric central pattern generator in crustacea: a set of conditional neuronal oscillators, *J. Comp. Physiol.* 163:715.

Bidaut, M., 1980, Pharmacological dissection of pyloric network of the lobster stomatogastric ganglion using picrotoxin, *J. Neurophysiol.* 44:1089.

Delcomyn, F., 1980, Neural basis of rhythmic behavior in animals, *Science* 210:492.

Dickinson, P.S., Mecsas, C., and Marder, E., 1990, Peptidergic modulation of a multioscillator system in the lobster. I. Activation of the cardiac sac motor pattern by the neuropeptides proctolin and red pigment concentrating hormone, *J. Neurophysiol.* 61:833.

Dickinson, P.S., and Nagy, F., 1983, Control of a central pattern generator by an identified modulatory interneurone in Crustacea. II. Induction and modification of plateau properties in pyloric neurones, *J. Exp. Biol.* 105:59.

Eisen, J.S., 1984, Separation and characterization of the synaptic actions of electrically coupled neurons in the stomatogastric ganglion of the lobster, *Panulirus interruptus*. Ph.D. thesis, Brandeis University.

Eisen, J.S., and Marder, E., 1982, Mechanisms underlying pattern generation in lobster stomatogastric ganglion as determined by selective inactivation of identified neurons. III. Synaptic connections of electrically coupled pyloric neurons, *J. Neurophysiol.* 48:1392.

Eisen, J.S., and Marder, E., 1984, A mechanism for production of phase shifts in a pattern generator, *J. Neurophysiol.* 51:1375.

Flamm, R.E., and Harris-Warrick, R.M., 1986a, Aminergic modulation in the lobster stomatogastric ganglion. I. Effects on the motor pattern and activity of neurons within the pyloric circuit, *J. Neurophysiol.* 55:847.

Flamm, R.E., and Harris-Warrick, R.M., 1986b, Aminergic modulation in the lobster stomatogastric ganglion. II. Target neurons of dopamine, octopamine and serotonin within the pyloric circuit, *J. Neurophysiol.* 55:866.

Gola, M., and Selverston, A.I., 1981, Ionic requirements for bursting activity in lobster stomatogastric neurons, *J. Comp. Physiol.* 145:191.

Graubard, K., 1978, Synaptic transmission without action potentials: input-output properties of a nonspiking presynaptic neuron, *J. Neurophysiol.* 41:1014.

Grillner, S., Wallén, P., Dale, N., Brodin, L., Buchanan, J., and Hill, R., 1987, Transmitters, membrane properties, and network circuitry in the control of locomotion in lamprey, *Trends Neurosci.* 10:34.

Harris-Warrick, R.M., Marder, E., Selverston, A.I., and Moulins, M., eds., 1992a, "Dynamic Biological Networks. The Stomatogastric Nervous System", MIT Press, Cambridge, MA.

Harris-Warrick, R.M., Nagy, F., and Nusbaum, M.P., 1992b, Neuromodulation of stomatogastric networks by identified neurons and transmitters, *in*: "Dynamic Biological Networks. The Stomatogastric Nervous System", R.M. Harris-Warrick, E. Marder, A.I. Selverston, and M. Moulins, eds., MIT Press, Boston, MA.

Hartline, D.K., 1979, Pattern generation in the lobster (*Panulirus*) stomatogastric ganglion. II. Pyloric network simulation, *Biol. Cybern.* 33:223.

Hartline, D.K., 1987, Modeling stomatogastric ganglion, *in* "The Crustacean Stomatogastric System", A.I. Selverston and M. Moulins, eds., Springer-Verlag, Berlin.

Hooper, S.L., and Marder, E., 1987, Modulation of the lobster pyloric rhythm by the peptide, proctolin, *J. Neurosci.* 7:2097.

Hooper, S.L., and Moulins, M., 1989, A neuron switches from one network to another by sensory induced changes in its membrane properties, *Science* 244:1587.

Hooper, S.L., and Moulins, M., 1990, Flexibility in the stomatogastric nervous system of the lobster: II. Synaptic and cellular mechanisms responsible for a long lasting restructuring of the pyloric network, *J. Neurophysiol.* 64:1574.

Hooper, S.L., Nonnotte, L., and Moulins, M., 1990, Flexibility in the stomatogastric nervous system of the lobster: I. Sensory input induces long lasting changes in the output of the pyloric network, *J. Neurophysiol.* 64:1555.

Johnson, B.R., and Hooper, S.L., 1992, Overview of the stomatogastric nervous system, *in* "Dynamic Biological Networks. The Stomatogastric Nervous System", R.M. Harris-Warrick, E. Marder, A.I. Selverston, and M. Moulins, eds., MIT Press, Boston, MA.

Kepler, T.B., Marder, E., and Abbott, L.F., 1990, The effect of electrical coupling on the frequency of model neuronal oscillators, *Science* 248:83.

Marder, E., 1984, Mechanisms underlying neurotransmitter modulation of a neuronal circuit, *Trends Neurosci.* 7:48.

Marder, E., Abbott, L.F., Buchholtz, F., Epstein, I., Golowasch, J., Hooper, S.L., and Kepler, T., 1993, Physiological insights from cellular and network models of the stomatogastric nervous system of lobsters and crabs, *American Zoologist* 33:29.

Marder, E., Abbott, L.F., Kepler, T.B., and Hooper, S.L., 1992, Modification of oscillator function by electrical coupling to non-oscillatory neurons, *in* "Induced Rhythms in the Brain", E. Baser and T.H. Bullock, eds., Birkhauser, Boston.

Marder, E., and Eisen, J.S., 1984a, Transmitter identification of pyloric neurons: electrically coupled neurons use different transmitters, *J. Neurophysiol.* 51:1362.

Marder, E., and Eisen, J.S., 1984b, Electrically coupled pacemaker neurons respond differently to same physiological inputs and neurotransmitters, *J. Neurophysiol.* 51:1345.

Marder, E., and Hooper, S.L., 1985, Neurotransmitter modulation of the stomatogastric ganglion of decapod crustaceans, *in* "Model Neural Networks and Behavior", A.I. Selverston, ed., Plenum Press, New York, NY.

Miller, J.P., 1987, Pyloric mechanisms, *in* "The Crustacean Stomatogastric System," A.I. Selverston and M. Moulins, eds., Springer-Verlag, Berlin.

Miller, J.P., and Selverston, A.I., 1979, Rapid killing of single neurons by irradiation of intracellularly injected dye, *Science* 206:702.

Miller, J.P., and Selverston, A.I., 1982a, Mechanisms underlying pattern generation in lobster stomatogastric ganglion as determined by selective inactivation of identified neurons. II. Oscillatory properties of pyloric neurons, *J. Neurophysiol.* 48:1378.

Miller, J.P., and Selverston, A.I., 1982b, Mechanisms underlying pattern generation in lobster stomatogastric ganglion as determined by selective inactivation of identified neurons. IV. Network properties of pyloric system, *J. Neurophysiol.* 48:1416.

Mulloney, B., Perkel, D.H., and Budelli, R.W., 1981, Motor-pattern production: interaction of chemical and electrical synapses, *Brain Res.* 229:25.

Nagy, F., and Dickinson, P.S., 1983, Control of a central pattern generator by an identified modulatory interneurone in crustacea. I. Modulation of the pyloric motor output, *J. Exp. Biol.* 105:33.

Nusbaum, M.P. and Marder, E., 1989, A modulatory proctolin-containing neuron (MPN). II. State-dependent modulation of rhythmic motor activity, *J. Neurosci.* 9:1600.

Pearce, R.A., and Friesen, W.O., 1985, Intersegmental coordination of the leech swimming rhythm. II. Comparison of long and short chains of ganglia. *J. Neurophysiol.* 54:1460.

Raper, J.A., 1979, Nonimpulse-mediated synaptic transmission during the generation of a cyclic motor pattern, *Science* 205:304.

Rezer, E., and Moulins, M., 1983, Expression of the crustacean pyloric pattern generator in the intact animal, *J. Comp. Physiol.* 153:17.

Russell, D.F., and Graubard, K., 1987, Cellular and synaptic properties, *in* "The Crustacean Stomatogastric System", A.I. Selverston and M. Moulins, eds., Springer-Verlag, Berlin.

Russell, D.F., and Hartline, D.K., 1978, Bursting neural networks: A reexamination, *Science* 200:453.

Russell, D.F., and Hartline, D.K., 1982, Slow active potentials and bursting motor patterns in pyloric network of the lobster, *Panulirus interruptus*, *J. Neurophysiol.* 48:914.

Selverston, A.I., and Miller, J.P., 1980, Mechanisms underlying pattern generation in lobster stomatogastric ganglion as determined by selective inactivation of identified neurons. I. Pyloric system, *J. Neurophysiol.* 44:1102.

Selverston, A.I., and Moulins, M., eds., 1987, "The Crustacean Stomatogastric System", Springer-Verlag, Berlin.

Selverston, A.I., Russell, D.F., Miller, J.P., and King, D.G., 1976, The stomatogastric nervous system: Structure and function of a small neural network, *Prog. Neurobiol.* 7:215.

THE ROLE OF PERCEPTION IN TIMING: FEEDBACK CONTROL IN MOTOR PROGRAMMING AND TASK DYNAMICS

Geoffrey P. Bingham

Department of Psychology
Indiana University
Bloomington, IN 47405

INTRODUCTION

The source of temporal order in human motor behavior is currently the subject of a debate between the advocates of two principal approaches, the motor programming approach and the task-dynamic approach (Meijer and Roth, 1988). Advocates of motor programming argue that timing is determined by a neural program that imposes temporal order on the components of the motor system (Gielen, 1991; Keele, Cohen, and Ivry, 1990; Schmidt, 1988; Viviani and Laissard, 1979). Advocates of the task-dynamic approach argue that timing is determined by a task specific dynamic assembled through interaction among various components of a perception/action system. The components of this system include the nervous system, the muscles, tendons, and joints, as well as circulatory and environmental elements (Beek, 1989; Bingham, 1988; Kelso and Kay, 1986; Kugler and Turvey, 1987). Although the two approaches differ in many respects, the key difference is in the relative importance assigned to perception in modifying ongoing actions.

CRITIQUE OF THE MOTOR PROGRAMMING APPROACH TO TIMING

In the motor programming model, perception is treated as having a small or indirect role in the modification of motor behaviors once they are initiated (Rosenbaum, 1991; Schmidt, 1982; Schmidt, 1988). There are two reasons for this assumption. First, only small modifications in the form of an ongoing behavior are thought to be possible due to the potential instability of feedback control. Substantial feedback delay combined with high corrective gain has been shown to yield unstable behavior (Franklin, Powell, and Emami-Naeini, 1994; Hogan, et al., 1987). Given the substantial feedback delays that must be incurred within the nervous system, stability can theoretically only be achieved by keeping corrective gain low. Thus, according to this analysis, only small modifications can be made to ongoing behaviors, implying that behaviors must be largely preprogrammed.

Second, assuming that changes to ongoing behaviors must be small, the only way a behavior can be strongly modified is through knowledge of results. Hypothetically, perception of the relative success or failure in achieving a goal after completion of a goal-directed behavior is used to re-parameterize or re-organize the behavior. The problem with this explanation lies in the indirect and ambiguous relation between perceptual variables and motor variables. According to the motor programming theory, perceptual error variables for complex behaviors can only be defined in the spatial domain. Motor control variables, on the other hand, are temporal and force related. Because the respective variables are incommensurate, mediating transformations are required. Determining the nature of the transformation from spatial inputs to temporally modulated outputs is called the problem of sensori-motor integration (Gielen, 1991; Rosenbaum, 1991).

Because the operative perceptual variables are assumed to be strictly spatial, perception cannot be a direct source of temporal order and timing must be derived elsewhere. In this approach, circuitry within the nervous system is assumed to impose timing on the motor apparatus (Gielen, 1991; Keele, Cohen, and Ivry, 1990; Schmidt, 1988; Viviani and Laissard, 1979). The problem with this approach is that non-neural components constrain the possible timing. These must be and, in fact, are reflected in the temporal order of functionally effective behaviors.

For instance, in forceful rhythmic behavior (e.g. hammering or marathon running), movements that are too slow or too fast do not allow for the storage and recovery of mechanical energy in passive elastic elements such as the elastic tendonous sheaths of muscles arranged in parallel and elastic connective tissue and tendons arranged in series (Alexander, 1984; Cavagna, 1977; Cavagna, Heglund, and Taylor, 1977; Cavanagh and Kram, 1985; Hasan, Enoka, and Stuart, 1985; Heglund and Cavagna, 1985; Komi, 1986). Similarly, movements that are too slow or too fast do not allow for optimal transfer of mechanical energy between elements that move linearly and elements that rotate, or for transfer of energy between limb segments (Aleshinsky, 1986; Bingham, Schmidt, and Rosenblum, 1989; Phillips, Roberts, and Huang, 1983). Furthermore, the frequency of muscle contraction should allow time for blood to perfuse a muscle after having been squeezed out by a contraction. If the frequency of contraction is too high (> .7-1 Hz), then the muscles will exhaust their supply of energy and oxygen thus limiting the duration of the rhythmic movement (Bingham et al., 1991; Laughlin, 1987; Laughlin and Armstrong, 1985; Morton, 1987). Considerable time will then be required to recover from the accumulated lactic acid and associated pain (Astrand and Rodahl, 1977; McMahon, 1984). Optimal frequencies of movement allow for energy conserving transformations that do not deplete metabolic energy sources. Optimal frequencies of rhythmic movement are generated reliably (Bingham et al., 1991; Emmerik et al., 1989; Kugler and Turvey, 1987; Schmidt and Turvey, 1992; Smoll, 1974; 1975a; 1975b; 1975c; Smoll and Schutz, 1982; Taguchi et al., 1981). In order to discover and maintain optimal frequencies of rhythmic activity, perceptual information about muscle and limb position, velocity, and metabolic states is required. However, motor programming theory assumes that corrections and guidance must be based primarily on perception of final states, implying that only positional information is used.

In maximally rapid actions, the force levels momentarily available depend on the immediately preceding states of the muscles involved (e.g. potentiation via pre-stretch or fatigue) and on their current states (e.g. muscle length and speed of shortening or lengthening) (Bell and Jacobs, 1986; Cavanagh and Komi, 1979; Komi, 1986; McMahon, 1984; Partridge, 1979). In rapid forceful acts (e.g. maximum distance throwing, speed skating or boxing), information about current and past muscle states is required to coordinate the activity in a given muscle with that in other muscles (Bingham et al., 1989). The high levels of energy developed in such actions would be impossible without the precise coordination which allows energy to be transferred from one body segment to another (Joris et al., 1985) and from one muscle to another (Bobbert, 1988). At high force levels, mistiming could yield spastic actions that would be both dangerous and painful. Sensory information about muscle and limb states is known to be available and it has been shown how such sensory information might be used to organize and coordinate muscular activity (Feldman, 1974; Feldman et al., 1990; Hasan et al., 1985; McMahon, 1984). These data argue against the assumption that only information about final states could be used to make substantial alterations in patterns of behavior.

Given the demonstrated need to coordinate neural with non-neural constraints on timing, and the insistence of motor programming theorists that the nervous system has sole responsibility for setting up timing, a motor programming model must somehow solve the problem of sensori-motor integration. For instance, a given spatiotemporal pattern of neural inputs to the various muscles involved in a rapid reach might result in a large spatial error in the final hand position due to interactive torques among the joints. The final perceived spatial error would then have to be converted to a new spatiotemporal pattern that would yield correct motor performance. A solution that has commonly been proffered in the motor programming account is trial and error.

The problem with a trial and error strategy is that there is no guarantee that successive trials will ever converge to accurate performance. Because the components of the action system are highly nonlinear and interactive, unidirectional changes in the timing of a given component can relate in a non-monotonic fashion to resulting spatial output. A small change in timing might initially reduce spatial error, but with subsequent similar changes, spatial error could increase. For instance, for a given level and timing of initial activity in shoulder flexors, continuous modulation of the timing and/or activity level of elbow flexors can change the resultant direction of motion at the joints and accordingly, at the hand (Martin and Prablanc, 1991). As the difference in initiation times is continuously reduced across trials, the resultant direction of hand motion can reverse (Hasan and Karst, 1989). The same can happen as the level of initial activity is continuously increased. This means that functionally effective timing cannot be found by linearly varying the timing at one joint or muscle group while holding the timing of others constant (Nelson, 1983). Simultaneous variations in initiation time and activity level in both joints is required in order to discover the combination that moves the hand in a given direction and at a given speed. This is one of many instances of the problem of context-conditioned variability (Turvey, 1977; Turvey, Fitch, and Tuller, 1982; Turvey, Shaw, and Mace, 1978). The limb movement that results from neural activation of a muscle is not unique and varies according to context. Factors that determine context are relative limb postures and movements, orientation with respect to gravity, current muscle length and velocity and preceding muscle states (Bingham, 1988). The large number of nonlinear factors combined with an assumed lack of an intrinsic relation between inputs and outputs means that the search for a viable organization of the inputs that control timing must be conducted in a rather haphazard fashion. A further complication lies in the large number of the units that must be controlled at each instant of time. Controllable units might be counted in terms of joints, which number in the hundreds, in terms of muscles, which number in the thousands, or in terms of motor units, that is, independently innervated packets of muscle fibers. These number in the tens of thousands. In all cases, the high numbers mean that effective organizations are not unique, and that degrees of freedom are redundant. When the problem of context conditioned variability is coupled with the degrees of freedom problem, trial and error search becomes entirely infeasible. The size of the space of possible inputs that must be searched in all of the possible contexts is astronomical and additional information or constraint is required to arrive reliably at a stable solution.

The motor programming approach has failed to recognize the extent to which factors other than neural pattern generators contribute to the timing of motor behaviors. Therefore, it has underestimated the contribution of ongoing perceptual information during behavior to the control and coordination of that behavior. The result has been an inherent inability to account for the stability and flexibility of observed behaviors.

This chapter will present an alternative approach to the motor programming theory. This task-dynamic theory assumes that behavior evolves from the interactions of nonlinear components about various joints, so that continuous information about both spatial and temporal dimensions of the evolving pattern are used to guide it to completion. A key feature of the feedback control approach is that it dispenses with the assumption that perceptual variables are incommensurate with control variables. If we can discover perceptual variables that are intrinsically related to control variables, then the problem of sensori-motor integration dissolves and the problem of coordination can be confronted directly.

Alternative Perceptual Variables and the Task-Dynamic Approach

There is experimental evidence that a number of perceptual variables can be used to control motor behavior. There are at least two ways in which these variables might be used to circumvent the inherent instability of feedback control of ongoing behaviors. Briefly, the two ways in which perceptual variables could function are as follows. First, if feedback control were based on variables that anticipate future states, then this would alter the equation which balances corrective gains against delay times. Second, if feedback and control variables were both temporal and therefore commensurate, then delay times could be anticipated and thus not destabilize the resulting behavior.

The search for new perceptual variables that could influence motor behavior has been motivated by a number of different hypotheses (Gibson, 1966; 1979; Reed, 1988; Reed and Jones, 1982; Bingham, in press). The first hypothesis is that of direct perception, or the idea that the sensory apparatus can self-organize to detect analytically complex patterns that correspond uniquely to properties of events. Specific event properties are perceived when corresponding patterns of sensory input are detected. In vision, for example, the relevant optic patterns are spatiotemporal patterns of optic flow corresponding to movements through the environment. The second hypothesis is that the objects of perception are events rather than static objects. Thus, perceivable properties are inherently distributed over time and must be discovered in the same way as the optical patterns that might provide information about those properties. The third hypothesis is that perception and action are interactive and therefore mutually determining. According to this view, actions are events that generate ongoing patterns of sensory input that can be detected by the visual, auditory or somatosensory systems. The perception of such events can be used, in turn, to guide and control actions.

The need to go beyond strictly spatial perceptual variables becomes apparent as soon as the control of motor behavior is considered in the context of events. The spatiotemporal structure of actions embedded within surrounding events must be coordinated with the spatiotemporal structure of those events. For instance, interceptive actions such as batting or catching a ball, and various learned motor skills such as jitterbugging, fencing, or performing on the trapeze, impose strict temporal as well as spatial requirements on motor organization. There are two reasons why direct perception of spatial dimensions alone cannot account for these abilities. These reasons will be described briefly and then elaborated. First, metric spatial scaling is lost in the transformation from event to optical pattern, whereas metric temporal scaling is preserved. This loss of spatial scaling means that appropriate scaling of actions would have to be achieved through temporal perceptual variables. Second, conversion of time dependent variables to another scale would result in a loss of the original timing precision. Thus, it would seem advantageous to preserve a direct measurement of temporal variables.

In the visual system, absolute spatial metrics are lost when an event is projected into an optical flow pattern (Bingham, 1987; Bingham, 1988; Bingham, 1993a; Bingham, 1994). Mathematically, the reason for this loss is that event velocities are projected into optical pattern by dividing by the distance over which they are projected. Dimensionally, an event velocity is length/time (L/T) while the projection distance is length (L). When the first is divided by the second, the length dimension cancels ($[L/T]/L = 1/T$).

As an illustration of how and why this happens, imagine a ball dropped to the floor from height of 1 m. Using a point source of light as shown in Figure 1, a shadow, or optical pattern, can be projected onto a translucent screen which is viewed from the opposite side. Projecting the shadow onto a screen at a short distance from the ball would yield a short distance of drop imaged on the screen. Projecting onto a screen at a larger distance from the ball would yield a longer imaged distance of drop. Thus, the dropping distance on the screen depends on the distance from the screen, not the absolute distance of the drop. Nevertheless, although the metric distance of the actual fall is lost, the optic flow remains the same throughout. Accordingly, given the loss of spatial metrics in optical flow, a strict dependence only on spatial variables would leave us at a terrible disadvantage when trying to understand how the problem of scaling is solved in visual perception. In contrast to spatial variables, the scaling of temporal variables is preserved in optical flow. The time for the 1 m drop in Earth's normal gravitational field is the same in the optics as in the event, regardless of the distance of the screen from the event and the light source. Thus, it would seem reasonable to seek a solution to the scaling problem through the time dimension. In fact, research has demonstrated that we are supremely sensitive to temporal variables that exist in optical flows (e.g. Todd, 1981; Regan and Hamstra, 1993).

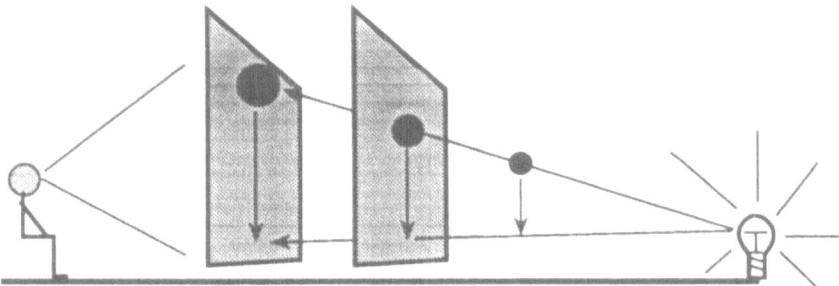

Figure 1. The shadow of a free fall event projected onto a rear projection screen using a point light source with the screen at two different distances from the event. The size of the image on the screen varies with this distance. This illustrates the loss of spatial metrics in the projection from events to optical pattern.

If, as motor programming advocates argue, the systems for motor control have direct access only to spatial variables, this would imply that access to time is indirect, that is, that time dependent variables must be derived via spatial measurements. Derivation of a time dependent variable such as velocity via measurements of position at internally clocked times would yield very poor estimates of velocity because the inevitable noise in position measurements would be amplified by differentiation. This inaccuracy would be exacerbated by any noise or instability in the internal clock. Psychophysical and modeling work in vision suggests that this problem is avoided because optical velocities are detected directly through dedicated, special purpose sensory elements, or correlation detectors (Borst and Egelhaaf, 1993; Lappin et al., 1975; Thompson, 1993). In vision, at least, there is evidence for a direct representation of time in perceived events. As will be shown later, the ability to discriminate optical velocities at particular optical positions yields a measurement of time.

Before describing temporal perceptual variables and the ways in which they might be used to address the issue of feedback control of motor activity, the next section describes the task-dynamic framework in which these variables have been identified and used.

TIMING VIA TASK DYNAMICS

Advocates of the task-dynamic approach argue that timing of motor activity is determined by organization that emerges from the interaction of components of the perception/action system. These include the central nervous system, muscles and tendons, joints and bones, blood circulation, and objects or conditions in the environment (Beek, 1989; Beek, Turvey, and Schmidt, 1992; Bingham, 1988; Feldman et al., 1990; Goldfield, Kay, and Warren, 1993; Kugler and Turvey, 1987; Latash, 1993; Riccio, 1993; Saltman and Kelso, 1986; Shaw and Kinsella-Shaw, 1988; Smith and Thelen, 1993; Thelen and Smith, 1994). According to this understanding, perception is used initially to discover useful forms of behavior and thereafter to guide their use. Constraints on potential forms of behavior are distributed across a great many heterogeneous nonlinear components listed above. The forms of behavior that emerge as these nonlinear components are allowed to interact cannot be predicted from the behavior of each component in isolation. See Bingham (1988) for an extended discussion of these points. Thus, as the resulting form of a new behavior emerges, it is perceived and modulated to an effective end.

As characterized in this approach, perception necessarily plays an important role in achieving stable goal-directed behavior via emerging organization. To gain some sense of how this occurs, consider two examples, one somewhat exotic and one mundane. The first example is learning to operate a lasso of the sort used in a rodeo show. The reader might have some experience with an effect that can be created with a short length of chain or rope. With the rope held at one end and allowed to hang downwards, as the hand is moved slowly in a small horizontal circle, the rope initially waggles about in an incoherent fashion. Eventually, as the frequency is increased and the appropriate higher frequency of hand motion is reached, the rope pops into a rotating S-curve carving out two somewhat flattened spheres in space. Once this coherent form is achieved, the performer may discover that the frequency of hand motion can now be reduced while retaining the pattern of motion of the rope. Less energy is required to maintain the stable form once it has been established. However, if the hand is moved at a still higher frequency, this will result in a transition to a new form of behavior, a new mode. An extra curve will appear in the S. With the right modulations of hand frequency, the behavior can be controlled and caused to oscillate between one mode and the other. Learning to operate a lasso is a similar exercise, with more modes possible because of the different length and shape of the rope. More patterns of motion are possible, but can only be discovered once the arm, hand and rope are assembled into a coherent form of behavior which can then be modulated.

This is not radically different from more mundane tasks such as brushing one's teeth, combing one's hair, wiping a counter, stirring gravy, scrambling eggs, polishing shoes, painting a wall, or patting a dog. To brush one's teeth, the hand and tooth brush are oscillated while controlling the locus of the equilibrium point of the oscillator as well as the direction or orientation of the oscillator. Small adjustments in the amplitude of the oscillator might also be possible. Careful control of the oscillator is desirable as known to anyone who has ever jammed his or her gums. Similarly, careful control is required to avoid spilling eggs in scrambling. A similar process takes place in learning to hammer, to swim, to jump rope or even learning to walk or to run. See the work of Esther Thelen for extensive examples of research on the development of walking and reaching from this perspective (e.g. Thelen and Smith, 1994; Thelen et al., 1993; Thelen and Ulrich, 1991). For related work see also various papers in Smith and Thelen (1993). In each case, the complex, interactive or nonlinear dynamics of the chained bone-segment system must be harnessed (together with the dynamics of any tools involved) via the nonlinear dynamics of the musculo-tendon (and circulatory) system to produce an emergent form of behavior that then can be parametrically controlled to effective ends. A related approach has been used in robotics to achieve stable and flexible solutions to the otherwise intractable problems posed by locomotor behavior in real time. A controllable hopping dynamic assembled from springs, dampers, drivers and sensory elements has been used by Raibert (1986) to construct robots that can run.

The strategy in the task-dynamic approach is to describe observable behaviors in terms of a task-dynamic model (Beek, 1989; Bingham, 1988; Saltzman and Kelso, 1986). Such models are developed and then elaborated on the basis of data from a number of different types of studies. In perturbation experiments, either dynamical parameters are altered or the initial conditions are changed and the resulting stability or change in the form of behavior is measured (e.g. Bingham et al., 1991; Kay, Saltzman, and Kelso, 1991; Kugler and Turvey, 1987). Inferences are drawn about the underlying dynamics. Dimensionality analysis and/or spectral analysis of movement data can yield information that constrains the order and type of components that are involved in the dynamic hypothesized to underlie the behaviors (e.g. Kay, 1989; Schmidt et al., 1991). Finally, studies that reveal dynamical properties of the underlying physiological components (e.g. of muscles, motor neurons and muscle afferents) support the formulation of task-dynamic models grounded in these component properties (e.g. Feldman et al., 1990). Although the dynamics in these models are often referred to as self-organizing, this does not mean that perception is excluded from the organizational process or that sensory elements are not among the basic components used to assemble the dynamic. On the contrary, dynamical properties are incorporated into stable organizations via perception.

The seminal example of a task-dynamic model is the λ-model developed by Anatol Feldman (Feldman, 1974; Feldman, 1980; Feldman, 1986; Latash, 1993). The model describes limb movement in terms of a mass-spring dynamic. The model incorporates both active and passive properties of the muscles spanning a joint together with local neural elements which sense muscle state and activate the muscle within specific regions of its state space depending on the length of the muscle and its rate of shortening or lengthening. The level of muscle activation is increased in proportion to the difference between the current length of the muscle and the length (λ) at which the muscle is initially activated. The two main control parameters in the model are the λ length, which determines the locus of the mass-spring's equilibrium point, and a proportionality constant which determines the stiffness, or level of muscle activation, as a function of the difference between λ and the current length.

Feldman has emphasized repeatedly that although muscles have spring-like properties, the mass-spring behavior at a joint cannot be attributed solely to the muscles. For instance, the musculo-tendon length-tension relation is so nonlinear as to be sometimes non-monotonic while the stiffness about a joint is approximately linear (Feldman, 1986; McMahon, 1984). Thus, peripheral afferent and efferent neural elements such as muscle spindles, α-, and γ-motor neurons, Renshaw cells, etc., must act together to modulate the nonlinear length-tension and force-velocity characteristics of opposing muscles in order to produce linear stiffness at a joint.

Why should the motor system be organized as a linear mass-spring? The likely reason is that the damped mass-spring dynamic exhibits a point stability (sometimes called a point attractor), such that motion stabilizes at an equilibrium point determined by the spring and the load. The stability of the system is determined by the stiffness and the damping.

In a system with a given linear stiffness and no damping, the linear mass-spring will oscillate about its equilibrium point with an amplitude determined by its initial state and at a frequency determined by the stiffness and mass values, independent of the amplitude. Of particular relevance in this case is that the mass will return towards the equilibrium point from any position to which it might be displaced.

When damping is added to the system, this yields the point stability which ensures rapid settling to the equilibrium position. The linear stiffness sends the mass towards the equilibrium point. The damping enables it to stay there by dissipating the mechanical energy of the spring and the motion. In the λ-model, damping is generated by both non-neural and neural components. Non-neural dissipation is produced by viscous damping in the muscles and in the joint itself. A neural contribution to damping is achieved via muscle spindle afferents which sense speed of shortening and adjust the length at which the muscle contraction is turned on or off. Damping and stiffness co-vary intrinsically in muscle, but the neural adjustment assures just slightly less than critical damping so that a stable posture is achieved in brief time.

The timing of movements produced by the λ-model is a function of the stiffness of the mass-spring system. Thus, in this model, timing is not prepackaged in the nervous system, but is produced actively by the interaction of neural and musculo-tendon elements modulated by sensory feedback. A natural by-product of a linear mass-spring organization is scaling of velocity with amplitude so as to preserve frequency or duration of movement. If the limb is momentarily perturbed to a different amplitude, the time of travel to the goal-point is preserved, that is, the temporal properties of the movement remain stable. As has often been observed, this temporal stability is a useful property for interlimb coordination where amplitude of each limb movement may vary due to noise in the system or as required by the particular task. Spatiotemporal stability is the result of three systems that keep delays in the feedback loop small and provide relatively high gain corrections. Each operates on a different time scale. First, non-neural components of the system provide an immediate response in which corrective force levels vary intrinsically with the state of the system. Second, neural elements in the periphery activate reflex circuits with short transmission distances and quick loop times. Third, sensory information about spatiotemporal properties of the behavior itself is used to modulate intrinsic muscle properties through continuous feedback.

The mass-spring model was originally developed to account for the timing and control of discrete movements about a single joint. More recently, the model has been elaborated for application to the problem of intra-limb coordination in multijoint reaching (Feldman et al., 1990; Flanagan, Ostry, and Feldman, 1990; Flanagan, Ostry, and Feldman, 1993). Mass-spring organization at the joints is combined in the elaborated model to yield mass-spring behavior at the hand. Control of the hand trajectory is achieved, according to the model, via an equilibrium point which follows a virtual trajectory determined by a superordinate control dynamic (Latash, 1993). The same approach can be used to apply the λ-model to the production of rhythmic movements in which case the equilibrium point would be moved back and forth between a pair of targets. In either case, we have stepped beyond the inherent stability of the linear mass-spring and once again must confront the problem of feedback control. How is the equilibrium point effectively guided along a virtual trajectory to a target without incurring the instability of feedback control? The control dynamic cannot be arbitrary.

TIMING AND THE PROBLEM OF FEEDBACK CONTROL
IN RHYTHMIC MOVEMENTS

A function that externally controls the behavior of the equilibrium point in the λ-model is referred to as a forcing function. Without such forcing, timing is determined by the stiffness, and the mass-spring dynamic is autonomous and stable. Mathematically, autonomous means that no term in the equation defining motion is a function of time explicitly, that is, none are of the form f(t). In an autonomous equation, all terms are of the form f(x, dx/dt, d^2x/dt^2). With inclusion of an external forcing function (i.e. f(t) = cos(t)), in the λ-model, extrinsically determined timing is imposed on the dynamic and the dynamic becomes nonautonomous. The imposition of extrinsic timing re-introduces the problem of feedback control. The stability of the resulting behavior depends on the relation between the timing characteristics of the linear mass-spring and the forcing function. Certain forcing frequencies can produce unbounded increases in the amplitude of oscillation. To ensure stable behavior, the forcing must be matched to the intrinsic timing properties of the linear mass-spring.

Before considering how to address this problem via the λ-model, we must take into account the observed nature of rhythmic movements. Recent investigations have shown that the properties exhibited by rhythmic movements do not correspond to those of a nonautonomous oscillator. Perturbation of rhythmic movements produces phase resetting, a property that could not be produced by an extrinsically forced linear mass-spring (Kay, et al., 1991). Autonomous nonlinear oscillators have been used to simulate this and other observed characteristics of rhythmic movements.

The linear mass-spring model for discrete movements was designed, in part, to capture the observed preservation of preferred movement duration over changes in target distance, that is, preferred frequency over changes in movement amplitude. Although frequency of movement does not change from preferred frequency with manipulations of amplitude, manipulation of the frequency of rhythmic movement has been found to produce regular changes in movement amplitude. As frequency is increased significantly above preferred rates, amplitude decreases. A task-dynamic model with nonlinear damping terms has been designed to simulate this relationship and the limit cycle stability of rhythmic limb movements (Kay, Kelso, Saltzman, and Schøner, 1987).

The limit cycle is the closed trajectory that results when limb velocity is plotted as a function of position. The limit cycle trajectory is the stable motion to which perturbed trajectories return at limit. It has been demonstrated experimentally that when a limb in oscillation is perturbed by a brief impulse, the limb returns rapidly to the limit cycle (Kay et al., 1991; Sternad, Beek, and Turvey, 1994). When perturbation momentarily changes the amplitude of motion, the peak velocity changes automatically to preserve the current frequency of oscillation as determined by the stiffness and mass of the oscillator. Thus, the timing of the oscillatory movement is preserved.

Amplitude stability is produced by nonlinear damping. Two nonlinear damping terms in the model describe the dissipation of energy and the input of energy into the system. The stable limit cycle is the set of states at which energy inflow balances energy outflow. Perturbations which send the movement off of the limit cycle, that is, speeding up or slowing down the limb, reflect energy imbalances. The excess energy is dissipated, or additional energy is supplied until the trajectory reaches the balanced energy flow of the limit cycle. Thus, the balanced energy flow within the limit cycle yields both spatial and temporal stability in the rhythmic movements.

The forced λ-model includes energy outflow and energy inflow components that could, in principle, yield limit cycle stability. The damping of the λ-model produces energy outflow. The forcing of the equilibrium point produces energy inflow. The question is how to make the forced λ-model equivalent to the autonomous nonlinear task-dynamic of Kay et al. (1987) and circumvent the problem of instability due to feedback control. In an autonomous system, the forced behavior of the equilibrium point must, by definition, be a function of system states (that is, f(x, dx/dt), not f(t)). In the autonomous task-dynamic model, the nonlinear damping terms are functions of limb position and velocity. This implies that sensory information about system states is used, although the model proposed by Kay et al. is not explicit about how this is achieved. Nevertheless, for an autonomous forced λ-model, sensory information about the state variables would be necessary to relate the forcing of the equilibrium point to the stiffness and damping of the linear mass-spring and thus produce a stable movement amplitude.

The problem of feedback control arises at this point and can be expressed as two specific questions. First, how should measurements of limb position and velocity be related to the timing of the forcing? Second, the equilibrium point is hypothetically controlled via the brain. This implies that a substantial delay would intervene between measurements of system state and generation of a forcing signal. How might this neural delay be incorporated into a stable dynamic?

The formulation of potential answers to these two questions originated with the application of the λ-model to account for the ability of people and animals to reliably perform rhythmic behaviors at preferred frequencies. Preferred frequencies are generated in various locomotory gaits including the quadruped gaits of horses (Hoyt and Taylor, 1981) and the bipedal gaits of humans (Holt, Hamill, and Andres, 1990; McMahon, 1984) as well as in many other rhythmic activities. The reliability with which preferred frequencies are produced means that performers are able to find the same preferred frequencies on different occasions. Rhythmic movements at preferred rates are more *stable* from cycle to cycle than movements at other frequencies (Kelso, 1984; Kelso and Kay, 1986; Kelso, Scholz, and Schøner, 1986; Kelso et al., 1987), *and they require a minimum of energy expenditure* (Cavagna et al., 1977; Hoyt and Taylor, 1981; McMahon, 1984). Kugler and Turvey (1987) argue at length that reliability entails a perceptual component in the organization of rhythmic behaviors, that is, a means by which preferred frequencies can be identified.

Using a paradigm intended to simulate essential elements of intralimb coordination in bipedal locomotion, Kugler and Turvey (1987) studied the timing and stability of rhythmic movements produced when people were asked to oscillate two hand-held pendulums simultaneously at the preferred rates. Different preferred frequencies were produced depending on the relative lengths of the pendulums. The two pendulums in a pair were often of distinctly unequal lengths. Participants were allowed to oscillate the pendulums in an exploratory fashion until a preferred mutual frequency was discovered. Rhythmic motions were then recorded while the participant oscillated the pendulum at a steady state. The finding was that a given individual homed in on the same frequency of oscillation for a given pair of pendulums on occasions that might be separated by weeks. See also Bingham et al. (1991), Rosenblum and Turvey (1988), Turvey et al. (1986), Turvey et al. (1988) and especially, Schmidt and Turvey (1992). Each time, the participant actively discovered a preferred frequency which was not the average of the preferred frequencies for each of the pendulums oscillated alone.

How do people reliably home in on preferred frequencies and why are they so stable? With the understanding of rhythmic limb movement as produced by a force driven, damped, harmonic oscillator, Hatsopoulos and Warren (in press) have suggested that preferred frequencies correspond to the resonant frequencies of the autonomous portion of the oscillator, that is, the linear mass-spring of the λ-model. Driving such an oscillator at its resonant frequency is energetically advantageous in the sense that minimum forcing is required to produce oscillation of a given amplitude. However, the problem remains that the resonant frequency must somehow be identified so that the driving frequency can be tuned to the resonant frequency. Hatsopoulos and Warren suggested that the required information is temporal and formulated in terms of relative phase. Phase is a purely temporal parameter, determined by time within an oscillatory cycle. A damped harmonic oscillator is driven at its resonant frequency when the relative phase between the driver and the oscillator is 90°. To the extent that relative phase can be perceived, rhythmic movement can be stably maintained at the preferred frequency. Phase (ϕ) is itself a function of system states (that is, f (ϕ)= f (x, dx/dt)). It corresponds to a nonlinear combination of an oscillator's position and velocity (that is, $\phi = \arctan[(dx/dt)/x]$). Thus, via perception of phase, the forcing can become a function of the system states and a fully autonomous organization can be achieved. The organization is nonlinear due to the arctan function. See Schmidt, Beek and Turvey (1992) for extended discussion of the role of perception in transforming dynamical systems from nonautonomous to autonomous organization.

The solution proposed by Hatsopoulos and Warren would account in part for the reliability and energy optimization observed in rhythmic movements at preferred rates. With the ability to perceive phase, a phase of 90° could reliably be established between a limb and controlled behavior of the equilibrium point. Oscillation at the resonant frequency would be energetically efficient. How stable is such a system? If determination of phase and control of forcing is cortically modulated, then we encounter the problem of transmission delays. In principle, the problem is to maintain a continuously available variable, the relative phase, at a single value, namely 90°. Initially, this would appear to be a standard feedback control problem in which the dynamic response of a system is brought to and maintained at a set point value. To preserve stability, the gain of adjustment would have to be modulated relative to the feedback delay time. However, because the controlled variable in this case is temporal, as long as the delay time is stable, it can simply be incorporated into the value assigned to the controlled variable, that is, the relative phase. Relevant delays have been variously estimated as between 50 ms and 200 ms (Carlton, 1992; Carlton, 1981; Flanagan et al., 1993). Estimates of synaptic transmission time from the arm to the head and back are closer to 50 ms (Evarts and Tanji, 1974). Transmission times of 50-200 ms are compatible with typical preferred movement times of about 1 second.

The organization suggested by Hatsopoulos and Warren can be formally modeled and studied via simulations. The linear damped mass-spring of the λ-model can be expressed as:

$$m\ddot{x} = -k(x - x_{ep}) - b\dot{x} \qquad (1)$$

where m is the mass, k is the stiffness, b is the damping, x_{ep} is the location of the equilibrium point, x is limb position, and the dot indicates differentiation with respect to time either once (that is, dx/dt) or twice (d^2x/dt^2). According to Hatsopoulos and Warren, rhythmic movements are generated by moving the equilibrium point according to some function which drives the oscillator. This function can be incorporated as follows:

$$m\ddot{x} = -kx - b\dot{x} + f[x, \dot{x}, t] \qquad (2)$$

where with $f = kx_{ep}$ we have equation (1). With $f = f(t) = kx_{ep}(t)$, the oscillator becomes nonautonomous, because the forcing is a function of an extrinsically imposed time. But, if we require $f = f(x, dx/dt)$, then the equation remains autonomous. Accordingly, as suggested by Hatsopoulous and Warren, if we set $f = f(\phi)$ where ϕ (the phase) is a function of x and dx/dt the following function is obtained:

$$m\ddot{x} = -kx - b\dot{x} + c\sin\left(tan^{-1}\left[\frac{\dot{x}}{|x| + \varepsilon}\right]\right) \qquad (3)$$

where c determines the strength of the forcing (and the amplitude of oscillation) and the expression in parentheses represents the phase angle ϕ ($\varepsilon \ll 1$ is used simply to avoid division by 0.)

This is an autonomous oscillator that exhibits a stable limit cycle behavior as shown in Figure 2. The form of the functional dependence on ϕ (that is, $\sin(\phi)$) guarantees that the relative phase between the forcing and the oscillator is 90° and that the damped, linear oscillator is driven at its resonant frequency. When k, b, and/or c are perturbed by Gaussian noise, the oscillator exhibits behavior typical of limb movements in which trajectories wander within a narrow band around the strictly deterministic limit cycle trajectory. Like the so-called hybrid model developed by Kay et al. (1987; 1991), this oscillator exhibits an inverse frequency-amplitude relationship when the frequency is manipulated by varying stiffness. At a given stiffness value, velocity and amplitude co-vary so that frequency is preserved during transient response to perturbations away from the limit cycle.

Subjects tested by Kay et al. (1991) exhibited a consistent phase advance in response to perturbations whether the perturbing impulse momentarily advanced or delayed the limb movement. The model reproduces this result when stiffness is taken to be a function of the amplitude of the oscillator. Under these conditions, when the amplitude exceeds a value determined by the strength of the forcing signal, stiffness increases until the transient response drops below the criterion. This can be treated as simply adding another sensory component that increases the stiffness in response to perturbation. Because the functional dependence of stiffness is on x, a state variable, the oscillator remains autonomous.

Finally, scaling the value of a single parameter in the model, namely the strength of the forcing signal, yields a transition from the point stable behavior characteristic of discrete movements (at c=0) to the stable limit cycle behavior characteristic of rhythmic movements (at c>0). In short, the model succeeds in extending the λ-model for discrete movements to the realm of rhythmic movements where it accounts for an impressive array of experimental results.

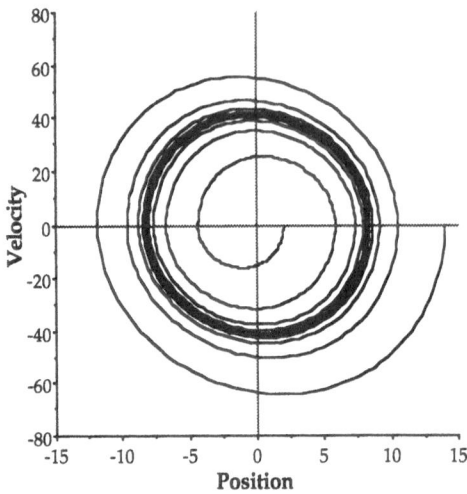

Figure 2. Phase portrait for the dynamic described by equation (3). Trajectories with initial conditions placed either inside or outside of the limit cycle converge to the limit cycle trajectory.

This phase modulated λ-model requires the assumption that performers are able to perceive relative phase. Unpublished studies performed by the author in collaboration with Richard C. Schmidt have shown that visual perception of relative phase is good. Somatosensory psychophysical studies that would be more relevant to the phase modulated λ-model have not been performed.

TIMING AND THE PROBLEM OF FEEDBACK CONTROL
IN DISCRETE MOVEMENTS

In addition to advances in task-dynamic models, recent work on central pattern generators has revealed that an integrated organization of neural networks, sensory inputs, and muscle and limb properties actually underlies behavioral timing that had previously been attributed to the action of neural pattern generators alone (Pearson, 1989; Rossignol, Lund, and Drew, 1988). With these recent developments in the understanding of rhythmic movement, leading advocates of motor programming have retrenched in the domain of discrete movements. Visually guided reaching would seem to require the use of spatial information in feedback. Accordingly, Schmidt (1992) has argued that two aspects of rapid targeted reaching require the imposition of preprogrammed timing patterns on the motor apparatus.

First, Schmidt argues that the spatial location of a target must be perceived and transformed into the appropriate temporal pattern of motor activity. If vision contributes exclusively spatial information to processes responsible for moving the hand to a target, then, it is argued, the motor timing required to accomplish the task must be determined in the CNS via preestablished relations between spatial perceptual variables and temporal motor variables. Once determined through trial and error, an appropriate pre-established timing program would be used to determine muscle activation patterns.

Second, Schmidt argues that the use of visual/spatial information entails significant neural transmission distances and hence, processing times. The long delays entailed by visual guidance would indeed prohibit the use of local sensori-motor neural organization and feedback control as in the λ-model. However, these hypotheses are called into question by a growing body of evidence for strong ongoing visual guidance during rapid targeted reaches and recent modeling efforts that indicate that feedback delays have been overestimated. In addition, there is evidence for the use of anticipatory visual variables that could alter the relation between delays and feedback gains so as to allow significant corrections to stable movements. An example of the use of an anticipatory variable is the use of temporal visual variables to guide the hand to a target. Just as for phase in the context of rhythmic movements, these temporal variables are equivalent to ratios of state variables of a task dynamic and are assembled to execute reaching movements. Like phase, these temporal visual variables can be related directly to motor timing in a control system. The use of temporal variables for visuo-motor control suggests that the relation between perception and action can be direct and need not be mediated by transformations that have been formulated on the basis of previous trial and error learning.

Visually Guided Reaching

Targeted reaching can be considered as having two phases. During the initial fast phase, the majority of the distance to a target is covered at high speed. During the subsequent slow phase, final adjustments are made for target acquisition. Using discontinuities in the velocity profiles of reaches as evidence for visual guidance, Keele, among others, has argued that the fast phase is ballistic while the slow phase is visually guided (Abrams, 1992; Keele and Posner, 1968). If large corrections in feedback control of reaching are unstable because visual feedback is too slow, then only small corrections to slow movements should be possible. However, a large body of evidence has accumulated over the last decade showing clearly that the fast phase of reaching is subject to continuous visual guidance as well.

Foremost are "double-step targeting" studies. In these, upon a signal, a participant reaches as rapidly as possible to a target. During some of the reaches, either the distance or the direction of the target is changed. In every case, regardless of whether changes in target location are introduced, participants rapidly adjust their reaches to acquire the target. In some studies, the target location is changed during the reaction time of the reach, before the hand actually begins to move. In these experiments where the second target appeared after a delay of 50 milliseconds or more after the first, the hand has been described as moving initially towards the first target location, and adjusting to move towards the second after about 200-300 ms (Georgopoulos, Kalaska, and Massey, 1981; Soechting and Lacquaniti, 1983). With shorter intervals of 25-50 ms between the presentation of targets 1 and 2, the hand has been described as moving in a direction half-way between the two target locations before adjusting (Sonderen, Gon, and Gielen, 1988). In other studies, the change in target location was introduced after the hand started to move. In this case, adjustments have been described as occurring 100 ms after the onset of the second target (Gielen, Heuvel, and Gon, 1984; Paulignan et al., 1990; Pellisson et al., 1986).

The problem with all of these descriptions and reported response times is in identifying when the response to the change occurs. Change of target distance or direction often results in smooth adjustments with no identifiable discontinuity in the velocity profile (Elliott, 1991; 1992; Pellisson et al., 1986). Although elaborate methods have been devised to identify a change in movement direction (Martin and Prablanc, 1991), the delay due to the need for muscular torques to overcome inertia of the limb can obscure the occurrence of control changes. Thus, response times are likely to be overestimated on the basis of kinematics alone.

To address this problem, Feldman and colleagues have extended the λ-model to simulate the control of multi-joint limb movements (Feldman et al., 1990; Flanagan et al., 1990; 1993). In the model, a virtual trajectory of the equilibrium point for the hand determines the hand trajectory. In all simulations, the virtual trajectory could be programmed to respond to perturbations within 65 ms. The variations in hand path curvature were determined by the dynamics of the arm and its control structure. Simulations generated using the model have reproduced most of the phenomena described in double-step targeting studies, including hand trajectories directed initially towards either the first target or half-way between the first and second target depending on the time of perturbation. The simulations have also reproduced smooth adjustments, and the variations in curvature of the hand path that have been used in other studies to estimate response times (Flanagan et al., 1993). The implication of these modeling results is that the response to changes in the visual location of a target is rapid and reflects continuous visual modulation of a virtual equilibrium-point trajectory.

Although some experimental studies suggest that visual guidance occurs throughout a reach, there is other evidence to suggest that visual guidance occurs primarily during the decelerative portion of a reach. Velocity profiles in targeted reaching are characteristically asymmetric, with more time spent in deceleration than acceleration (Zalaznik, Schmidt, and Gielen, 1986). Also, as the accuracy required in a reach increases, the accelerative portion of the trajectories remains unchanged while the decelerative portion increases in duration (Marteniuk et al., 1987). Finally, removal of visual information during the act of reaching has been found to have little effect on the accelerative portion of reaches but distinct effects on the decelerative portion (Carlton, 1981; 1992; Paillard, 1982). Because the target is foveated during a targeted reach, Sivak and Mackenzie (1992) have suggested that visual guidance occurs primarily while the hand is in central vision during the latter portion of a reach.

While continued visibility of the target and hand during a reach are not required to bring the hand within the vicinity of a stationary target (Goodale, Pelisson, and Prablanc, 1986; Pelisson et al., 1986), visibility of both the target and the hand with free head movement is required for accurate target acquisition. A large number of studies have shown that when either the target or the hand alone is visible, performance is inferior to that when both are visible. Further, a number of studies indicate that vision may not improve performance without head movement (Biguer, Prablanc, and Jeannerod, 1984; Carlton, 1992; Carnaham, 1992; Carson et al., 1993; Elliott and Allard, 1985; Jeannerod and Prablanc, 1983; Marteniuk, 1978; Prablanc et al., 1979a; Prablanc et al., 1979b; Proteau and Cournoyer, 1990; Sivak and Mackenzie, 1992). The importance of visual information for accurate reaching does not change with extensive practice in a given task. On the contrary, the availability of visual information becomes increasingly important with increasing skill because improved skill does not transfer to performance without vision (Proteau, 1992; Proteau et al., 1987). Clearly, visual information is needed and used to guide targeted reaching. The implication of the collective evidence is that while visual information is obtained before movement is initiated, either the information itself or the use of the information in movement execution is subject to errors that require subsequent fine tuning and guidance during deceleration. These findings raise the question of how the combined problems of sensori-motor integration and feedback delays are overcome to allow for stable visual guidance. The λ-model would seem to solve the problem of transformation from spatial to temporal variables because only spatial coordinates would be required if one could simply place the equilibrium point at the perceptually determined location of a target. The timing would then be produced by the autonomous control structure of the λ-model. Unfortunately, its not so simple.

First, accuracy in either estimating target position or in bringing the hand to that position is sometimes insufficient to acquire a target and adjustments must be made during the reach. As already mentioned, double-step targeting experiments have shown that large adjustments can be made smoothly and efficiently during the high velocity portion of a reach. This implies that large adjustments would have to be made in the location of the equilibrium point. Second, one must often circumvent obstacles on route to a target. For instance, a seated person reaching from below the hip to a target just above the knee must move the hand around the leg along the way. To accomplish this, the equilibrium point must be moved along a curved path and the timing of the movement of the equilibrium point relative to the movement of the limb becomes an issue. Third, some experiments indicate that the equilibrium point is indeed moved along a trajectory preceding the limb. In deafferented monkeys, if the limb is kicked ahead of the equilibrium point, then the limb moves backwards briefly, presumably back behind the equilibrium point, before continuing along the reach (Bizzi et al., 1984).

Together, the nature of reaching tasks and the experimental evidence require that the equilibrium point be moved and adjusted with respect to the hand. Thus, the feedback control problem remains in two respects. First, even if the feedback delay can be shortened to about 65 ms by the rapid response of the λ-model or its neural equivalent, this still limits the strength of corrections and threatens the stability of response. While smaller than previous estimates by half, 65 ms still represents a significant delay.

Second, the visual scaling problem remains. How is the spatial location of both target and hand determined rapidly, reliably and accurately during the course of a reach? Binocular vision does not solve the scaling problem uniquely because monocular vision allows apprehension of spatial scale at levels of resolution that may be comparable to those obtained with binocular vision (Bingham and Stassen, 1994; Lappin and Love, 1992). However, in monocular vision, absolute spatial scale is available only with accompanying head movement, so there is the need to coordinate head and hand movements. See Bingham and Stassen (1994) for a review. The amplitudes and durations of head movement required for given levels of spatial resolution and accuracy remain to be determined, but if these are significant, then sufficiently rapid apprehension of spatial locations of the hand during rapid reaches may not be possible. The need to determine the velocity, i.e., speed and direction of hand movement, introduces additional difficulties.

Temporal scaling is preserved in the projection from events to optical flow and time dimensioned optical variables can be used to help solve these problems of visual guidance. Although a stationary target might not seem to entail use of temporal perceptual variables, the need to guide the hand once it appears in motion within the visual field does introduce this need.

Temporal Perceptual Variables

τ is a temporal optical variable. It is the relative rate of expansion of the optical image of an approaching object, that is, the ratio of image size to image expansion rate. τ is equivalent to the ratio of object distance to object velocity. Before the discovery of τ, interceptive actions such as catching or hitting a projectile were hypothesized to require visual apprehension of object distance and object velocity to form an estimate of arrival time. However, Lee (1974) showed that τ specifies the time to contact with the eye for an object traveling towards an observer at constant velocity. For an object that is accelerating or decelerating, the momentary value of τ under- or overestimates the time to contact. Although τ corresponds to a ratio of position and velocity, its value provides no information about either object distance or velocity. An object at large distance traveling at high velocity may yield the same time to contact and τ value as an object at a small distance traveling at lower velocity. Thus, τ is purely temporal in its dimensions and does not distinguish between these situations. Todd (1981) has shown that human observers can reliably detect and use τ differences of about 100 ms. Further, Regan and Hamstra (1993) have shown that τ is can be resolved with a Weber constant of 10%. A number of investigations have shown that τ is used to coordinate interceptive actions (Bootsma and Oudejans, 1993; Bootsma and Peper,

1992a; Bootsma and Peper, 1992b; Bootsma and van Wieringen, 1990; Lee, 1980; Lee and Thomson, 1982; Lee et al., 1983; Savelsbergh,Whiting, and Bootsma, 1991; Schiff and Detwiler, 1979; Schiff and Oldak, 1990; Sideway, McNitt-Gray, and Davis, 1989). However, in all of these instances, the object to be intercepted was moving towards the observer. The next section shows how temporal optical variables might be used to guide the hand away from an observer and towards a target.

Temporal Perceptual Variables for Guiding the Hand to a Target

Information that becomes available when the hand comes into view during a reach can be described in terms of two time dimensioned optical variables. Figure 3 illustrates the geometry of the hand and target with respect to the eye, expressed in polar coordinates. The first optical variable, τ_ϕ, is directional and corresponds to the time to close the angle ϕ or the optical gap between the image of the target and the image of the hand (that is, the time until $\phi = 0$). The second, τ_R, represents the time for the hand to travel to a distance from the eye equal to the distance of the target from the eye.

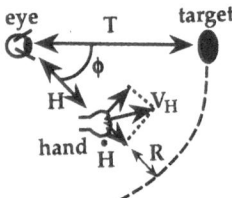

Figure 3. The optical geometry in reaching to a target. The relations are isolated in a plane containing the eye, the hand, and the target ignoring rotations of this plane about an axis from eye to target. T is the distance from eye to target. H is the distance from eye to hand. R is the distance from the hand to a sphere at target distance from the eye. V_H is the tangential velocity of the hand. ϕ is the visual angle between the hand and target.

As shown in Figure 3, ϕ is an optical variable (that is, a visual angle) so that τ_ϕ is derived as:

$$\tau_\phi = \tau_\phi(t) = \frac{\phi(t)}{\dot{\phi}(t)} \tag{4}$$

Derivation of a radial time to contact, τ_R is more complex. τ_R specifies the time before the hand contacts the surface of a sphere, centered at the eye, with a radius (T) equal to the radial distance of the target from the eye. The distance of the hand from this surface along a radial line extended from the eye is $R = T - H$. Thus, τ_R can be derived as

$$\tau_R = \tau_R(t) = \frac{Rt}{\dot{Rt}} = \frac{T(t) - H(t)}{\dot{T}(t) - \dot{H}(t)}. \tag{5}$$

Because T and H are geometrical variables, not optical variables, the optical relations corresponding to τ_R must be derived from equation 5 in terms of two components, one for hand motion and one for head motion. For the hand motion component, we assume that the head does not move so that T is constant and dT/dt = 0. From equation 5, this yields

$$\tau_{R_H}(t) = \frac{H(t) - T}{\dot{H}(t)} = \frac{H(t)}{\dot{H}(t)}\left[1 - \frac{T}{H(t)}\right] = -\frac{I_H(t)}{\dot{I}_H(t)}\left[1 - \frac{I_H(t)}{I_T}\frac{S_T}{S_H}\right] \tag{6}$$

where

$$I_H(t) = \frac{S_H}{H(t)}$$

is hand image size with the hand being approximated as a sphere of diameter S_H,

$$\dot{I}_H(t) = \frac{-\dot{H}(t)S_H}{H^2(t)}$$

is hand image expansion rate, and I_T is target image size. Note that a ratio of two non-optical variables appears in the rightmost term of equation 5, namely, the ratio of target size to hand size. This means that observers must have an appreciation of target size relative to hand size to be able to use τ_R. Because head motion is small in comparison to hand motion (that is, dT/dt is small in comparison to dH/dt) τ_R in full form is closely approximated by equation 5, especially as R grows small. In our studies, we have found that the head movement contribution to τ_R is negligible. Thus, we use τ_R as approximated by equation 5.

Hypothesized Strategies for the Use of τ Variables to Control Approach Behaviors

How might τ variables be used to guide a hand to a target? Different strategies have been proposed for the use of τ variables to control approach behaviors. The use of a "τ-Margin Strategy" has been proposed in tasks where the approach is not controlled by the observer (Bootsma and Peper, 1992; Bootsma and Wieringen, 1990; Lee, et al., 1983; Savelsberg, Whiting and Bootsma, 1991; Sideway, McNitt-Gray and Davis, 1989). In this strategy, certain τ values (i.e. margin values) are used to initiate behavior in preparation for contact. The hypothetical margin values are specific to the task and allow for errors due to non-constant approach velocities.

Alternatively, when the approach is controlled by the observer, the use of a "Constant dτ/dt Strategy" has been hypothesized, with dτ/dt equal to -.05 (Kim, Turvey, and Carello, 1993; Lee, 1976; 1980). There are three problems with this strategy: (1) it requires that observers perceive and maintain the value of a 2nd order optical variable;(2) the dynamics of the system responsible for dissipating energy are not taken into account (Kaiser and Phatak, 1993); and (3) no optic dτ/dt variable has been derived for guidance of the hand in 3-space, although Bootsma has derived and investigated the use of a dτ/dt variable under the assumption that the angle of approach to the target is constant (Bootsma and Peper, 1992; Bootsma and Oudejans, 1993). However, this assumption is inappropriate for the control of unconstrained targeted reaching.

A possible two part strategy for using τ_ϕ and τ_R to guide the hand would be as follows: First, use a Constant-τ_ϕ Strategy. Second, keep $\tau_R \geq \tau_\phi$, with τ_R converging on τ_ϕ as ϕ approaches 0, to prevent the hand from passing the target distance before ϕ is closed.

Mathematically, using constant τ would mean reaching a target in infinite time which is presumably why this strategy had not been considered previously. However, a constant τ approach entails a linear decrease in velocity with diminishing target distance and rapid convergence. The convergence time, T, is determined as follows: $T = \tau_c \ln P$, where P is the proportion of the initial distance remaining and τ_c is the constant value of τ. For instance, approaching from an initial distance of 20 cm (a typical value for the distance at which the peak hand velocity might occur) to within 5 mm of a target would take 370 ms with $\tau_c = 100$ ms. Placing the original aimpoint a small distance (e.g. 5 mm) beyond the target would result in contact in finite time at a velocity proportional to that distance (e.g. at 5 cm/s) yielding a gentle tap. Alternatively, the Constant-τ_ϕ Strategy could be abandoned when ϕ is sufficiently small, for instance, when the hand begins to occlude the target.

The advantages of this strategy are: (1) it requires only that an observer perceive and maintain the value of a 1st order optical variable; (2) the optical variables are derived explicitly; and (3) the τ_c value can be determined with respect to the controlled approach dynamics. Values of τ_ϕ expected in maximally rapid targeted reaches would be slightly greater than the perception-action loop time, estimated to be 60-100 ms. Values 10% greater would allow sufficient resolution of the τ value. A similar margin would be expected between τ_R and τ_ϕ.

Evidence that human observers use this strategy was found in studies of adjustment to changes in visual target direction in rapid visually guided reaching (Bingham, 1993b; Bingham, Muchisky, and Romack, 1991; Bingham and Romack, 1994; Bingham, Romack, and Stassen, 1993a; 1993b; Romack, Buss, and Bingham, 1992; Zhang and Bingham, 1993).

Reaching with Changes in Visual Target Direction

In these studies, participants performed rapid reaches from a launch platform next to their hip to place a stylus in a target hole located just above and beyond their knee. The cartesian coordinates of LEDs placed on the thumb and head were sampled at 100 Hz via a 2 camera WATSMART system. Maximally rapid reaches were performed with the restriction that the stylus not collide with the face of the target at high speed. The head was free to move. First, a set of reaches was performed by participants who wore a set of goggles that admitted a 45° visual field for the right eye and blocked the left eye. We call this condition "clear goggle". The right goggle was fitted with 3 LEDs, above, below and to the right of the eye. We performed measurements to establish the border of the field of view relative to these LEDs. In a set of reaches immediately following those in the clear goggle condition, participants performed rapid reaches while wearing a 10° prism mounted over the right goggle. The prism was mounted base left so that the visual field was displaced 10° to the right. In all reaches, vision was occluded until just before the reach was initiated.

Representative paths for the last clear goggle reach and the first prism reach are shown in Figure 4a projected on the horizontal x-y plane. The initial prism trajectories appear as if two double-step targeting changes had taken place. First, the participants reached towards the original, actual, target location, or halfway between the actual and displaced positions. Then, they moved towards the apparently displaced target. Finally, with detection of the perturbation to visual direction, participants adjusted to bring the hand to the actual target location.

Turning to an examination of the visual information that may have been used by participants, τ_ϕ and τ_R were derived at each sampling interval using measured 3-D positions and velocities of the eye, hand, and target. Spatial means were computed for each quantity in terms of the distance to the target. Representative mean τ_ϕ and τ_R trajectories for the clear goggle set of reaches to fixed targets are shown in Figure 4b together with the mean percentage of the hand's tangential velocity directed towards the target.

Associated coefficients of variation are shown in Figure 4c. As ϕ became small, all participants maintained constant τ_ϕ trajectories at values close to 100 ms with variability less than 10% of the mean. Also, as ϕ became small, τ_R approached and attained a value of 200-300 ms with variability less than 20% of the mean. The $d\tau_R/dt$ and $d\tau_\phi/dt$ trajectories are shown in Figure 4d. $d\tau/dt$ is also shown although we know of no optical correlate for 200-300 ms with variability less than 20% of the mean. The $d\tau_R/dt$ and $d\tau_\phi/dt$ trajectories are shown in Figure 4d. $d\tau/dt$ is also shown although we know of no optical correlate for this variable. All

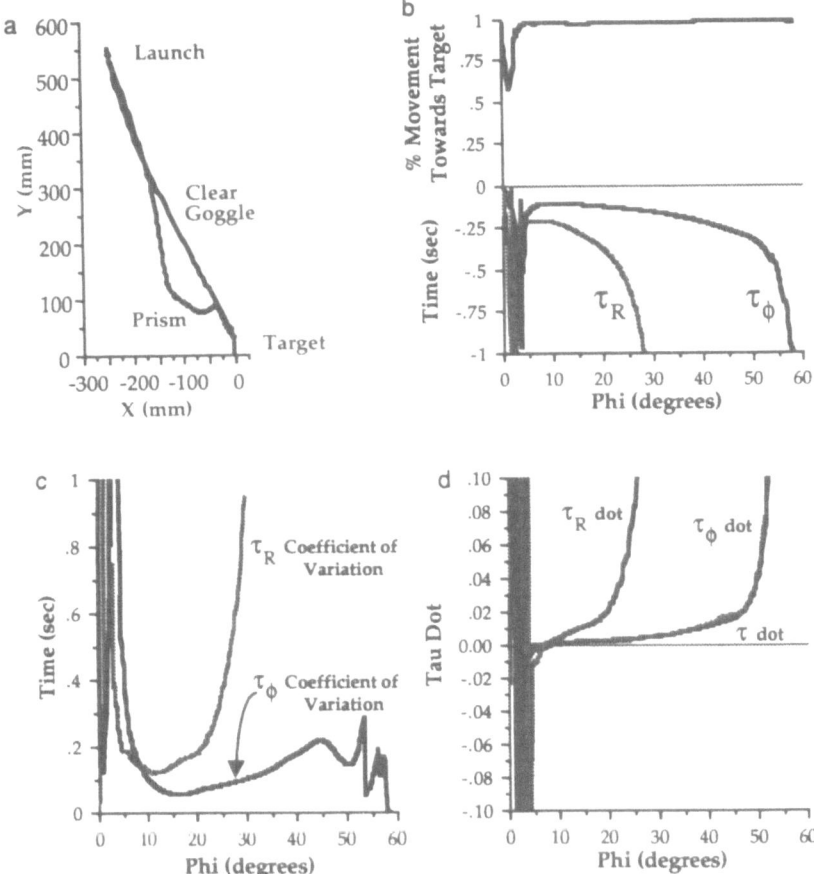

Figure 4. **A.** Representative paths of hand movement from launch platform (at upper left) to target (at (0,0), lower right) projected onto the horizontal X-Y plane. Compare the first reach with a prism (Prism) to the last reach without a prism (Clear Goggle). **B.** Representative τ_ϕ and τ_R trajectories for a reach to a fixed target as a function of ϕ, that is, the visual angle subtended between hand and target. Also shown is a trajectory for the percent of hand tangential velocity directed towards the target. The target is at $\phi = 0$ to the left and the starting point of the hand is on the right. Negative values on the time axis are used only for convenience in plotting. **C.** Representative trajectories for the coefficients of variation associated with τ_ϕ and τ_R for a reach to a fixed target as a function of ϕ. The target is at $\phi = 0$. (d) Representative trajectories for $d\tau_\phi/dt$, $d\tau_R/dt$ and dv/dt for a reach to a fixed target as a function of ϕ. The target is at $\phi = 0$.

dτ/dt variables changed rapidly until the end of the fast phase where the values approached a constant value near 0 as predicted by the constant-τ strategy, not near -.05 as predicted by the constant dτ/dt strategy. Finally, in the majority of trials the participants did not follow a constant-τ trajectory all of the way to the target, but only to within a few millimeters of the target as shown in Figure 4b.

How did participants detect the change in the visual location of the target in prism trials? Responses directing the hand to the actual target location occurred well before the hand occluded the apparent location of the target 10° to the right of the actual location. As shown in Figure 5, participants made a correction about 100 ms after detecting the violation of the $\tau_R \geq \tau_\phi$ condition, so that τ_R became less than τ_ϕ. These results provide provisional evidence that people may indeed be using a constant-τ_ϕ strategy with the additional condition that $\tau_R < \tau_\phi$.

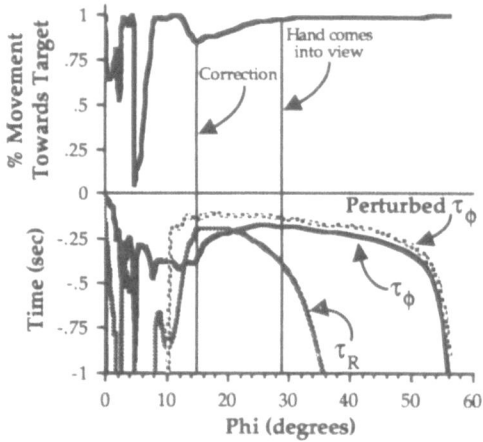

Figure 5. Representative τ_ϕ, τ_R and perturbed-τ_ϕ trajectories plotted versus ϕ for the first reach perturbed by prismatic viewing. Also shown is a trajectory for the percent of hand tangential velocity directed towards the target. The target is at $\phi = 0$. Perturbed-τ_ϕ was computed using a target location at 10° to the right of the actual target location. Note that if τ_R is gauged with respect to perturbed-τ_ϕ, the hand could pass the target distance before the gap between the hand and the actual target direction is closed. This becomes more likely when τ_R is smaller than τ_ϕ. Negative values on the time axis are used only for convenience in plotting.

Using τ in a λ-Model of Visually Guided Reaching

If reaching is achieved via control of an equilibrium point for the hand as predicted by a multi-joint λ-model, then a means of using τ variables to guide the behavior of the equilibrium point must be discovered. The use of both τ_ϕ and τ_R in the full multi-joint model has not yet been explored. However, I have investigated how a τ variable could be used to drive a uni-dimensional damped mass-spring to a target. A function is required for a forcing term, $x_{ep}(\tau)$, which determines the behavior of the equilibrium point of the mass-spring as a function of perceived τ :

$$m\ddot{x} = -kx - b\dot{x} + x_{ep}(\tau) \tag{7}$$

where τ is

$$\tau = \frac{x - x_{target}}{\dot{x}}$$

To achieve the tendency to preserve a specified constant-τ value, $x_{ep}(\tau)$ was determined as the solution to a first order equation which describes the control dynamic:

$$\dot{x}_{ep} = sgn[\tau^* - \tau_c]\gamma(\tau^*)$$

where the sgn function determines the sign of the gain, γ, and thus the direction of motion of the equilibrium point, τ_c is the intended constant-τ value, and τ* is τ delayed by 65 ms. The latter was used to simulate visual feedback delay. In this case, it was not possible to incorporate the delay time into the temporal control value itself as was done in the phase driven λ-model for rhythmic movement. Nevertheless, in this case, the stable delay could be incorporated into τ_c together with a margin for error yielding a value of 100 ms for τ_c. The sign of the gain, γ, is determined by whether τ* is greater than or less than τ_c. This model acts to preserve τ_c. For simulations, x_{ep} was set to x_{target} once x fell within less than a millimeter of the target. In actual reaching, this might be achieved via physiological tremor and a final haptically mediated adjustment. Also, to simulate the effect of the hands only coming fully into view during the decelerative portion of the movement, the feedback portion of the model was not turned on until the limb reached a position midway between x(t = 0) and x_{ep}(t = 0). Note that the autonomy of the task dynamic was preserved because feedback was turned on as a function of a state variable, the position of the hand. As shown in Figure 6a, the model proved not only to be quite stable, but to produce simulated trajectories similar to those observed in our experiments on reaching. Compare Figure 6a with Figure 4b.

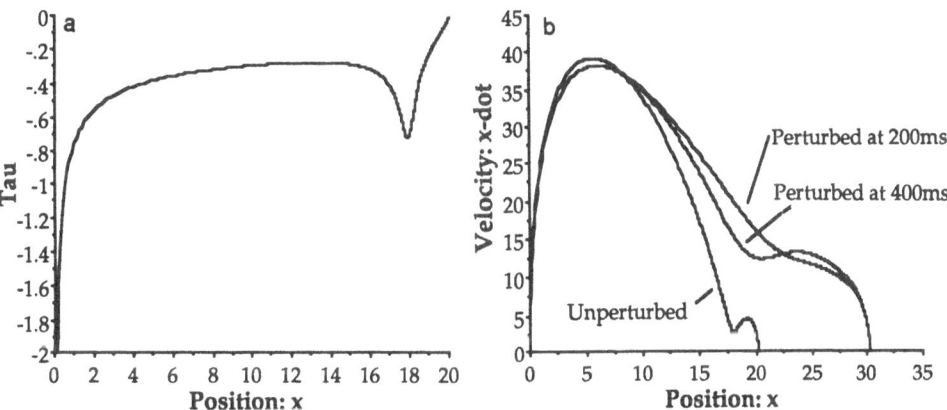

Figure 6. A. τ trajectory simulated by the dynamic model for approach to a target at position 20. In this case, the simulated hand starts at x = 0 and travels to x = 20. **B.** Simulated phase plane trajectories generated by the dynamic model for approach to a target at position 20 or perturbed target location at 30.

In simulations of double-step targeting, shifts of the target from 20 cm to 30 cm early in the reach produced relatively smooth changes in velocity profiles; when the same shifts occurred late in the movement, they tended to produce a distinct deceleration followed by re-acceleration as shown in Figure 6b. This is a typical result in double step targeting experiments. In all cases, large adjustments were successfully achieved in brief time.

There are 3 aspects of this model that should be emphasized. First, the model allows large corrections despite significant feedback delays. Use of anticipatory temporal variables for visual feedback produces stable behavior despite significant feedback gain. Second, time is incorporated into the perceptual variable through direct measurement of optical velocity. Special purpose sensory apparatus, such as correlation detectors, yields direct measurement of optical velocity, a spatial-temporal quantity. Third, the temporal perceptual variable τ is mathematically equivalent to state variables of the task dynamic, that is, position and velocity of the hand as controlled via the forced λ-model dynamic. If biologically correct, this model resolves the problem of sensori-motor integration. Also, because the forcing of the equilibrium point is a function of the state variables, the forced λ-model remains autonomous. The timing behavior arises from the interaction of the various components assembled into the task dynamic. There is no imposition of predetermined timing by the nervous system.

Nevertheless, the visual scaling problem has been incorporated into the model by assuming that initial visual information about the target direction, distance, and size was available. To successfully acquire a target, the model requires that the initial position of the equilibrium point, $x_{ep}(t = 0)$, be in the direction of the target relative to the initial position of the limb, $x(t = 0)$ and at an appropriate distance. Furthermore, to produce sufficiently rapid movements by taking advantage of the mass-spring portion of the model, the initial position of the equilibrium point was placed at a position 60 – 70% of the distance to the target. Finally, the multi-joint model incorporating use of both τ_ϕ and τ_R requires information about target size because it is used in the derivation of τ_R.

Information About Target Distance and Direction Available Before Reach Initiation via Rhythmic Movement of the Head

Information about the direction of a target is available in optical flow generated by head movement. When one moves the head towards a target in the surround, the optical pattern exhibits a radial outflow from a node which lies in the target image. Warren has shown that observers can use this information to judge the direction of movement with about $1°$ accuracy (Warren and Hannon, 1990; Warren, et al., 1991; Warren, Morris, and Kalish, 1988).

The problem of scaling distance is more difficult. The evidence indicates that head motion is required for monocular apprehension of absolute distance. See Bingham and Stassen (1994) for review and discussion. Analyses of optic flow have shown that if information about the velocity of head motion is available, then optic flows can be scaled to yield measurements of absolute distance. Unfortunately, reliable information about head velocity is not known to be available and is unlikely. On the other hand, somatosensory information about the amplitude of head motion is known to be available (Clark and Horch, 1986; McCloskey, 1978; McCloskey, 1980). Accordingly, we assumed that somatosensory information about the amplitude of head movement would be used to scale information about target distance.

Head motion was modeled as if it were generated by a simple physical oscillator. The resulting harmonic motion is an approximation to the limit cycle trajectory of a nonlinear oscillator such as the phase driven λ-model. The information about distance was derived in terms of time-dimensioned variables using the symmetry of the oscillatory trajectories. When the head moves towards an object, then τ at the peak velocity (τ_{pv}) is mid-way in time or in position between the endpoints of the oscillatory motion and thus can be easily found and used. Alternatively, τ_{pv} could be determined through input from the vestibular system which would signal the occurrence of peak velocity and a change from acceleration to deceleration. The value of τ_{pv} divided by the period of oscillation (T) is equivalent to the distance of the object in units of the amplitude of head oscillation scaled by 2π.

As shown in Figure 7, the head is assumed to move along the line of sight directed toward a target surface. Head movement is modeled via a harmonic mass-spring dynamic. The head moves with an amplitude, A, about the equilibrium point of the mass-spring which is located at a distance D from the target. The equation of motion for the mass-spring is

$$\ddot{x}(t) = -\frac{k}{m}[x(t) - D] \qquad (8)$$

where k is a linear stiffness and m is the mass, both assumed constant. Using this equation and its solution together with

$$\tau = \frac{x}{\dot{x}}$$

we obtained

$$\frac{\tau_{pv}}{T} = \frac{1}{2\pi}\frac{D}{A} \qquad (9)$$

Thus, as the head oscillates towards an object, $2\pi\tau_{pv}/T$ specifies that the object is twice as far as the distance that the head moves.

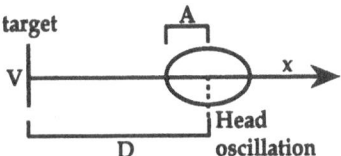

Figure 7. Schematization of head movement. A harmonic phase trajectory for head oscillation is shown at distance D from a target surface. Head movement of amplitude A along the x direction toward and away from the target. Variation in position, x(t), is shown along the abscissa and variation in velocity, V, along the ordinate. The target is located at x = 0.

Assuming that τ_{pv} is available optically and that A is available kinesthetically, the only problem that remains in principle is the determination of T, the period of oscillation. Because the period of voluntary movement at preferred rates has been found to be stable and reliably reproduced, a possible solution is to assume head movement at a preferred rate and to incorporate the value into a scaling constant, $2\pi/T_p$, (where T_p is a constant equivalent to the individuals preferred period of head motion). Head movement at preferred rates would be optimal for measurement of distance for another reason. Voluntary movement to an endpoint at preferred rates of motion has been found to be optimal for yielding the most accurate judgments of position via kinesthesis (Clark and Horch, 1986; Eklund, 1972; Paillard and Brouchon, 1968; 1974). Use of T_p in lieu of some direct measurement of period would introduce some noise into the resulting distance estimates, but the results would be sufficient to initialize the task-dynamic control scheme for visually guided reaching.

CONCLUSION

The advantage of the autonomous dynamical organization hypothesized under the task-dynamic approach is that the temporal pattern depends on the intrinsic timing characteristics of the various components of the motor system including the muscles, tendons, limbs and joints, as well as the nervous system. These constraints on timing must be accommodated in any model of motor organization. To be used in motor organization, the intrinsic timing characteristics of non-neural components must be perceived. To track the behaviors of the muscles and limbs, the system must be able to measure the relevant state variables, that is, positions and velocities. Velocity is a spatiotemporal quantity while position is spatial, so the ratio of velocity to position yields a quantity that is dimensionally equivalent to frequency (1/time), that is, phase. Perception of phase is thus equivalent to the simultaneous measurement of velocity and position and yields an effective measure of time. Phase perception can be used to overcome the instability of feedback control and to produce stable timing in rhythmic movements at preferred frequencies. Phase perception would, of course, also be invaluable in establishing other forms of coordination in motor behavior, in particular, intralimb coordination.

Similarly, a ratio of position to velocity yields time-to-contact, or τ, which we used to solve the distance perception problem. With this solution, we come full circle using the task-dynamic for perceptually driven rhythmic movements to obtain an initial estimate of target distance and direction that can be used, in turn, to parameterize the perceptually driven task-dynamic for targeted reaching. In no instance would timing be imposed by the nervous system on the remaining motor components. Rather, in all cases, timing would emerge from the interaction of the various components of the action system assembled into a perceptually modulated task-dynamic. Perceptual control would be achieved via time dimensioned variables, either phase or τ, that are equivalent to ratios of task-dynamic state variables, that is, positions and velocities. The use of anticipatory time dimensioned perceptual variables, provide a solution to the problem of sensori-motor integration, and overcomes the limitations of feedback control.

ACKNOWLEDGMENTS

The research described in this paper was supported in part by NSF Grant BNS-9020590 and by the Institute for the Study of Human Capabilities at Indiana University. An earlier draft of this paper was circulated in the summer of 1994. I am very grateful to the following colleagues for helpful comments on that draft: Reinoud Bootsma, Ellen Covey, Thomas Fikes, Christopher Pagano, Richard C. Schmidt, Robert Port, and Esther Thelen.

REFERENCES

Abrams, R. A., 1992, Coordination of eye and hand for aimed limb movements, *in*: "Vision and Motor Control", L. Proteau and D. Elliot, eds., Elsevier, Amsterdam.

Aleshinsky, S. Y., 1986, An energy "sources" and "fractions" approach to the energy expenditure problem IV. *J. Biomechanics*, 19:287.

Alexander, R. M., 1984, Elastic energy storage in running vertebrates, *Am. Zool.*, 24:85.

Astrand, P.-O., and Rodahl, K., 1977, "Textbook of Work Physiology", McGraw-Hill, New York, NY.

Beek, P. J., 1989, "Juggling Dynamics", Free University Press, Amsterdam .

Beek, P. J., Turvey, M. T., and Schmidt, R. C., 1992, Autonomous and nonautonomous dynamics of coordinated rhythmic movements, *Ecol. Psychol.*, 4:65.

Bell, D. G., and Jacobs, I., 1986, Electro-mechanical response times and rate of force development in males and females, *Medicine Science Sports Exercise*, 18:31.

Biguer, B., Prablanc, C., and Jeannerod, M., 1984, The contribution of coordinated eye and head movements in hand pointing accuracy, *Exp. Brain Res.*, 55:462.

Bingham, G. P., 1987, Dynamical systems and event perception: a working paper parts I-III, *Perception-Action Workshop Review*, 2:4.

Bingham, G. P. , 1988, Task-specific devices and the perceptual bottleneck, *Human Movement Sci.*, 7:225.

Bingham, G. P., 1993a, Perceiving the size of trees: Form as information about scale, *J. Exp. Psychol.: Human Percept. Perform.*, 19:1.

Bingham, G. P. (1993b). "Spatio-temporal Information in Visually Guided Reaching," Paper presented at the ONR Conference on Neural Representations of Temporal Patterns, Duke University, Durham, NC.

Bingham, G. P., Dynamics and the problem of visual event recognition, in: "Mind as Motion: Dynamics, Behavior and Cognition", R. Port and T. v. Gelder, eds., MIT Press, Cambridge, MA, in press.

Bingham, G. P., Muchisky, M. M., and Romack, J., 1991, "'Adaptation' to Displacement Prisms is Sensorimotor Skill Acquisition", Paper presented at a meeting of the Psychonomics Society, San Fransisco, CA.

Bingham, G. P., and Romack, J. L., 1995, Adaptation to displacement prisms is skill acquisition: Analysis of reaching trajectories, submitted.

Bingham, G. P., Romack, J. L., and Stassen, M. G., 1993a, Optical information in visually guided reaching with perturbation of visual direction, in: "Studies in Perception and Action", S. Valenti and J. B. Pittenger, eds., Vancouver, Ontario, Canada.

Bingham, G. P., Romack, J. L., and Stassen, M. G., 1993b, "Targeted Reaching with Perturbation of Visual Direction." Paper presented at the 18th Interdisciplinary Conference, Jackson Hole, WY.

Bingham, G. P., Schmidt, R. C., and Rosenblum, L. D., 1989, Hefting for a maximum distance throw: A smart perceptual mechanism, J. Exp. Psychol.: Human Percept. Perform. 15:507.

Bingham, G. P., Schmidt, R. C., Turvey, M. T., and Rosenblum, L. D., 1991, Task-dynamics and resource dynamics in the assembly of a coordinated rhythmic activity, J. Exp. Psychol.: Human Percept. Perform., 17:359.

Bingham, G. P., and Stassen, M. G., 1994, Monocular egocentric distance information generated by head movement, Ecol. Psychol., 6:219.

Bizzi, E., Acornero, N. Chapple, W., and Hogan, N., 1984, Posture control and trajectory formation during arm movement, J. Neurosci, 4:2738.

Bobbert, M. F. , 1988, "Vertical Jumping: A Study of Muscle Functioning and Coordination", Free University Press, Amsterdam..

Bootsma, R. J., and Oudejans, R. R. D., 1993, Visual information about time-to-collision between two objects, J. Exp. Psychol.: Human Percept. Perform., 19:1041.

Bootsma, R. J., and Peper, C. E., 1992a, Predictive visual information sources for the regulation of action with special emphasis on catching and hitting, in: "Vision and Motor Control", L. Proteau and D. Elliot, eds., Elsevier, Amsterdam.

Bootsma, R. J., and van Wieringen, P. C., 1990, Timing an attacking forehand drive in table tennis, J. Exp. Psychol.: Human Percept. Perform., 16:21.

Borst, A., and Egelhaaf, M., 1993, Detecting visual motion: Theory and models, in: "Visual Motion and Its Role in the Stabilization of Gaze", F. A. Miles and J. Wallman, eds., Elsevier, Amsterdam.

Carlton, L. G., 1981, Processing visual feedback information for movement control, J. Exp. Psychol.: Human Percept. Perform., 7:1019.

Carlton, L. G., 1992, Visual processing time and the control of movement, in: "Vision and Motor Control", L. Proteau and D. Elliot, eds., Elsevier, Amsterdam.

Carnaham, H., 1992, Eye, head and hand coordination during manual aiming, in: "Vision and Motor Control", L. Proteau and D. Elliot, eds., Elsevier, Amsterdam.

Carson, R. G., Goodman, D., Roneo, C., and Elliot, D., 1993, Asymmetries in the regulation of visually guided aiming, J. Motor Behav., 25:21.

Cavagna, G. A., 1977, Storage and utilization of elastic energy in skeletal muscle, in: "Exercise and Sport Sciences Reviews", R. S. Hutton, ed., Journal Publishing Affiliates, Santa Barbara, CA.

Cavagna, G. A., Heglund, N. C., and Taylor, C. R., 1977, Mechanical work in terrestrial locomotion: Two basic mechanisms for minimizing energy expenditure, Am. J. Physiol., 233:R242.

Cavanagh, P. R., and Komi, P. V., 1979, Electromechanical delay in human skeletal muscle under concentric and eccentric contraction, Eur. J. App. Physiol., 42:159.

Cavanagh, P. R., and Kram, R., 1985, Mechanical and muscular factors affecting the efficiency of human movement, Medicine Science Sports Exercise, 17:326.

Clark, F. J., and Horch, K. W., 1986, Kinesthesia, in: "Handbook of Perception and Human Performance V. 1 Sensory Processes and Perception", K. R. Boff, L. Kauffman, and J. P. Thomas, eds., Wiley, New York, NY.

Eklund, G., 1972, Position sense, and state of contraction. The effects of vibration, J. Neurol. Neurosurg. Psychiatry, 35:606.

Elliot, D., 1991, Discrete versus continuous visual control of manual aiming, Human Movement Sci., 10:393.

Elliot, D., 1992, Intermittent versus continuous control of manual aiming movements, in: "Vision and Motor Control", L. Proteau and D. Elliot, eds., Elsevier, Amsterdam.

Elliot, D., and Allard, F., 1985, The utilization of visual feedback information during rapid pointing movements, Quart. J. Exp. Psychol., 37A:407.

Emmerik, R. E. A., Brinker, B. P. L. M., Vereijken, B., and Whiting, H. T. A., 1989, Preferred tempo in the learning of a gross cyclical action, Quart. J. Exp. Psychol., 41A:251.

Evarts, E. V., and Tanji, J., 1974, Gating of motor cortex reflexes by prior instruction, *Brain Res.*, 71:479.

Feldman, A. G., 1974, Control of the length of a muscle, *Biophysics*, 19:771.

Feldman, A. G., 1980, Superposition of motor programs: I. Rhythmic forearm movements in man, *Neurosci.*, 5:81.

Feldman, A. G., 1986, Once more on the equilibrium-point hypothesis (lambda model) for motor control, *J. Motor Behav.*, 18:17.

Feldman, A. G., Adamovich, S. V., Ostry, D. J., and Flanagan, J. R., 1990, The origin of electromyograms: Explanation based on the equilibrium point hypothesis, *in*: "Multiple Muscle Systems: Biomechanics and Movement Organization", J. M. Winters and S. L. Woo, eds., Springer-Verlag, New York, NY.

Flanagan, J. R., Ostry, D. J., and Feldman, A. G., 1990, Control of human jaw and multi-joint arm movements, *in*: "Cerebral control of speech and limb movements", G. E. Hammond, ed., Elsevier, Amsterdam.

Flanagan, J. R., Ostry, D. J., and Feldman, A. G., 1993, Control of trajectory modifications in target-directed reaching, *J. Motor Behav.*, 25:140.

Franklin, G. F., Powell, J. D., and Emami-Naeini, A., 1994, "Feedback Control of Dynamic Systems", Addison-Wesley, Reading, MA.

Georgopoulos, A. P., Kalaska, J. F., and Massey, J. T., 1981, Spatial trajectories and reaction times of aimed movements: Effects of practice, uncertainty, and change in target location, *J.Neurophysiol.*, 46:725.

Gibson, J.J., 1968, "The Senses Considered as Perceptual Systems", Allen and Unwin, London.

Gibson, J.J., 1979, "The Ecological Approach to Visual Perception", Houghton Mifflin, Boston, MA.

Gielen, C. C. A. M., 1991, The role of internal representations in the control of fast arm and eye movements, *in*: "Movement Control", R. Jacobs and W. E. I. Rikkert, eds., Free University Press, Amsterdam.

Gielen, C. C. A. M., Heuvel, P. J. M. v. d., and Gon, J. J. D. v. d., 1984, Modification of muscle activation patterns during fast goal-directed arm movements, *J. Motor Behav.*, 16:2.

Goldfield, E. C., Kay, B. A., and Warren, W. H., Jr., 1993, Infant bouncing: The assembly and tuning of action systems, *Child Dev.*, 64:1128.

Goodale, M. A., Pelisson, D., and Prablanc, C., 1986, Large adjustments in visually guided reaching do not depend on vision of the hand or perception of target displacement, *Nature*, 320:748.

Hasan, Z., Enoka, R. M., and Stuart, D. G., 1985, The interface between biomechanics and neurophysiology in the study of movement: Some recent approaches, *Exercise Sport Sci. Rev.*, 13:169.

Hasan, Z., and Karst, G. M., 1989, Muscle activity for initiation of planar, two-joint arm movements in different directions, *Exp. Brain Res.*, 76:651.

Hatsopoulos, N. G., and Warren, W. H., Jr., Resonance tuning and stiffness control in arm swinging, *J. Motor Behav.*, in press.

Heglund, N. C., and Cavagna, G. A., 1985, Efficiency of vertebrate locomotory muscles, *J. Exp. Biol.*

Henneman, A., Somjen, G. and Carpenter, D., 1965, Excitability and inhibitability of motoneurons of different sizes, *J. Neurophysiol.*, 28:599.

Hogan, N., Bizzi, E., Mussa-Ivaldi, F. A., and Flash, T., 1987, Controlling multi-joint motor behavior, *Exercise Sport Sci. Rev.*, 15:153.

Holt, K. G., Hamill, J., and Andres, R. O., 1990, The force-driven harmonic oscillator as a model for human locomotion, *Human Movement Sci.*, 9:55.

Hoyt, D. F., and Taylor, C. R., 1981, Gait and the energetics of locomotion in horses, *Nature*, 292:239.

Jeannerod, M., and Prablanc, C., 1983, The visual control of reaching movements, *in*: "Motor Control Mechanisms in Man", J. E. Desmedt, ed., Raven Press, New York, NY.

Joris, H. J. J., van Muyen, A. J. E., van Ingen Schenau, G. J., and Kemper, H. C. G., 1985, Force, velocity, and energy flow during the overarm throw in female handball players, *J. Biomech.*, 18:409.

Kaiser, M. K., and Phatak, A. V., 1993, Things that go bump in the light: On the optical specification of contact severity, *J. Exp. Psychol.: Human Percept. Perform.*, 19:194.

Kay, B. A., 1989, The dimensionality of movement trajectories and the degrees of freedom problem: A tutorial, *in*: "Self-Organization in Biological Work Spaces", P. N. Kugler, ed., North-Holland, Amsterdam.

Kay, B. A., Kelso, J. A. S., Saltzman, E., and Schøner, G., 1987, The space-time behavior of single and bimanual rhythmical movements:data and limit cycle mode, *J. Exp. Psychol.: Human Percept. Perform*, 13:564.

Kay, B. A., Saltzman, E., and Kelso, J. A. S., 1991, Steady-state and perturbed rhythmical movements: A dynamical analysis, *J. Exp. Psychol.: Human Percept. Perform*, 17:183.

Keele, S. W., Cohen, A., and Ivry, R., 1990, Motor programs: Concepts and issues, *in*: "Attention and Performance XIII: Motor Representation and Control", M. Jeannerod, ed., Erlbaum, Hillsdale, N.J.

Keele, S. W., and Posner, M. I., 1968, Processing of visual feedback in rapid movements, *J. Exp. Psychol.*, 77:155.

Kelso, J. A. S., 1984, Phase transitions and critical behavior in human bimanual coordination, *Am. J. Physiol.*, 246:R1000.

Kelso, J. A. S., and Kay, B. A., 1986, Information and control: A macroscopic analysis of perception-action couplings, *in*: "Perspectives on Perception and Action", H. Heuer and A. F. Sanders, eds, Erlbaum, Hillsdale, N.J.

Kelso, J. A. S., Scholz, J. P., and Schøner, G., 1986, Nonequilibrium phase transitions in coordinated biological motion: Critical fluctuations, *Physics Lett. A*, 118:279.

Kelso, J. A. S., Schøner, G., Scholz, J. P., and Haken, H., 1987, Phase-locked modes, phase transitions and component oscillators in biological motion, *Physica Scripta*, 35:79.

Kim, N., Turvey, M. T., and Carello, C., 1993, Optical information about the severity of upcoming contacts, *J. Exp. Psychol.: Human Percept. Perform.*, 19:179.

Komi, P. V., 1986, The stretch-shortening cycle and human power output, *in*: "Human Muscle Power", N. L. Jones, N. McCartney, and A. J. McComas, eds, Human Kinetics, Champaign, IL.

Kugler, P. N., and Turvey, M. T., 1987, "Information, Natural Law, and the Self-Assembly of Rhythmic Movement", Erlbaum, Hillsdale, N.J.

Lappin, J. S., Bell, H. H., Harm, O. J., and Kottas, B., 1975, On the relation between time and space in the visual discrimination of velocity, *J. Exp. Psychol.: Human Percept. Perform.*, 1:384.

Lappin, J. S., and Love, S. R., 1992, Planar motion permits perception of metric structure in stereopsis, *Percept. Psychophysics*, 51:86.

Latash, M. L., 1993, "Control of Human Movement", Human Kinetics, Champaign, IL.

Laughlin, M. H., 1987, Skeletal muscle blood flow capacity: Role of the muscle pump in exercise hyperemia, *Am. J. Physiol.*, 253:H993.

Laughlin, M. H., and Armstrong, R. B., 1985, Muscle blood flow during locomotory exercise, *Exercise Sport Sci. Rev.*, 13:95.

Lee, D. N., 1976, A theory of the visual control of braking based on information about time to collision, *Perception*, 5:437.

Lee, D. N., 1980, The optic flow field: The foundation of vision, *Philos. Trans. Royal Soc. London B*, 290:169.

Lee, D. N., and Thomson, J. A., 1982, Vision in action: The control of locomotion, *in*: "Analysis of Visual Behavior", D. J. Ingle, M. A. Goodale, and R. J. Mansfield, eds., MIT Press, London.

Lee, D. N., Young, D. S., Reddish, P. E., Lough, S., and Clayton, T. M. H., 1983, Visual timing in hitting an accelerating ball, *Quart. J. Exp. Psychol.*, 35A:333.

Marteniuk, R. G., 1978, The role of eye and head positions in slow movement execution, *in*: "Information Processing in Motor Learning and Control", G. E. Stelmach, ed., Academic Press, New York, NY.

Marteniuk, R. G., MacKenzie, C. L., Jeannerod, M., Athenes, S., and Dugas, C., 1987, Constraints on human arm movement trajectories, *Canadian J. Psychol.*, 41:365.

Martin, O., and Prablanc, C., 1991, Two-dimensional control of trajectories towards unconsciously detected double step targets, *in*: "Tutorials in Motor Neuroscience", J. Requin and G. E. Stelmach, eds., Kluwer Academic Publishers, Netherlands.

McCloskey, D. I., 1978, Kinesthetic sensibility, *Psychol. Rev.*, 58:763.

McCloskey, D. I., 1980, Kinaesthetic sensations and motor commands in man, *in*: "Spinal and Supraspinal Mechanisms of Voluntary Motor Control and Locomotion", J. E. Desmedt, eds, Karger, Basel.

McMahon, T. A., 1984, "Muscles, Reflexes, and Locomotion", Princeton University Press, Princeton, NJ.

Meijer, O. G., and Roth, K., 1988, "Complex Movement Behavior: The Motor-Action Controversy", Elsevier, Amsterdam.

Mochon, S., and McMahon, T. A., 1980, Ballistic walking, *J. Biomech.*, 13:49.

Mochon, S., and McMahon, T. A., 1981, Ballistic walking: an improved model, *Math. Biosci.*, 52:241.

Morton, R. H., 1987, A simple model to link hemodynamics, fatigue, and endurance in static work, *J. Biomech.*, 20:641.

Nelson, W. L., 1983, Physical principles for economies of skilled movements, *Biol. Cyber.*, 46:135.

Paillard, J., 1982, The contribution of peripheral and central vision to visually guided reaching, *in*: "Analysis of Visual Behavior", D. J. Ingle, M. A. Goodale, and R. J. Mansfield, eds., MIT Press., London.

Paillard, J., and Brouchon, M., 1968, Active and passive movements in the calibration of position sense, *in*: "The Neuropsychology of Spatially Oriented Behavior", S. J. Freedman, ed., Dorsey, Homewood, IL.

Paillard, J., and Brouchon, M., 1974, A proprioceptive contribution to the spatial encoding of position cues for ballistic movements, *Brain Res.*, 71:273.

Partridge, L. D., 1979, Muscle properties: A problem for the motor controller physiologist, *in*: "Posture and Movement", R. E. Talbott and D. R. Humphrey, eds., Raven Press, New York, NY.

Paulignan, Y., MacKenzie, C. L., Marteniuk, R. G., and Jeannerod, M., 1990, The coupling of arm and finger movements during prehension, *Exp. Brain Res.*, 79:431.

Pearson, K.G., 1989, Sensory elements in the pattern generating networks, *in*: "Workshop on the Mechanics, Control, and Animation of Articulated Figures", Cambridge, MA.

Pellisson, D., Prablanc, C., Goodale, M. A., and Jeannerod, M., 1986, Visual control of reaching movements without vision of the limb, *Exp. Brain Res.*, 62:303.

Pellisson, D., Prablanc, C., Goodale, M. A., and Jeannerod, M., 1986, Visual control of reaching movements without vision of the limb. II. Evidence of fast unconscious processes correcting the trajectories of the hand to the final position of a double-step stimulus, *Exp. Brain Res.*, 62:303.

Phillips, S. J., Roberts, E. M., and Huang, T. C., 1983, Quantification of intersegmental reactions during rapid swing motion, *J. Biomech.*, 16:411.

Prablanc, C., Echalier, J. F., Jeannerod, M., and Komilis, E., 1979a, Optimal response of eye and hand motor systems in pointing at a visual target. II. Static and dynamic visual cues in the control of hand movement, *Biol. Cybern.*, 35:183.

Prablanc, C., Echalier, J. F., Komilis, E., and Jeannerod, M., 1979b, Optimal response of eye and hand motor systems in pointing at a visual target. I. Spatio-temporal characteristics of eye and hand movements and their relationships when varying the amount of visual information, *Biol. Cybern.*, 35:113.

Proteau, L., 1992, On the specificity of learning and the role of visual information for movement control, *in*: "Vision and Motor Control", L. Proteau and D. Elliot, eds., Elsevier, Amsterdam.

Proteau, L., and Cournoyer, J., 1990, Vision of the stylus in a manual aiming task: The effects of practice, *Quart. J. Exp. Psychol.*, 42B:811.

Proteau, L., Marteniuk, R. G., Girouard, Y., and Dugas, C., 1987, On the type of information used to control and learn an aiming movement after moderate and extensive training, *Human Movement Sci.*, 6:181.

Raibert, M. H., 1986, "Legged Robots That Balance", MIT Press, Cambridge, MA.

Regan, D. and Hamstra, S.J., 1993, Dissociation of discrimination thresholds for time to contact and for rate of angular expansion, *Vision Res.*, 33:5/6.

Reed, E. S., 1988, "James J. Gibson and the Psychology of Perception", Yale University Press, New Haven, CT.

Reed, E. S., and Jones, R. K., eds., 1982, "Reasons for Realism: Selected essays of James J. Gibson", Erlbaum, Hillsdale, N.J.

Riccio, G. E., 1993, Information in movement variability about the qualitative dynamics of posture and orientation, *in*: "Variability and Motor Control", K. M. Newell and D. M. Corcoj, eds., Human Kinetics, Champaign, IL.

Romack, J. L., Buss, R. A., and Bingham, G. P., 1992, "Adaptation" to displacement prisms is sensorimotor learning, *in*: "Proceedings of the 14th Annual Conference of Cognitive Science Society", Erlbaum, Bloomington, IN.

Rosenbaum, D. A., 1991, "Human Motor Control", Academic Press, New York, NY.

Rosenblum, L. D., and Turvey, M. T., 1988, Maintenance tendency in coordinated rhythmic movements: Relative fluctuations and phase, *Neurosci.*, 27:289.

Rossignol, S., Lund, J.P. and Drew, T., 1988, The role of sensory inputs in regulating patterns of rhythmical movements in higher vertebrates, *in*: "Neural Control of Rhythmic Movements in Vertebrates", A.H. Cohen, S. Rossignol and S. Grillner, eds., Wiley, New York, NY.

Saltman, E., and Kelso, J. A. S., 1986, Skilled actions: A task dynamic approach, *Psychol. Rev.*, 94:1.

Savelsbergh, G. J. P., Whiting, H. T. A., and Bootsma, R. J., 1991, Grasping tau, *J. Exp. Psychol.: Human Percept. Perform.*, 17:315.

Schiff, W., and Detwiler, M. L., 1979, Information used in judging impending collision, *Perception*, 8:647.

Schiff, W., and Oldak, R., 1990, Accuracy of judging time to arrival: Effects of modality, trajectory, and gender, *J. Exp. Psychol.: Human Percept. Perform.*, 16:303.

Schmidt, R. A., 1982, "Motor Control and Learning: A Behavioral Emphasis", Human Kinetics, Champaign, IL.

Schmidt, R. A., 1988, Motor and action perspectives on motor behavior, *in*: "Complex Movement Behavior: The Motor-Action Controversy", O. G. Meijer and K. Roth, eds., Elsevier, Amsterdam.

Schmidt, R. C., Beek, P. J., Treffner, P. J., and Turvey, M. T., 1991, Dynamical substructure of coordinated rhythmic movements, *J. Exp. Psychol.: Human Percept. Perform.*, 17:635.

Schmidt, R. C., and Turvey, M. T., 1992, Long term consistencies in assembling coordinated rhythmic movements, *Human Movement Sci.*, 11:349.

Shaw, R. E., and Kinsella-Shaw, J., 1988, Ecological mechanics: A physical geometry for intentional constraints, *Human Movement Sci.*, 7:155.

Sideway, B., McNitt-Gray, J., and Davis, G., 1989, Visual timing of muscle preactivation in preparation for landing, *Ecol. Psychol.*, 1:253.

Sivak, B., and Mackenzie, C. L., 1992, The contribution of peripheral vision and central vision to prehension, *in*: "Vision and Motor Control", L. Proteau and D. Elliot, eds., Elsevier, Amsterdam.

Smith, L. B., and Thelen, E., eds., 1993, "A Dynamic Systems Approach to Development", MIT Press, Cambridge, MA.

Smoll, F. L., 1974, Development of rhythmic ability in response to selected tempos, *Perceptual Motor Skills*, 39:767.

Smoll, F. L., 1975a, Between-days consistency in personal tempo, *Perceptual Motor Skills*, 41:731.

Smoll, F. L., 1975b, Variability in development of spatial and temporal elements of rhythmic ability, *Perceptual Motor Skills*, 40:140.

Smoll, F. L., 1975c, Preferred tempo in performance of repetitive movements, *Perceptual Motor Skills*, 40:439.

Smoll, F. L., and Schutz, R. W., 1982, Accuracy of rhythmic motor behavior in response to preferred and nonpreferred tempos, *J. Human Movement Stud.*, 8:123.

Soechting, J. E., and Lacquaniti, F., 1983, Modification of trajectory of a pointing movement in response to a change of target location, *J. Neurophysiol.*, 49:548.

Sonderen, J. F. v., Gon, J. J. D. v. d., and Gielen, C. C. A. M., 1988, Conditions determining early modification of motor programmes in response to changes in target location, *Exp. Brain Res.*, 71:320.

Stein, R. B., 1982, What muscle variable(s) does the nervous system control in limb movements? *Behav. Brain Sci.*, 5:535.

Sternad, D., Beek, P. J., and Turvey, M. T., 1994, "Mechanical Perturbations of Rhythmic Forearm Movements", Unpublished manuscript.

Taguchi, S., Gliner, J. A., Horvath, S. M., and Makamura, E., 1981, Preferred tempo, work intensity, and mechanical efficiency, *Perceptual Motor Skills*, 52:443.

Thelen, E., Corbetta, D., Kamm, K., Spencer, J., Schneider, K., and Zernicke, R. F., 1993, The transition to reaching: Mapping intention and intrinsic dynamics, *Child Dev.*, 64:1058.

Thelen, E., and Smith, L. B., 1994, "A Dynamic Systems Approach to the Development of Cognition and Action", MIT Press, Cambridge, MA.

Thelen, E., and Ulrich, B. D., 1991, Hidden skills: A dynamic systems analysis of treadmill stepping during the first year, *Monographs Soc. Res. Child Dev.*, Serial No. 223:56.

Thompson, P., 1993, Motion psychophysics, *in*: "Visual Motion and Its Role in the Stabilization of Gaze", F. A. Miles and J. Wallman, eds., Elsevier, Amsterdam.

Todd, J. T., 1981, Visual information about moving objects, *J. Exp. Psychol.: Human Percept. Perform.* 7:795.

Turvey, M. T., 1977, Preliminaries to a theory of action with reference to vision, *in*: "Perceiving, Acting and Knowing", R. Shaw and J. Bransford, eds., Erlbaum, Hillsdale, N.J.

Turvey, M. T., Fitch, H. L., and Tuller, B., 1982, The Bernstein perspective. I. The problems of degrees of freedom and context-conditioned variability, *in*: "Human Motor Behavior: An Introduction", J. A. S. Kelso, ed., Erlbaum, Hillsdale, N.J.

Turvey, M. T., Rosenblum, L. D., Schmidt, R. C., and Kugler, P. N., 1986, Fluctuations and phase symmetry in coordinated rhythmic movements, *J. Exp. Psychol.: Human Percept. Perform.*, 12:564.

Turvey, M. T., Schmidt, R. C., Rosenblum, D. L., and Kugler, P. N., 1988, On the time allometry of coordinated rhythmic movements, *J. Theor. Biol.*, 130:285.

Turvey, M. T., Shaw, R. E., and Mace, M., 1978, Issues in the theory of action: Degrees of freedom, coordinative structures and coalitions, *in*: "Attention and Performance Vol 7", J. Requin, ed., Erlbaum, Hillsdale, N.J.

Viviani, P., and Laissard, G., 1979, Timing control in motor sequences, *in*: "The Development of Timing Control and Temporal Organization in Coordinated Action", J. Fagard and P. H. Wolff, eds., Elsevier, Amsterdam.

Warren, W. H., and Hannon, D. J., 1990, Eye movements and optical flow, *J. Optical Soc. Am. A*, 7:160.

Warren, W. H., Mestre, D. R., Blackwell, A. W., and Morris, M. W., 1991, Perception of circular heading from optical flow, *J. Exp. Psychol.: Human Percept. Perform.*, 17:28.

Warren, W. H., Morris, M. W., and Kalish, M., 1988, Perception of translation heading from optical flow, *J. Exp. Psychol.: Human Percept. Perform.*, 14:646.

Zalaznik, H. N., Schmidt, R. A., and Gielen, C. C. A. M., 1986, Kinematic properties of rapid aimed hand movements, *J. Motor Behav.*, 18:353.

Zhang, H.-N., and Bingham, G. P., 1993, Head movement and optic flow with displacement prisms in visually guided reaching. Paper presented at the North American Meeting of the International Society for Ecological Psychology, Smith College, Northampton, MA.

NEURAL MODELS OF TEMPORALLY ORGANIZED BEHAVIORS: HANDWRITING PRODUCTION AND WORKING MEMORY

Stephen Grossberg

Center for Adaptive Systems and
Department of Cognitive and Neural Systems
Boston University
111 Cummington Street
Boston, MA 02215
USA

INTRODUCTION: NEURAL CONTROL OF MOVEMENT AND MEMORY

The brain uses several different types of mechanisms to control the temporal organization of behavior. This chapter summarizes biological neural networks which model two types of temporal control. The first model is the VITEWRITE model of handwriting production (Bullock, Grossberg, and Mannes, 1993). The second model is the STORE model for encoding sequences of events in working memory (Bradski, Carpenter, and Grossberg, 1992). Both models have arisen from a computational analysis of relevant behavioral and neural data bases. In both models, the temporal properties of the behavior are not explicitly represented in the network, but instead are emergent properties of multicellular interactions. This fact raises the issue of what organizational principles enable the networks to generate goal-oriented temporal relationships despite the fact that these relationships are not explicitly represented in the model mechanisms.

THE VITEWRITE MODEL OF HANDWRITING PRODUCTION

The VITEWRITE model addresses a number of key issues concerning the skilled performance of sequential actions: What is a motor program? How can a complex movement be flexibly performed at will with variable speed, size, and style without requiring new learning? How does the brain control a redundant manipulator that possesses more degrees of freedom than the space in which it moves? How can smooth curvilinear movements be organized by such a redundant manipulator? In particular, how is the timed launching of different groups, or synergies, of muscles achieved so that the desired directions, distances, and curvatures of movement are achieved? How, moreover, can "acts of will" that vary the speed and size of movements achieve their goal, thereby changing distances and curvatures of movement, without disrupting the correct directions of movement that preserve its overall form through time?

The VITEWRITE model, summarized in Figure 1, introduces a new concept of how a "motor program" can control skilled sequential movements. This motor program is not explicitly represented in the model. Rather, it is an emergent property of feedback interactions between a working memory representation of desired movement directions (called a Vector Plan), and a trajectory generator for moving the limb (called a VITE circuit). The VITEWRITE model also provides a new analysis of how the use of a redundant manipulator can simplify the problem of motor planning. The VITEWRITE model demonstrates how a working memory can control writing movements that exhibit many properties of human handwriting when it interacts reciprocally with a suitably defined trajectory generator coupled

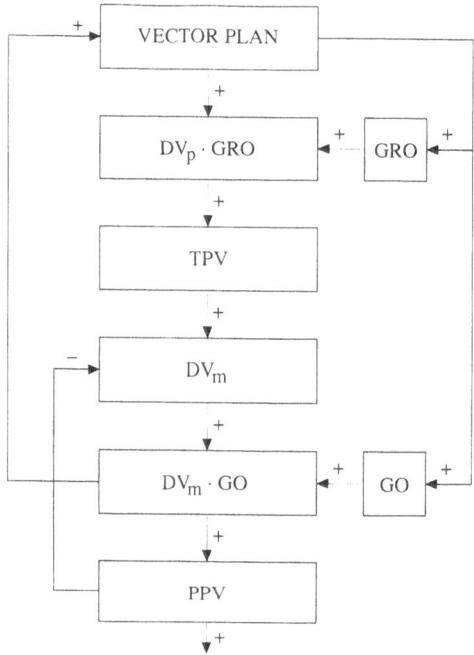

Figure 1. Schematic of the VITEWRITE model: A vector plan functions as a motor program that stores discrete planning Difference Vectors DV_p in a working memory. A GRO signal determines the size of script and a GO signal its speed of execution. After the vector plan and these will-to-act signals are activated, the circuit generates script automatically. Size-scaled planning vectors $DV_p \cdot$ GRO are read into a Target Position Vector (TPV). An outflow representation of present position, the Present Position Vector (PPV), is subtracted from the TPV to define a movement Difference Vector (DV_m). The DV_m is multiplied by the GO signal. The net signal $DV_m \cdot$ GO is integrated by the PPV until it equals the TPV. The signal $DV_m \cdot$ GO is thus an outflow representation of movement speed. It is used to automatically trigger read-out of the next planning vector DV_p. See text for details. Reprinted with permission from Bullock, Grossberg, and Mannes (1993).

to a model hand with redundant degrees of freedom. These results extend the applicability of the VITE model from the control of reaching behaviors (Bullock and Grossberg, 1988, 1991) to the control of complex curvilinear trajectories.

Using a hand with redundant degrees of freedom, here taken to be three (Figure 2), simplifies the motor program, or plan, in at least three ways, that will be explained in subsequent sections. First, each of the three motor synergies, or coordinated muscle groups, of such a hand can be controlled with unimodal velocity profiles. Second, the Vector Plan working memory consists of a discrete set of difference vectors that are read into a VITE circuit at prescribed times. These difference vectors, called *planning* vectors, represent the direction and desired amount of contraction of a motor synergy. They are denoted by DV_p below. Third, the motor program automatically launches transient directional commands to the hand synergies at only two phases in a movement – when the hand begins to move, or when a peak velocity in one of the synergies is achieved.

Such a motor program can be utilized with a VITE model because the VITE model contains a processing stage at which an outflow representation of intended movement velocity is represented. This stage computes the product $DV_m \cdot GO$ of difference vectors DV_m multiplied by a GO signal. The continuously changing DV_m vectors are called *movement* vectors. They are not the discrete planning vectors DV_p. The GO signals that multiply the movement vectors are "will to act", or analog speed, signals that cause movement of a motor synergy if its DV_m is not equal to zero. The $DV_m \cdot GO$ outflow commands continuously move the synergy towards a desired target configuration until its DV_m equals zero. The maxima in time of these $DV_m \cdot GO$ outflow commands in the VITE trajectory generator are used as

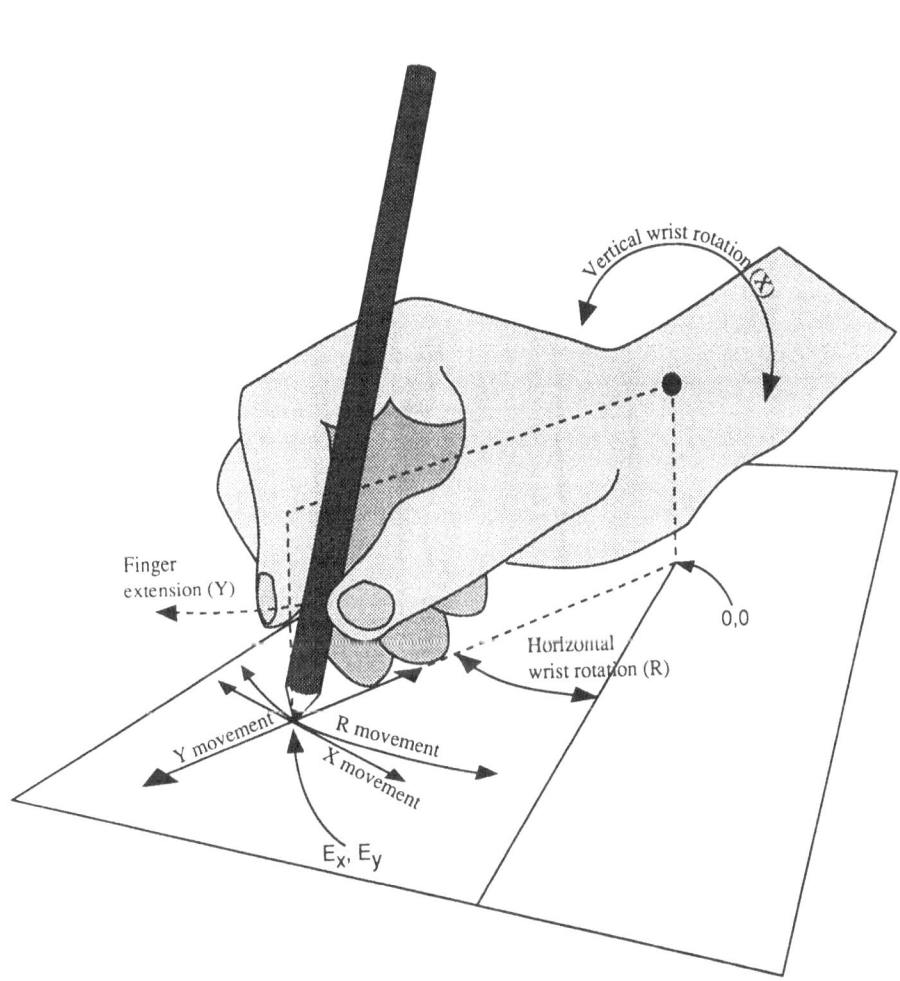

Figure 2. The geometric model of the hand to be controlled, with three degrees of freedom: finger extension/retraction, which moves the pen along the up-down (Y) axis, vertical wrist rotation (supination/pronation), which has the effect of moving the pen along the left-right (X) axis, and horizontal wrist rotation (R), which has two effects: rotating the other two axes, and moving the pen left-right. Reprinted with permission from Bullock, Grossberg, and Mannes (1993).

feedback control signals to read-out the next planning vector DV_p. Using this type of internal feedback loop, an increase in the GO signal can speed up a handwriting movement without changing its form. In a similar way, a GRO signal (defined below) can multiply the planning vectors DV_p before the net signals $DV_p \cdot GRO$ arrive at the VITE model, resulting in a handwritten movement of different size but the same form.

In summary, the VITEWRITE model converts the Vector Plan's temporally discrete and disjoint planning vectors $DV_p \cdot GRO$ into smooth curvilinear trajectories among temporally overlapping synergetic movements. The unimodal temporal shapes of the $DV_m \cdot GO$ outflow velocity commands to the motor synergies are an emergent property of the entire VITEWRITE circuit. When a peak in one synergy's $DV_m \cdot GO$ function is attained, it can activate read-out of a planning vector from the motor program to the VITE circuitry that controls other synergies. These properties enable the VITEWRITE model to avoid explicit storage of within-stroke time lags, to use few memory resources to store the planning vectors, to employ activity-based $DV_m \cdot GO$ decisions to automatically read out the planning vectors, and to thereby achieve speed and size rescaling in response to scalar GRO (size) and GO (speed) acts-of-will, while effortlessly concatenating letter shapes into words.

The VITEWRITE model builds upon desirable properties of the VITE model that have been described in previous studies of VITE-controlled reaching. Indeed, a role for the VITE model in handwriting control was noticed soon after its introduction in Bullock and Grossberg (1988), since the VITE circuit generates emergent properties that mimic key properties of handwriting data. These include the isochrony principle, namely the tendency for strokes of different size to be completed with approximately equal duration (Schomaker, Thomassen, and Teulings, 1989; Viviani and Terzuolo, 1983); skewed velocity profiles (Wann, Nimmo-Smith, and Wing, 1988), typically with faster rise and slower fall in velocity; the synthesis of continuous complex movements from unit segments (Soechting and Terzuolo, 1987); and the tendency of maximal curvatures of a trajectory to occur at locations of minimum velocity (Abend, Bizzi, and Morasso, 1982; Fetters and Todd, 1987; Viviani and Terzuolo, 1980).

The three main components of the VITEWRITE model are: (1) a geometrical model of the hand, (2) a VITE neural trajectory generator, and (3) a Vector Plan working memory. By combining these elements, precise extrinsic control of onset and offset timing is unnecessary. Instead, the times at which subsequent movement commands are read out from the working memory are automatically determined by events in the trajectory generator, which, in turn, are sensitive to previous working memory and volitional signals.

GEOMETRY OF THE HAND

The number of motor segments used in handwriting is large, involving every joint from the shoulder to the fingers. The analysis here is restricted to the hand only, which still has a total of seven degrees of freedom from the wrist to the fingertip. Most of these hand joints operate in concert during handwriting to control three main sets of motor synergists, or muscle groups. Accordingly, the hand model in Bullock, Grossberg, and Mannes (1993) has three degrees of freedom: vertical wrist rotation (supination/pronation, called X), finger extension or retraction (called Y), and horizontal wrist rotation (called R), as in Figure 2.

The extra, third degree of freedom, R, can be used to reduce the complexity of both the motor program and the neural trajectory generator. As an example, consider the simple stroke depicted in Figure 3. In Cartesian space, this stroke can be generated by a mix of unimodal and bimodal velocity profiles with unequal component movement durations, as shown in Figure 3a. By adding a third degree of freedom, which, at least in this example, acts in much the same way as the horizontal component, the same stroke can now be generated using only unimodal, bell-shaped velocity profiles with equal durations. Because of this simplification, there is a unique maximum outflow velocity during each synergetic movement that can be used to trigger readout of the next working memory command. In this way, a redundant degree of freedom can be used to reduce the complexity of both trajectory generation and motor planning.

A further simplification is made by considering the relative amplitudes of synergetic movement that are characteristic of skilled handwriting. Both the effects of finger extension and vertical wrist rotation in handwriting are small in relation to the total range (cf. Lacquaniti et al., 1987), and the radius of horizontal wrist rotation is rather large in relation to finger extension and vertical wrist rotation. The trajectories of each of these components are thus good approximations to straight lines. Therefore, we further simplify the geometrical hand model by modelling both X (vertical wrist rotation) and Y (finger extension) as an orthogonal system of spatially straight lines. However, since these axes of movement are mounted on the hand (and not fixed with respect to the drawing surface), this coordinate system can be rotated by horizontal wrist motion.

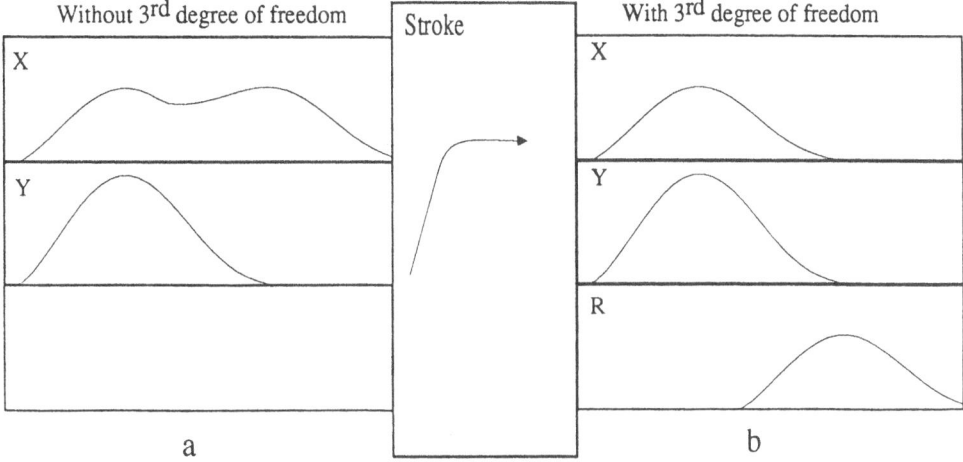

Figure 3. A stroke that is greatly simplified by use of three degrees of freedom. Left: With two degrees of freedom, the stroke shown in the middle can only be obtained by a mix of bimodal and unimodal velocity profiles, since the horizontal component is non-zero before and after the bend. Right: Using a third degree of freedom (R), which acts much like X, allows production of the same shape with only unimodal velocity profiles. This presumably simplifies neural control. Reprinted with permission from Bullock, Grossberg, and Mannes (1993).

Under these assumptions, if the wrist is located at spatial location (0,0), then the pen tip, or end effector location (E_x, E_y) can be found by

$$E_x = (l + y)\sin r + x\cos r \qquad (1)$$

$$E_y = (l + y)\cos r - x\sin r \qquad (2)$$

where x and y denote the X and Y excursions, respectively, and r stands for the horizontal angle of the hand with respect to the arm. The length of the hand from the wrist to the knuckles, denoted as l, is large relative to the X, Y and R excursions.

SYNCHRONOUS TRAJECTORY FORMATION BY THE VECTOR INTEGRATION TO ENDPOINT MODEL

The Vector Integration To Endpoint, or VITE, model of Bullock and Grossberg (1988, 1991) is a neural model of a trajectory generator whose outflow commands control multi-joint motor trajectories. The model shows how a group of muscles may be dynamically bound into a motor synergy, and once bound, how the synergy can perform synchronous movements at variable speeds. We therefore often call synergy, synchrony, and speed the "3 S's" of trajectory formation. The VITE model outputs are the inputs to a second neural model called the FLETE model for Factorization of LEngth and TEnsion. The FLETE model suggests how outflow commands from a VITE circuit may be transformed into positionally accurate movements of an arm that is subjected to variable external forces and stiffness levels (Bullock and Grossberg, 1991; Bullock, Contreras-Vidal, and Grossberg, 1992). In other words, the VITE model forms part of the Platonic trajectory planning apparatus of a larger movement control system, whereas the FLETE model controls Newtonian force and motor related factors to ensure that the arm closely tracks VITE outflow movement commands. To accomplish this, the FLETE model compares VITE outflow velocity DV·GO and positional PPV signals with dynamic and static inflow signals from the muscles themselves to trigger either reactive responses to positional errors or adaptively timed gain changes that serve to predictively preempt errors before they can occur on future movement trials. In the original references, the VITE model is interpreted in terms of neural data about brain regions such as parietal cortex, motor cortex, and basal ganglia. The FLETE model is interpreted in terms of neural data about the spinal cord and cerebellum. These spinocerebellar interactions will not be further discussed here.

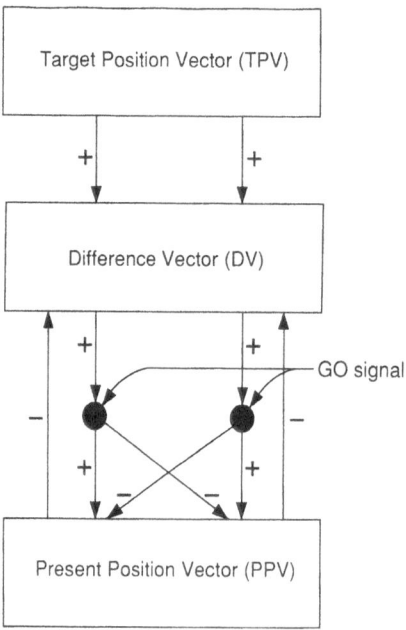

Figure 4. The VITE circuit, the neural controller of each component agonist-antagonist pair of the hand. Reprinted with permission from Bullock, Grossberg, and Mannes (1993).

The VITE circuit consists of four neural processing stages that are depicted in Figure 4. The first stage computes a Target Position Vector (TPV) that encodes desired limb positions. As in the VITEWRITE model (Figure 1), these target locations are derived from signals coded in terms of muscle lengths from higher processing stages. The Present Position Vector (PPV) stage integrates its inputs over time to generate outflow movement signals to spinal neuron pools, which in turn act on muscles capable of moving the arm. The Difference Vector (DV) stage continuously computes the difference between PPV and TPV using excitatory outflow signals from the TPV and inhibitory corollary discharge, or efference copy, signals from the PPV. This DV is denoted by DV_m in Figure 1.

Outflow from the DV to PPV is multiplied, or gated, by a nonspecific GO signal. Before any movement begins, a desired position command may be loaded into the TPV and relayed to the DV. This operation is called motor priming (Georgopoulos et al., 1984). Until the GO signal grows positive, however, no change in PPV can occur. Once the GO signal becomes positive, the PPV can start integrating its input signals at the rate DV· GO (see Figure 4). This multiplicative interaction maintains the direction coded by DV while using the GO signal to modulate the speed of movement in this direction. The size of the GO signal is assumed to grow monotonically once a movement is initiated. Since the PPV integrates DV· GO, the rate of change of the outflow PPV signal, namely $dPPV/dt$, tracks DV· GO. Thus DV· GO provides an internal measure of the commanded movement velocity $dPPV/dt$. The DV is driven to zero by inhibitory feedback from PPV to DV as the PPV approaches the TPV. The system thus equilibrates when the PPV equals the TPV. If a single GO signal multiplies all outflow commands from the DV equally, all components of a given motor synergy tend to complete their movement synchronously, regardless of GO signal magnitude or component movement amplitude (Bullock and Grossberg, 1988).

COORDINATION OF MULTIPLE MOTOR SYNERGIES WITH ASYNCHRONOUS ONSETS AND OFFSETS

The production of curved trajectories during handwriting requires that distinct movement components have distinct but overlapping velocity profiles. These phase lags suggest that the several hand synergies (finger extension, horizontal wrist rotation, and vertical wrist rotation) cannot be grouped into one TPV with a single GO signal, since the VITE circuit would work towards making all component movements terminate at the same time. Instead, each of the three synergies of the hand model is controlled by its own VITE circuit, with separately initiated GO signals. These GO signals are reset before the onset of a new movement by each synergy. The assumption of multiple GO signal channels is consistent with data on the proposed anatomical site of GO signal generation, namely the basal ganglia (Bullock and Grossberg, 1991, Horak and Anderson, 1984a, 1984b). Recent reports indicate that pathways through the basal ganglia maintain somatotopy, or motor-channel specificity (Parent, 1990), and work summarized by Golani (1992) implicates the basal ganglia in gating the degrees of freedom that are incorporated into different movements.

MODEL EQUATIONS

The equations that govern the dynamics of the multi-channel VITE circuit are as follows. The TPV is denoted by $T = (T_1, T_2, \ldots, T_n)$, the PPV by $P = (P_1, P_2, \ldots, P_n)$, the movement vector DV_m by $V = (V_1, V_2, \ldots, V_n)$, the planning vector DV_p by $D = (D_1, D_2, \ldots, D_n)$, the GRO signal by $S = (S_1, S_2, \ldots, S_n)$, and the GO signal by $G = (G_1, G_2, \ldots, G_n)$, where index i denotes the ith motor synergy.

Target Position Vector

$$T_i(t_{i,j+1}) = T_i(t_{ij}) + S_i D_i(t_{ij}). \tag{3}$$

The TPV receives planning vector inputs $D_i(t_{ij})$ from higher processing stages. These inputs embody directional commands whose size, scaled by S_i, determines the distance travelled by each synergy. At launch times t_{ij}, $j = 1, \ldots, n$, the jth planning vector $(D_1(t_{1j}), D_2(t_{2j}), \ldots, D_n(t_{nj}))$, scaled by the GRO signals $(S_1, S_2, \ldots, S_n)$, is added to the TPV, as in (3).

Difference Vector

$$\frac{dV_i}{dt} = \alpha(-V_i + T_i - P_i). \tag{4}$$

By equation (4), the movement vector V tracks the difference $T - P$ at rate α. Equation (4) simplifies the original VITE equations (Bullock and Grossberg, 1988), which used an opponent push-pull mechanism to avoid negative values for V_i (see Figure 4). Here, agonist and antagonist activations are lumped into one variable by allowing negative values.

GO Signal

$$G_i(t) = G_0(t - t_{ij})^n, \qquad t_{ij} \leq t < t^*_{ij}, j = 1, ..., n, \tag{5}$$

where G_0 is a constant and t_{ij} is the jth time at which synergy i is launched. The GO signal grows monotonically until time t^*_{ij}, when it is reset to zero. This stereotyped and repetitive GO signal rule is capable of generating arbitrary shapes of cursive script letters shapes. In all simulations, $n = 1.4$, which produces nearly symmetrical bell-shaped velocity profiles. Equation (5) for the growth of the GO signal is one of many that could be used, and is chosen solely for convenience. Bullock and Grossberg (1988) showed that many psychophysical properties of arm movements could be fit by a wide variety of GO signal shapes. In particular, they showed that a physically plausible GO signal could be generated by two or more neurons activated in series by a step function input. In the VITE model, using a cascade to generate a GO signal is one of two determinants of the velocity profile – the DV being the other. In the Plamondon (1989) handwriting model, a much longer cascade is used to generate the entire velocity profile.

Present Position Vector

$$\frac{dP_i}{dt} = V_i G_i. \tag{6}$$

The PPV integrates its input signals at the rate $V_i G_i$. Since V_i tracks $T_i - P_i$ and P_i increases if $T_i > P_i$ or decreases if $T_i < P_i$, the process continues until all $P_i = T_i$, $i = 1, 2, \ldots, n$.

FEEDBACK CONTROL OF SEQUENTIAL MOVEMENT COMMANDS

To produce the smooth, curved trajectories of script, synergy DV_p directions and GO signal onsets need to be appropriately timed. The onset timing for the next stroke in a motor program could be determined in two ways. In one way, the time of launching the next stroke is a parameter of the motor program (cf. Schomaker et al., 1989). In the second way, some event in the controller itself, or even downstream from the controller, triggers execution of the next stroke. The first possibility faces the difficulty that the motor program may not be able to compensate for changes in stroke size and speed of execution. For example, the shape of a trajectory could be very different at different execution speeds, unless the timing of successive onsets could automatically compensate for such motor variability.

If triggering a successive stroke is contingent on a characteristic event in the velocity vector of the movement, this problem is avoided, since onset lags then shift automatically with speed of execution. An outflow representation of each synergy's velocity is encoded in the VITE model by the model neurons whose activities represent the DV_m· GO processing stage. Such an outflow representation avoids the instability problems that could otherwise occur if delayed inflow signals from the muscles themselves were used. Simulation studies by Bullock, Grossberg, and Mannes (1993) have shown that two events are suitable to launch a stroke. These are: (1) times when there is a match between TPV and PPV, and consequently all velocities are close to zero. This event is called a *postural launch*; (2) times at the peak of one or more velocity traces. This event is called a *dynamic launch*. In a dynamic launch, a peak in one of the velocity profiles initiates movement of new synergies by triggering read-in of new targets and reset of their respective GO signals. The new targets may be zero for some

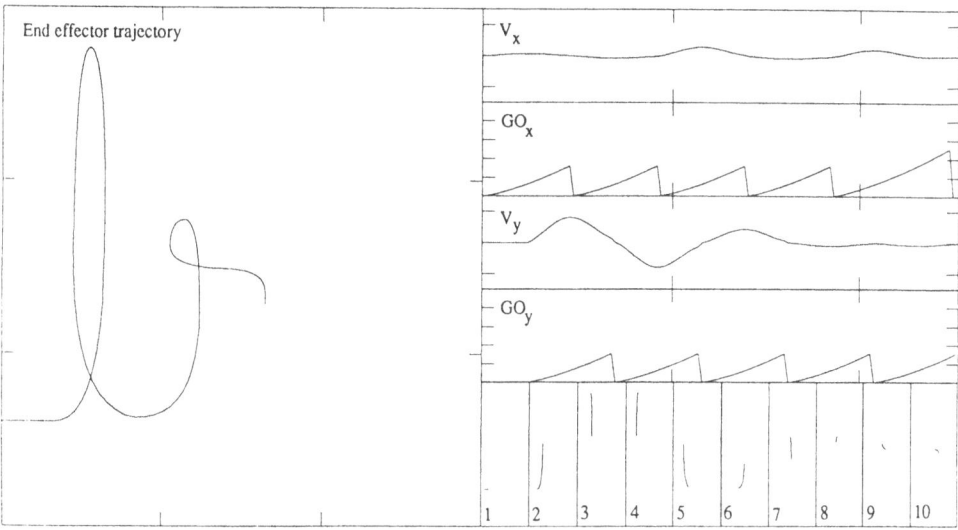

Figure 5. An example showing how to generate the end-effector trajectory drawn in the left panel. V_x, V_y denote X and Y velocities, respectively. GO-signal values for each of these components are plotted below the velocity profiles. The smaller panels labeled 1-10 show the end-effector trajectory during the time interval along the axis which the panels touch above. Reprinted with permission from Bullock, Grossberg, and Mannes (1993).

or all components. In the postural launch, a point of zero velocity can also trigger a new movement. Thus the launch times t_{ij} in (5) and Figure 5 occur either when a synergy is at rest or when the outflow speed command DV· GO of another synergy reaches a maximal size. Reset occurs at times t^*_{ij} when the PPV of the synergy equals its TPV. This control scheme is robust with respect to changes in command timing. Perturbing onset timing results in rounder shapes if a dynamic launch occurs before the peak of another velocity profile, and more angular shapes if the launch occurs after the peak. If a new target is launched only when one of these two types of events occurs, then the phase relations between any two component velocity traces are limited to either 0 or 90 degrees. Each peak and zero in the outflow velocity trace DV_m· GO can activate read-out of the next planning vector DV_p from the working memory, as in Figure 5. Such a DV_p reads a new directional movement command into the

TPV of the VITE circuit. Each DV_p also activates the GO signal of the corresponding synergy. These TPV commands point in the independent X, Y, and R directions. Their amplitudes equal the maximal excursion of the letter in that direction. The order, timing, and size of these synergy commands determine the curvature of the movement. All the stored commands in the vector plan that characterizes a letter in this scheme are generated at discrete times in independent directions. The VITE model automatically converts these temporally discrete commands into continuously curved trajectories of appropriate shape. Such a controller affords a huge compression of the premotor and motor commands needed to generate cursive script. Some key properties of movement generated by the VITEWRITE model are summarized below.

SIMULATIONS OF CURSIVE SCRIPT: FLEXIBLE CONTROL OF SIZE, SPEED, AND STYLE

An example of a script letter "b" is shown in Figure 5. The motor program – that is the sequence of directional targets for the controller – is summarized in Table 1. Each row in Table 1 corresponds to a stroke segment shown in the small panels on the lower right side of Figure 5.

Table 1. Notation for a motor program, characterizing the letter shape shown in Figure 6. X is launched first, with a target of 10 length units (corresponding to about 5 mm). The next half cycle, which is launched at the velocity peak of the X motion, executes an upward (Y) motion of 110 units. At the Y velocity peak, an X motion in the other direction is triggered. This temporally overlapping succession of X and Y is continued until the last pattern of the motor program, which launches no component, and so movement comes to a halt. Numbers in round brackets denote the TPV_i during the second half-cycle, i.e. the decreasing part of the velocity profile. Reprinted with permission from Bullock, Grossberg, and Mannes (1993).

Half cycle	X	Y	R
1	10	0	0
2	(10)	110	0
3	-10	(110)	0
4	(-10)	-110	0
5	40	(-110)	0
6	(40)	60	0
7	-10	(60)	0
8	(-10)	-15	0
9	30	(-15)	0
10	(30)	-10	0

To start with, an X motion to the right is launched (stroke segment 1 in Figure 5 and half cycle 1 in Table 1). At the time when X reaches maximum velocity, a Y motion upwards is launched (stroke 2). At the peak of this Y motion, a small X motion to the left is launched (stroke 3), and so forth. The letter "b" is a relatively simple example because the trajectory of this letter is a variation of a circle, but with different amplitudes for X and Y in every stroke. The similarity to a circular trajectory can also easily be seen by the up-down alternation of the

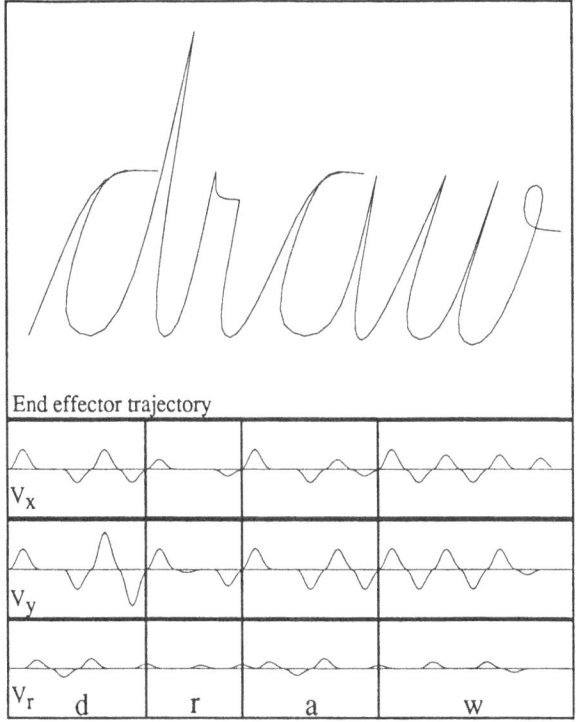

Figure 6. An example of connecting letters by concatenating individual motor programs. Reprinted with permission from Bullock, Grossberg, and Mannes (1993).

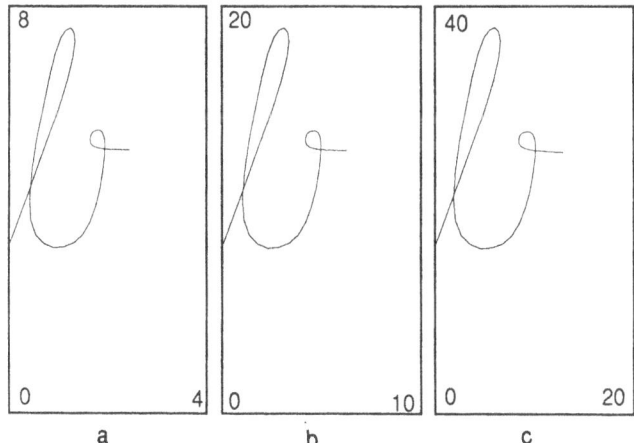

Figure 7. Shape invariance with two different hand geometries: Panels **A** through **C** show perfect shape invariance of the letter "b", scaled to three different sizes by choosing three different values for the GRO parameter S. The trajectories were reduced to fit in the panels. The numbers in the right and top corners of each panel indicate the panel's size in mm prior to reduction. The end effector position was calculated by equations (1) and (2). Reprinted with permission from Bullock, Grossberg, and Mannes (1993).

velocity profiles. Bullock, Grossberg, and Mannes (1993) discuss how use of a consistent stylistic strategy for each letter enables letters to be effortlessly connected into word shapes, an example of which is depicted in Figure 6. Some aspects of the kinematics of handwriting trajectories are invariant with respect to variations in starting point, slant, and size (Viviani and Terzuolo, 1980; Morasso, 1981). These invariances are also exhibited by the model. Figure 7 displays variations of a "b" letter trajectory achieved by rescaling the volitional GRO command. This "b" shape is created by a different combination of synergetic commands than in Figure 5, thereby illustrating the flexibility that can be achieved using a redundant manipulator. In each panel of Figure 7, all the components D_i of TVP_i are multiplied by the same GRO scalar S, but S varies across the panels. This variable GRO command modifies the size of the letters produced, but leaves the trajectory shape invariant. The simplified geometrical model defined in equations (1) and (2) produces perfect shape invariance under size scaling. If a more elaborate geometrical model of the hand is used, extreme finger angles at the border of the workspace produce distortions; see Bullock, Grossberg, and Mannes (1993). In addition, changes in writing style can be achieved by multiplying each component D_i of TPV_i by a different scalar S_i.

Shape invariance under speed rescaling is demonstrated in Figure 8, which shows the same letter performed at a given speed and at double that speed. This is achieved by rescaling the GO signal via parameter G_0 in equation (5). This simulation assumes that new synergies are instantaneously launched at the velocity maxima of other synergies. If a small but finite reaction time is needed to launch, invariance would not be substantially influenced until speeds were attained at which the duration of each synergy was not much greater than the reaction time. Then the smooth curvature of the letter shape would begin to deteriorate, leading to straighter trajectories followed by more sudden changes of curvature.

The ease with which size and speed invariance are demonstrated in the VITEWRITE model derives from the model's use of DVs to control updating of the TPV in equation (3) and updating of the PPV in equation (6). Once DV directional control is available, scalar GRO and GO volitional signals can transform a stereotyped series of DVs into motor performances whose sizes and speeds can be adjusted to match variable environmental conditions. Models which utilize DVs for their spatial and trajectory control have generically been called Vector Associative Maps, or VAMs (Gaudiano and Grossberg, 1991).

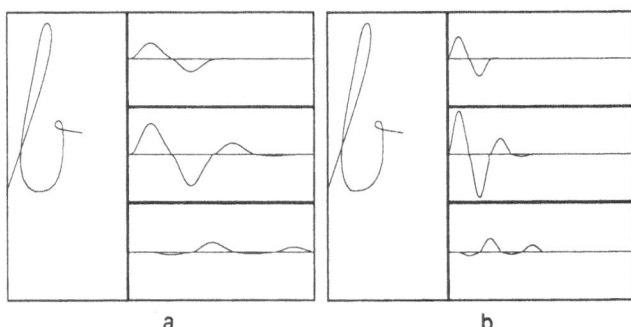

a   b

Figure 8. Shape invariance under speed rescaling: The same motor program is executed at two different speeds, simulated by scaling the magnitude of the GO signal. Panel **A** shows the letter "b" executed at a "normal" speed ($G_0 = 1$), panel **B** at twice that speed ($G_0 = 2$). Reprinted with permission from Bullock, Grossberg, and Mannes (1993).

THE TWO-THIRDS POWER LAW RELATING CURVATURE AND VELOCITY

Another widely observed invariant of movement is the strong coupling between velocity and curvature (Morasso, 1981; Abend, Bizzi, and Morasso, 1982). In general, peaks in the curvature profile occur at troughs in the velocity profile. Lacquaniti, Terzuolo, and Viviani (1983) formulated a "two-thirds power law" to describe the empirical relation between curvature and velocity. This law relates angular velocity $A(t)$ to curvature $C(t)$ as $A(t) = kC(t)^{2/3}$, which can be rewritten for tangential velocity $V(t)$ as $V(t) = kR(t)^{1/3}$, where $R(t) = 1/C(t)$ denotes the radius of curvature. Figure 9a plots model curvature and model tangential velocity for the letter "b"; Figure 9b plots model tangential velocity alongside the tangential velocity predicted from model curvature by the two-thirds power law. The agreement is close

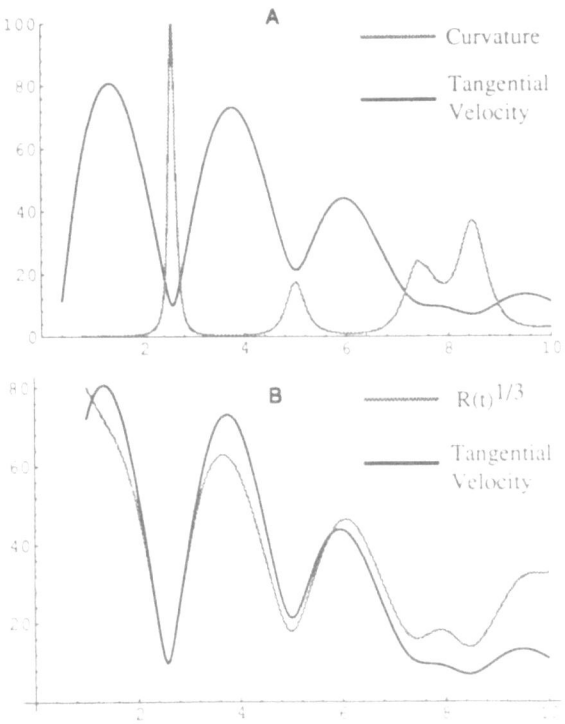

Figure 9. Relationship between pen tip (tangential) velocity $V(t)$ and curvature for the letter "b." The simulated pen tip trajectory $x(t),y(t)$ was least-squares fitted to a polynomial. Velocity was computed as $V(t) = ((x')^2 + (y')^2)^{1/2}$, and curvature by the formula $C(t) = (x'y'' - y'x'')/V(t)^3$. Plot **A** plots curvature and velocity, which show the expected inverse relationship. Plot **B** compares the velocity $V(t)$ with the predicted curvature $kR(t)^{1/3}$, $k = 10$ according to the two-thirds power law (Wann, Nimmo-Smith, and Wing, 1988). Reprinted with permission from Bullock, Grossberg, and Mannes (1993).

but not perfect. Indeed, the two-thirds power function is itself an imperfect descriptor of human performance. Wann, Nimmo-Smith, and Wing (1988) have noted that one basis for the discrepancy is that human velocity profiles are not perfectly symmetrical about the peak velocity value. VITE velocity profiles show the same duration dependent deviation from perfect symmetry that is seen in human handwriting movements (Bullock and Grossberg, 1988, 1991; Nagasaki, 1989).

COMPLEX SKILLED MOVEMENT AS AN EMERGENT PROPERTY

The VITEWRITE model demonstrates how a multi-channel VITE trajectory generator, controlling a suitably designed hand with redundant degrees of freedom, enables a simple motor program to generate complex curvilinear movements that have many of the properties that humans exhibit when they produce cursive script. In particular, the existence of a $DV_m \cdot GO$ processing stage enables the VITE model to trigger readout of new motor commands at peak values of a synergy's outflow velocity profile. Using this trigger, the DV_ms that update the TPV and the PPV processing stages may be modulated by volitional GO signals that rescale the speed of handwriting without changing its form. Likewise, the use of a motor program that consists of planning vectors DV_p released by velocity-sensitive events in the trajectory generator enable volitional GRO signals to rescale the size of handwriting without changing its form. From a higher computational viewpoint, the use of difference vectors such as DV_p and DV_m in Figure 1, gated by volitional commands such as GRO and GO, and integrated to yield positional commands such as TPV and PPV, provide a computational framework for analyzing how many goal-oriented complex movements are made and flexibly modified under variable task conditions. Neural network architectures in which these directional, volitional, and positional commands are interactively repeated have been called VAM Cascades (Gaudiano and Grossberg,1991). Accumulating theoretical and empirical evidence points to VAM Cascades as a computational framework for the control of planned biological movements. See Bullock, Grossberg, and Guenther (1993), Grossberg et al. (1993), and Guenther et al. (1994) for further discussion.

The second part of this chapter addresses the question of how a working memory, such as the Vector Plan in Figure 1, may be organized. What is a working memory, and what are the organizational principles that govern its design?

WHAT IS A WORKING MEMORY?

Working memory is the type of memory in which a novel temporally ordered sequence of events, such as a telephone number, can temporarily be stored and then performed (Baddeley,1976). Working memory is a kind of short-term memory (STM) and, unlike long-term memory (LTM), it can quickly be erased by a distracting event. There is a large experimental literature on the topic of working memory, as well as a variety of models (Atkinson and Shiffrin, 1971; Cohen and Grossberg, 1987; Cohen, Grossberg, and Stork, 1987; Elman, 1990; Grossberg, 1970, 1978a, 1978b; Grossberg and Pepe, 1971; Grossberg and Stone, 1986; Gutfreund and Mezard, 1988; Guyon, Personnaz, Nadal, and Dreyfus, 1988; Jordan, 1986; Reeves and Sperling, 1986; Schreter and Pfeifer, 1989; Seibert, 1991; Seibert and Waxman, 1990a, 1990b; Wang and Arbib, 1990).

The present class of models, called STORE (Sustained Temporal Order REcurrent) models, exhibit properties that have not previously been available in a dynamically defined working memory (Bradski, Carpenter, and Grossberg, 1992, 1994). In particular, STORE working memories are designed to encode the invariant temporal order of sequential events, or items, that may be presented with widely differing growth rates, amplitudes, durations, and interstimulus intervals. The STORE model is also designed to enable all possible groupings of the events stored in STM to be stably learned and remembered in LTM, even when new events are perturbing the system. In other words, these working memories enable chunks (also called compressed, categorical, or unitized representations) of a stored list to be encoded in LTM in a manner that is not erased by the continuous barrage of new inputs to the working memory.

Working memories with these properties are important in many applications which require properties of behavioral self-organization. One application is the VITEWRITE model described above or, more generally, working memories for sensorimotor planning whose distributed representations can be unitized during learning and subsequently read out on demand during performance. In the case of the VITEWRITE model, the working memory reads out planning vectors DV_p that are generated during sensory-motor imitation and learning. Then volitional commands, such as the GO and GRO signals of the VITEWRITE model, can flexibly modify the invariant contents of the working memory to generate movements that can rapidly adapt to variable task conditions. Other applications include working memories for eye movement control (Grossberg and Kuperstein, 1989), for variable-rate speech perception (Cohen and Grossberg, 1986; Cohen, Grossberg, and Stork, 1987), and for 3-D visual object recognition (Bradski, Carpenter, and Grossberg, 1992).

INVARIANCE PRINCIPLE AND NORMALIZATION RULE

The STORE neural network working memories are based upon algebraically characterized working memories that were introduced in Grossberg (1978a, 1978b). These algebraic working memories were designed to explain psychological data concerning working memory storage and recall. In these models, individual events are stored in working memory in such a way that the pattern of STM activity across event representations encodes both the events that have occurred and the temporal order in which they have occurred. In the cognitive literature, such a working memory is often said to store both *item* information and *order* information (Healy, 1975; Lee and Estes,1981; Ratcliff, 1978). Item information is encoded by what nodes are active. Order information is represented by their relative activation, with the most active nodes performed first. The models also include a mechanism for reading out events in the stored temporal order. A rehearsal wave, or nonspecific arousal input, causes the most active node to be read-out first, as it self-inhibits its own activation via negative feedback to enable the next-most-active node to be read-out. An event sequence can hereby be performed from STM even if it is not yet incorporated through learning into LTM, much as a new telephone number can be repeated the first time that it is heard.

The large data base on working memory shows that storage and performance of temporal order information from working memory is not always veridical (Atkinson and Shiffrin, 1971; Baddeley, 1976; Reeves and Sperling, 1986). The observed deviations from veridical temporal order in STM were shown to follow from two postulates of the algebraic working memory model that have clear adaptive value. These principles are called the Invariance Principle and the Normalization Rule (Grossberg, 1978b).

Invariance Principle. The spatial patterns of STM activation across the event representations of a working memory are stored and updated in response to sequentially presented events in such away as to leave invariant the temporal order of all groupings of previously presented events. In particular, a temporal list of events is encoded in STM in a way that preserves the stability of previously learned LTM codes for familiar sublists of the list.

For example, suppose that the word *MY* has previously been stored in STM and has, through learning, established a learned chunk in LTM. Suppose that the word *MYSELF* is then stored for the first time in STM. The word *MY* is a syllable of *MYSELF*. The STM encoding of *MY* as a syllable of *MYSELF* may not be the same as its STM encoding as a word in its own right. On the other hand, STM encoding of *MY* as part of *MYSELF* should not be allowed to force forgetting of the LTM code for *MY* as a word in its own right. If it did, familiar words, such as *MY*, could not be learned as parts of larger words, such as *MYSELF*, without eliminating the smaller words from the lexicon. More generally, new wholes could not be built from familiar parts without erasing LTM of the parts.

The Invariance Principle can be realized algebraically as follows, provided that no list items are repeated. Assume for simplicity that the ith list item is preprocessed by a winner-take-all network. Each list item then activates a single output node of the preprocessor network. (Properties of the working memory also obtain if a finite set of output nodes is activated for each item.) In the winner-take-all case, the winning node that is activated by the ith item sends a binary input I_i to the first working memory level F_1 (Figure 10). Let x_i denote the activity of the ith item representation of F_1. Suppose that I_i is registered in working memory at time t_i. At time t_i, the activity pattern $(x_1(t_i), x_2(t_i), \ldots, x_n(t_i))$ across F_1 stores the effects of the list $I_1, I_2, \ldots, I_i$ of previous inputs. The input I_i updates the activity values $x_k(t_{i-1})$ to new values $x_k(t_i)$ for all nodes $k = 1, 2, \ldots, i$ according to the following rule:

$$x_k(t_i) = \begin{cases} 0 & \text{if } k > i, \\ \mu_i & \text{if } k = i, \\ \omega_i x_k(t_{i-1}) & \text{if } k < i \end{cases} \qquad (7)$$

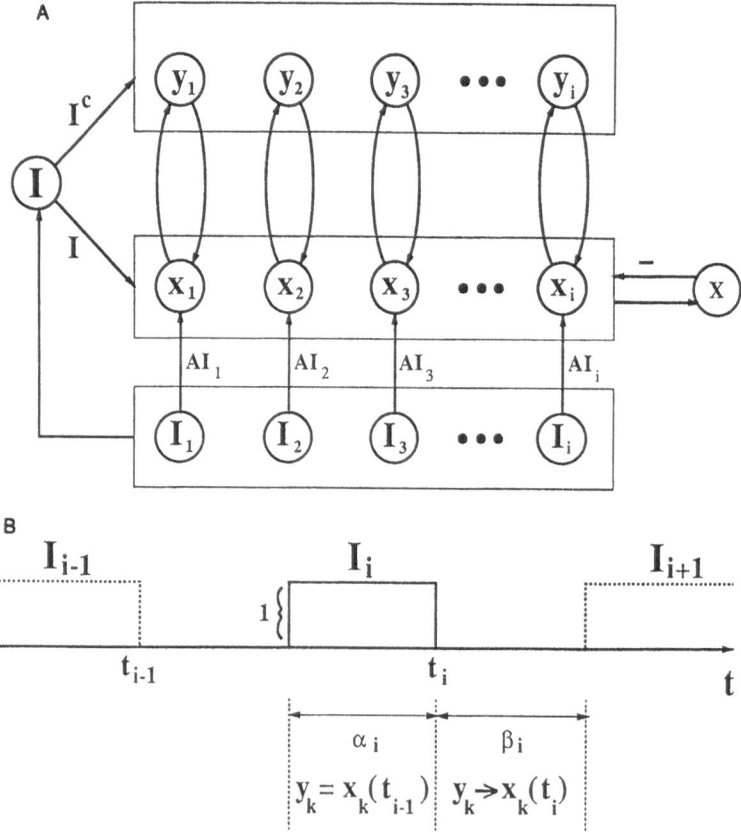

Figure 10. A. Elementary STORE model: STM activity x_i at level 1 registers the item input I_i, nonspecific shunting inhibition x, and level 2 STM y_i. STM activity y_i at level 2 registers x_i. Complementary input-driven gain signals I and I^c control STM processing at levels 1 and 2. B. Input $I_i(t)$ equals 1 for $t_i - \alpha_i < t \le t_i$. When all inputs are off $(t_i < t \le t_i + \beta_i)$ level 2 variables y_k relax to level 1 values $x_k(t_i)$. Reprinted with permission from Bradski, Carpenter, and Grossberg (1992).

174

By (7), at time t_i, the pattern $(x_1(t_{i-1}), x_2(t_{i-1}), \ldots, x_{i-1}(t_{i-1}))$ of previously stored STM activities is multiplied by a common factor ω_i as the ith item is stored with initial activity μ_i.

The storage rule (7) satisfies the Invariance Principle for the following reason. Suppose that F_1 is the first level of a two-level competitive learning, or self-organizing feature map network (Grossberg, 1976). Then F_1 sends signals to the second level F_2 via an adaptive filter. The total input to the jth F_2 node is $\Sigma_k x_k z_{kj}$, where z_{kj} denotes the LTM trace, or adaptive weight, in the path from the kth F_1 node to the jth F_2 node. In psychological terms, each active F_2 node represents a chunk of the F_1 activity pattern. When the jth F_2 node is active, the LTM weights z_{kj} converge toward x_k; in other words, the vector of LTM traces z_{kj} becomes parallel to the F_1 activity vector. When a new item i is added to the list, the Invariance Principle implies that the previously active items in the list will simply be multiplied by a common factor ω_i, thereby maintaining a constant ratio between the STM activities of previously active items. Constant STM activity ratios imply that the STM activity vector at F_1 remains parallel to its LTM weight vectors as the magnitudes of the STM activities change under new inputs. Hence, adding new list items does not invalidate the STM and LTM codes for sublists. In particular, the temporal order of items in each sublist, encoded as relative sizes of both the STM and the LTM variables, remains invariant.

Normalization Rule. The Normalization Rule algebraically formulates the classical property of the limited capacity of STM (Atkinson and Shiffrin, 1971). A convenient statement of this property is given by the equation

$$S_i = \sum_k x_k(t_i) = \mu_1 \theta_i + S(1 - \theta_i), \tag{8}$$

where $\theta_1 = 1$ and θ_i decreases towards 0 as i increases. For example, let $\theta_i = \theta^{i-1}$, with $0 < \theta < 1$. Total activity S_i then increases toward an asymptote, S, as new items are presented. Parameter S characterizes the "limited capacity" of STM. In human subjects, this parameter is determined by biological constraints. In an artificial neural network, parameter S can be set at any finite value.

Using (7) and (8), it was proved in Grossberg (1978a) that the rate at which S_i approaches its asymptote S helps to determine the form of the STM activity pattern. The pattern $(x_1, \ldots, x_i)$ can exhibit primacy (all $x_{k-1} > x_k$), recency (all $x_{k-1} < x_k$), or bowing, which combines primacy for early items with recency for later items (Grossberg, 1978a). These patterns correspond to properties of STM storage by human subjects. Model parameters are typically set so that the STM activity pattern exhibits a primacy gradient in response to a short list. Since more active nodes are read out of STM before less active nodes during performance trials, primacy storage leads to the correct order of recall in response to a short list. Using the same parameters, the STM activity pattern exhibits a bow in response to longer lists, and approaches a recency gradient in response to still longer lists. An STM bow leads to performance of items near the list beginning and end before items near the middle of the list. A larger STM activity at a node also leads to a higher probability of recall from that node when the network is perturbed by noise. An STM bow thus leads to earlier recall and to a higher probability of recall from items at the beginning and the end of a list.

These formal network properties are seen in experiments that test working memory in human subjects, such as free recall experiments during which subjects are asked to recall the items in a list after being exposed to them once in a prescribed order (Atkinson and Shiffrin, 1971; Healy, 1975; Lee and Estes, 1981). Effects of LTM on free recall data have also been analyzed by the theory (Grossberg, 1978a; 1978b).

The multiplicative gating in equation (7) and the normalization rule in (8) are algebraic versions of the general types of properties which are found in shunting competitive feedback networks (Grossberg, 1973). A task of STORE model research was to design specialized shunting networks which realize equations (7) and (8) of Bradski, Carpenter, and Grossberg (1992) as emergent properties of their real-time dynamics. The STORE model is a real-time shunting network, defined below, which exhibits the desired emergent properties. In particular, the STORE system moves from primacy to bowing to recency as a single model parameter is increased.

WORKING MEMORY INVARIANCE UNDER VARIABLE INPUT SPEED, DURATION, AND INTERSTIMULUS INTERVAL

Two types of real-time working memory models, *transient* models and *sustained* models, can realize the invariance and normalization properties. In a transient model, presentation of items with different durations can alter the temporal order of previously stored items. Transient memory models can still accurately represent temporal order if input durations are transformed in a preprocessing stage so that they have a constant duration. Sustained models allow input durations and interitem intervals to be essentially arbitrary: so long as these intervals are not too short, variations in input speed, duration, and interstimulus interval have no effect on the temporal order that is stored in STM. A sustained neural network model is defined below. This two-level STORE model codes lists of distinct items. A variant of the STORE model design can encode the invariant temporal order of lists in which each item may occur multiple times (Bradski, Carpenter, and Grossberg, 1994). Each item may also be represented by multiple nodes.

The first level of the STORE model (Figure 10) consists of nodes with STM activity x_i. The ith item is assumed to send a unit input I_i to the ith node for a time interval of length α_i. After an interstimulus interval of length β_i, the next item sends an input to the $(i + 1)^{st}$ node, and so on. Each STM node also receives shunting inhibition via a nonspecific feedback signal that is proportional to the total STM activity x. The second STORE level consists of excitatory interneurons whose activity y_i tracks x_i. A critical additional factor in the model is gain control that enables changes in x_i to occur only when an input is present and enables changes in y_i to occur only when no input is present.

This alternating gain control allows feedback from y_k to x_k $(k < i)$ to preserve previously stored patterns even when a new input I_i is present for a long time interval. These processes are defined below in the simplest way possible to permit complete analysis and understanding of the model's emergent properties.

The STORE model is defined by the dimensionless equations

$$\frac{dx_i}{dt} = [AI_i + y_i - x_i x]I \tag{9}$$

and

$$\frac{dy_i}{dt} = [x_i - y_i]I^c, \tag{10}$$

where

$$x \equiv \sum_k x_k, \tag{11}$$

$$I \equiv \sum_k I_k, \tag{12}$$

$$I^c \equiv 1 - I, \tag{13}$$

176

and

$$x_i(0) = y_i(0) = 0. \tag{14}$$

Parameter A in (9) scales the relative size of bottom-up inputs I_i to top-down feedback signals y_i. In (13), the notation "c" in I^c designates that the values of I^c are complementary to those of I; when I is on in (9), I^c is off in (10), and conversely. The input sequence I_i is given by:

$$I_i(t) = \begin{cases} 1 & \text{if } t_i - \alpha_i < t < t_i \\ 0 & \text{otherwise} \end{cases} \tag{15}$$

The input durations (α_i) and the interstimulus intervals ($\beta_i = t_i - \alpha_i - t_{i-1}$) are assumed to be large relative to the dimensionless relaxation times of x_i and y_i, set equal to 1 in equations (9) and (10). Thus each x_i reaches steady state when inputs are on ($I = 1$) and each y_i reaches x_i when inputs are off ($I^c = 1$). Otherwise, t_i and α_i can be arbitrary, and their values have no effect on patterns of memory storage.

The following properties of the STORE model are proved in Bradski, Carpenter and Grossberg (1992). The relative sizes of the activities in pattern $(x_1, \ldots, x_{i-1})$ are preserved when x_i becomes active. For large values of A, the total STM activity is approximately normalized at all times, whereas for small A, it grows rapidly as more inputs perturb the network. Since the size of parameter A in (9) reflects the degree to which the input I_i influences the STM pattern, recency for large A (present input dominates) and primacy for small A (past activities dominate) would be intuitively predicted. In fact, for large A, the pattern of STM activity $(x_1, \ldots, x_i)$ always shows a recency gradient. For small A, the STM patterns in response to short lists show a primacy gradient. In particular, if $x_1(t_1) > A$, $(x_1 \ldots x_i)$ shows a primacy gradient until $x_i(t_i) \leq A$. Presenting additional inputs $I_{i+1}, I_{i+2}, \ldots$ causes the STM pattern to bow. If $x_1(t_1) \leq A$, the STM pattern always exhibits recency. Recency occurs for all list lengths whenever $A \geq 1$, while small A values allow relatively long lists to be stored by primacy gradients. The position at which the STM pattern bows can be calculated iteratively. For example, the bow occurs at position $i = 2$ if $1 > A \geq .5(3 - \sqrt{5}) \cong 0.382$.

These properties of the STORE model are illustrated by the computer simulations summarized in Figure 11. Each row depicts STM storage of a list at a fixed value of A. In the left column, the STM vector $(x_1, x_2, \ldots, x_7)$ is depicted at times $t_1, t_2, \ldots, t_7$ when successive inputs $I_1, I_2, \ldots, I_7$ are stored. Each activity x_i is represented by the height of a vertical bar. The top row depicts a recency gradient, the seventh row a primacy gradient, and intermediate columns represent bows at each successive list position. The middle column graphs the ratios x_i / x_{i+1} through time. The horizontal graphs mean that the Invariance Principle is obeyed as soon as both items in each ratio are stored. The third column graphs the growth of total activity $x(t)$ to its capacity S. The input durations α_i in equation (15) varied randomly between 10 and 40. Such variations in input parameters had no discernible effect on the stored STM patterns.

CONCLUDING REMARKS

The VITEWRITE and STORE models suggest several general conclusions about the temporal organization of planned behaviors. First, the temporal properties of behavior may not be explicitly encoded in their controlling neural circuits. Rather, they may be implicitly coded in distributed properties that emerge as a result of interactions across multiple neurons and processing stages. For example, in the VITEWRITE model, feedback interactions between a Vector Plan working memory and a VITE trajectory generator determine the form and timing of handwriting production. These feedback interactions enable simple GO and GRO

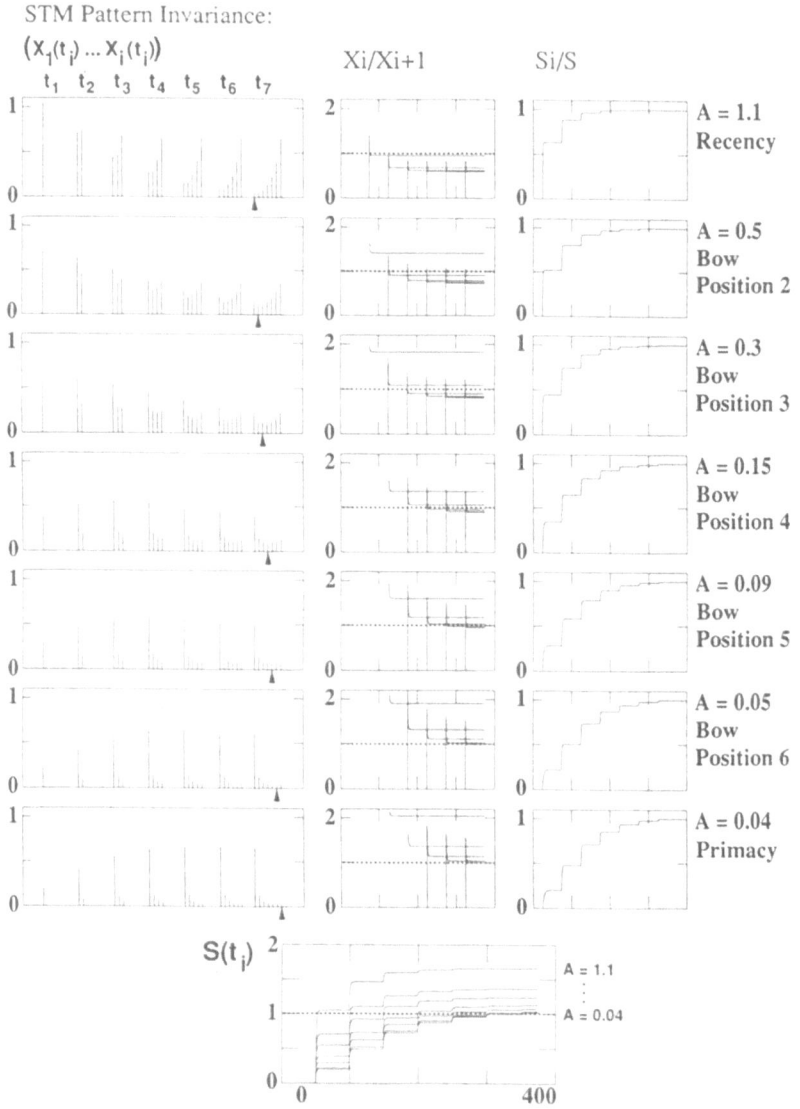

Figure 11. STORE model simulations for different values of the input parameter A. Left column: Each row codes the stored STM pattern after each of 7 items is presented; bar height codes stored STM activity of each item, with activities of later items to the right of earlier item activities. The pointer marks the list position of minimum activity in each of the 7 activity patterns. Middle column: Ratios of successive activities x_i / x_{i+1} through time, with horizontal graphs designating LTM invariance. Right column: Normalized total STM activity S_i/S through time. The STM patterns $(x_1(t_i) \ldots x_i(t_i))$ show recency for large A, bowing for intermediate values of A, and primacy for small values of A. Total activity $x(t_i) \equiv S_i$ grows toward the asymptote S as i increases for each value of A. When a new input I_i is stored, the previous pattern vector $(x_1 . \ldots x_{i-1})$ is amplified if $S_i \equiv x(t_i) < 1$; or depressed if $S_i \equiv x(t_i) > 1$; but the pattern of relative activities is preserved. For these simulations, input durations α_i were varied randomly between 10 and 40, with the intervals $(t_i - t_{i-1})$ set equal to 50. Reprinted with permission from Bradski, Carpenter, and Grossberg (1992).

volitional commands to flexibly alter the speed, size, and style of the handwriting representation that is invariantly coded in the Vector Plan. Likewise, the temporal order of the item representations that are remembered by a STORE working memory is implicitly coded by the relative sizes of the item representation activities. A nonspecific rehearsal wave, like the feedback signal from the $DV_m \cdot$ GO stage in Figure 1, translates these relative activities into the order of recall, with the largest activities being read-out earliest while self-inhibiting themselves in order to prevent perseverative read-out of the same items over and over again. The relative activities, in turn, are stored in a way that enables all possible groupings of stored items to be stably learned and recalled by unitized representations, or chunks, with which the working memory reciprocally interacts.

The second general conclusion is that different organizational principles may govern the design of neural circuits that control different aspects of timed behavior. In particular, the principles governing the VITEWRITE model concern how movement synergies can be controlled in such a way that individual synergies may be synchronously performed at variable speeds, yet different synergies may be flexibly reorganized by volitional commands that govern their individual times, speeds, and sizes of production. In contrast, principles governing the STORE model concern how sequences of events may be represented in STM in such a way that they may be stably unitized in and recalled from LTM. Different brain regions are, moreover, implicated in the control of these different neural designs. The VITE model has been used to explain data about parietal cortex, motor cortex and basal ganglia, whereas the STORE model may be used to interpret data about the frontal cortex and its interactions with other brain regions.

The principles that govern the VITEWRITE and STORE models do not exhaust the model neural designs that have been used to explain behavioral and neural data about temporally organized behaviors. Some examples of neural models constructed by our research group are listed below as well as references that discuss the work of many other authors. For example, conditioning of individual motor acts during reinforcement learning is adaptively timed. This process is modelled by a Spectral Timing model that is compared with data from hippocampus and cerebellum (Bullock, Fiala, and Grossberg, 1994; Grossberg and Merrill, 1992, 1995; Grossberg and Schmajuk, 1989). Circadian rhythms that control mammalian sleep and wake cycles are regulated on a much slower time scale. This oscillatory process is modelled by a Gated Pacemaker model that is compared with data from the suprachiasmatic nuclei of the hypothalamus (Carpenter and Grossberg, 1983, 1984, 1985). Oscillatory movement gaits and gait transitions, such as the cat gaits (walk, trot, pace, gallop), and the human gaits (walk, run) operate on a much faster time scale than circadian rhythms. These oscillatory movements are modelled by a GO Gait Generator model that is compared with data about spinal cord, basal ganglia, and motor cortex, among other neural structures (Cohen, Grossberg, and Pribe, 1992, 1993a, 1993b, 1993c; Grossberg, Pribe, and Cohen, 1994; Pribe, Grossberg, and Cohen, 1984). A central task of computational neuroscience is to further develop neural models of these various types of temporally organized behavior and to show how they may be integrated into a unified neural architecture for the real-time control of intelligent adaptive behavior.

ACKNOWLEDGEMENTS

The author wishes to thank Diana Meyers and Robin Locke for their valuable assistance in the preparation of the manuscript. The preparation of this paper, Technical Report CAS/CNS-TR-93-057, was supported in part by ARPA (ONR N00014-92-J-4015) and the Office of Naval Research (ONR N00014-91-J-4100 and ONR N00014-92-J-1309).

REFERENCES

Abend, W., Bizzi, E., and Morasso, P., 1982, Human arm trajectory formation, *Brain*, 105:331.

Atkinson, R.C., and Shiffrin, R.M., 1971, The control of short term memory, *Scientific American*, August:82.

Baddeley, A.D., 1976, "The Psychology of Memory", Basic Books, New York, NY.

Bradski, G., Carpenter, G.A., and Grossberg, S., 1992, Working memory networks for learning temporal order with application to 3-D visual object recognition, *Neural Computat.*, 4:270.

Bradski, G., Carpenter, G.A., and Grossberg, S., 1994, Store working memory networks for storage and recall of arbitrary temporal sequences, *Biol. Cybern.*, 71:469.

Bullock, D., Contreras-Vidal, J.L., and Grossberg, S., 1992, Equilibria and dynamics of a neural network model for opponent muscle control, *in:* "Neural Networks in Robotics", G. Bekey and K. Goldberg, eds., Kluwer Academic, Norwell, MA.

Bullock, D., Fiala, J.C., and Grossberg, S., 1994, A neural model of timed response learning in the cerebellum, *Neural Networks*, 7:1101.

Bullock, D. and Grossberg, S., 1988, Neural dynamics of planned arm movements: Emergent invariants and speed-accuracy properties during trajectory formation, *Psychol. Rev.*, 95:49.

Bullock, D. and Grossberg, S., 1991, Adaptive neural networks for control of movement trajectories invariant under speed and force rescaling, *Human Movement Sci.*, 10:3.

Bullock, D., Grossberg, S., and Guenther, F.H., 1993, A self-organizing neural model of motor equivalent reaching and tool use by a multijoint arm, *J. Cognit. Neurosci.*, 5:408.

Bullock, D., Grossberg, S., and Mannes, C., 1993, A neural network model for cursive script production, *Biol. Cybern.*, 70:15.

Carpenter, G.A. and Grossberg, S., 1983, A neural theory of circadian rhythms: The gated pacemaker, *Biol. Cybern.*, 48:35.

Carpenter, G.A. and Grossberg, S., 1984, A neural theory of circadian rhythms: Aschoff's rule in diurnal and nocturnal mammals, *Am. J. Physiol. (Regulatory, Integrative and Comp. Physiol.)*, 247:R1067.

Carpenter, G.A. and Grossberg, S., 1985, A neural theory of circadian rhythms: Split rhythms, after-effects, and motivational interactions, *J. Theor. Biol.*, 113:163.

Cohen, M. and Grossberg, S., 1986, Neural dynamics of speech and language coding: Developmental programs, perceptual grouping, and competition for short term memory, *Human Neurobiol.*, 5:1.

Cohen, M. and Grossberg, S., 1987, Masking fields: A massively parallel neural architecture for learning, recognizing, and predicting multiple groupings of patterned data, *App. Optics*, 26:1866.

Cohen, M.A., Grossberg, S., and Pribe, C., 1992, A neural pattern generator that exhibits frequency-dependent in-phase and anti-phase oscillations, *in:* "Proceedings of the International Joint Conference on Neural Networks, IV", IEEE Service Center, Piscataway, NJ.

Cohen, M.A., Grossberg, S., and Pribe, C., 1993a, A neural pattern generator that exhibits arousal-dependent human gait transitions, *in:* "Proceedings of the World Congress on Neural Networks, IV", Erlbaum Associates, Hillsdale NJ.

Cohen, M.A., Grossberg, S., and Pribe, C., 1993b, Frequency-dependent phase transitions in the coordination of human bimanual tasks, *in:* "Proceedings of the World Congress on Neural Networks, IV", IEEE Service Center, Piscataway, NJ. Technical Report CAS/CNS-T-93-018, Boston University, Boston, MA.

Cohen, M.A., Grossberg, S., and Stork, D., 1987, Recent developments in a neural model of real-time speech analysis and synthesis, *in:* "Proceedings of the IEEE International Conference on Neural Networks, IV", M. Caudill and C. Butler, eds., San Diego, CA.

Elman, J.L., 1990, Finding structure in time, *Cognit. Sci.*, 14:179.

Fetters, L. and Todd, J., 1987, Quantitative assessment of infant reaching movements, *J. Motor Behav.*, 19:147.

Gaudiano, P. and Grossberg, S., 1991, Vector associative maps: Unsupervised real-time error-based learning and control of movement trajectories, *Neural Networks*, 4:147.

Georgopoulos, A.P., Kalaska, J.F., Caminiti, R., and Massey, J.T., 1984, The representation of movement direction in the motor cortex: Single cell and population studies, *in:* "Dynamic Aspects of Neocortical Function", G.M. Edelman, W.E. Gall, and W.M. Cowan, eds., Wiley, New York, NY.

Golani, I., 1992, A mobility gradient in the organization of vertebrate movement: The perception of movement through symbolic language, *Behav. Brain Sci.*, 15:249.

Grossberg, S., 1970, Some networks that can learn, remember, and reproduce any number of complicated space-time patterns, II, *Stud. App. Math.*, 49:135.

Grossberg, S., 1973, Contour enhancement, short-term memory and constancies in reverberating neural networks, *Stud. App. Math.*, 52:217.

Grossberg, S., 1976, Adaptive pattern classification and universal recoding, I: Parallel development and coding of neural feature detectors, *Biol. Cybern.*, 23:121.

Grossberg, S., 1978a, Behavioral contrast in short-term memory: Serial binary memory models or parallel continuous memory models? *J. Math. Psychol.*, 17:199.

Grossberg, S., 1978b, A theory of human memory: Self-organization and performance of sensory-motor codes, maps, and plans, *in*: "Progress in Theoretical Biology", 5, R. Rosen and F. Snell, eds., Academic Press, Academic Press, New York. Reprinted *in*: "Studies of Mind and Brain", Grossberg, S. ed., Reidel Press, Boston, MA.

Grossberg, S., Guenther, F.H., Bullock, D., and Greve, D., 1993, Neural representations for sensory-motor control, II: Learning a head-centered visuomotor representation of 3-D target position, *Neural Networks*, 6:43.

Grossberg, S. and Kuperstein, M., 1989, "Neural Dynamics of Adaptive Sensory-Motor Control: Expanded edition", Pergamon Press, Elmsford, NY.

Grossberg, S. and Merrill, J.W.L., 1992, A neural network model of adaptively timed reinforcement learning and hippocampal dynamics, *Cognitive Brain Res.*, 1:3.

Grossberg, S. and Pepe, J., 1971, Spiking threshold and overarousal effects in serial learning, *J. Statistical Physics*, 3:95.

Grossberg, S. and Schmajuk, N.A., 1989, Neural dynamics of adaptive timing and temporal discrimination during associative learning, *Neural Networks*, 2:79.

Grossberg, S. and Stone, G.O., 1986, Neural dynamics of attention switching and temporal order information in short term memory, *Memory and Cognition*, 14:451.

Guenther, F.H., Bullock, D., Greve, D., and Grossberg, S., 1994, Neural representations for sensory-motor control, III: Learning a body-centered representation of 3-D target position, *J. Cognit. Neurosci.*, in press.

Gutfreund, H. and Mezard, M., 1988, Processing of temporal sequences in neural networks. *Physiol. Rev. Lett.*, 61:235.

Guyon, I., Personnaz, L., Nadal, J.P., and Dreyfus, G., 1988, Storage retrieval of complex sequences in neural networks, *Physiol. Rev. A*, 38:6365.

Healy, A.F., 1975, Separating item from order information in short-term memory, *J. Verbal Learning Verbal Behav.*, 13:644.

Horak, F.B. and Anderson, M.E., 1984a, Influence of globus pallidus on arm movements in monkeys. I. Effects of kainic acid-induced lesions, *J. Neurophysiol.*, 52:290.

Horak, F.B. and Anderson, M.E., 1984b, Influence of globus pallidus on arm movements in monkeys. II. Effects of stimulation, *J. Neurophysiol.*, 52:305.

Jordan, M.I., 1986, "Serial Order: A Parallel Distributed Processing Approach", Institute for Cognitive Science, Report 8604, University of California, San Diego, CA.

Lacquaniti, F., Ferrigno, G., Pedotti, A., Soechting, J.F., and Terzuolo, C., 1987, Changes in spatial scale in drawing and handwriting: Kinematic contributions by proximal and distal joints, *J. Neurosci.*, 7:819.

Lacquaniti, F., Terzuolo, C., and Viviani, P., 1983, The law relating kinematic and figural aspects of drawing movements, *Acta Psychol.*, 54:115.

Lee, C. and Estes, W.K., 1981, Item and order information in short-term memory: Evidence for multilevel perturbation processes, *J. Exp. Psychol.: Human Learning and Memory*, 1:149.

Morasso, P., 1981, Spatial control of arm movements, *Exp. Brain Res.*, 42:223.

Nagasaki, H., 1989, Asymmetric velocity profiles and acceleration profiles of human arm movements, *Exp. Brain Res.*, 74:319.

Parent, A., 1990, Extrinsic connections of the basal ganglia, *Trends Neurosci.*, 13:254.

Plamondon, R., 1989, Handwriting control: A functional model, *in*: "Models of Brain Function", R.M.J. Cotterill, ed., Cambridge University Press, Cambridge, UK.

Pribe, C., Grossberg, S., and Cohen, M. A., 1994, Neural control of interlimb oscillations, II: Biped and quadruped gaits and bifurcations, *Biol. Cybern.*, in press.

Ratcliff, R., 1978, A theory of memory retrieval, *Psychol. Rev.*, 85:59.

Reeves, A. and Sperling, G., 1986, Attention gating in short-term visual memory, *Psychol. Rev.*, 93:180.

Schomaker, L., Thomassen, A., and Teulings, H.L., 1989, A computational model of cursive handwriting, *in*: "Computer Recognition and Human Production of Handwriting", R. Plamondon, C.Y. Suen, and M.L. Simner, eds., World Scientific, Singapore.

Schreter, Z. and Pfeifer, R., 1989, Short-term memory/long-term memory interactions in connectionist simulations of psychological experiments on list learning, *in*: "Neural Networks from Models to Applications", L. Personnaz and G. Dreyfus, eds., I.D.S.E.T., Paris.

Seibert, M.C., 1991, "Neural Networks for Machine Vision: Learning Three-Dimensional Object Recognition", Ph.D. Thesis, Boston University, Boston, MA.

Seibert, M.C. and Waxman, A.M., 1990a, Learning aspect graph representations from view sequences, *in*: "Advances in Neural Information Processing Systems", Vol. 2, D.S. Touretzky, ed., Morgan Kaufmann Publishing, San Mateo, CA.

Seibert, M.C. and Waxman, A.M. , 1990b, Learning aspect graph representations of 3D objects in a neural network, *in*: "Proceedings of the International Joint Conference on Neural Networks (IJCNN-90) Washington, D.C.", M. Caudill, ed., Erlbaum, Hillsdale, NJ.

Soechting, J.F. and Terzuolo, C.A., 1987, Organization of arm movements is segmented, *Neuroscience*, 23:39.

Viviani, P. and Terzuolo, C.A., 1980, Space-time invariance in learned motor skills, *in*: "Tutorials in Motor Behaviour", G.E. Stelmach and J. Requin, eds., North-Holland, Amsterdam.

Viviani, P. and Terzuolo, C.A., 1983, The organization of movement in handwriting and typing, *in*: "Language Production", 2, B. Butterworth, ed., Academic Press, New York, NY.

Wang, D. and Arbib, M.A., 1990, Complex temporal sequence learning based on short-term memory, *Proc. IEEE.*, 78:1536.

Wann, J., Nimmo-Smith, I., and Wing, A.M., 1988, Relation between velocity and curvature in movement: Equivalence and divergence between a power law and a minimum-jerk model, *J. Exp. Psychol.: Human Percept. Perform.*, 14:622.

SPECIAL PURPOSE TEMPORAL PROCESSING IN HIPPOCAMPAL FIELDS CA1 AND CA3

Richard Granger, Makoto Taketani and Gary Lynch

ICS Department
Center for the Neurobiology of Learning and Memory
University of California
Irvine, CA 92717-3425
USA

INTRODUCTION: TIME AND LEARNING

The kind of learning you are doing now is characterized by rapid acquisition (the speed of reading), enormous capacity (further learning does not force out or interfere with prior learning) and persistence: what you learn from a page can potentially last decades. These features are extremely unusual for any biological mechanism, for which homeostasis is the norm. The only known mechanism having these properties is synaptic long-term potentiation, or LTP, and it is therefore the leading, and currently the only, candidate mechanism for rapid, persistent, high-capacity learning. In contrast, slowly accreted learning of the kind that occurs when one learns motor skills such as riding a bicycle or playing racquetball, may be subserved by a different set of mechanisms.

Different time courses of potentiation in different locales within the brain suggest memories of different durations, and possibly provide different contributions to the creation of memories. The embedding of these differently-enduring memories in different anatomical architectures gives rise to differential emergent function. Analysis of neural networks that might underlie memory formation has led to hypotheses about the separable functional roles of these networks in the formation of memories of different types and durations.

Many cortical processing pathways funnel information into the hippocampus and back out to cortex again. For this reason, among others, the hippocampus has been a focus of investigation in memory research for some time. It is known, for instance, that animals with hippocampal lesions have difficulty in rapid learning tasks, and humans with damage to the cortico-hippocampal system have similarly been found to have problems in encoding and retrieving memories. Each network in the cortico-hippocampal pathway has a drastically different anatomical architecture made up of very different cell types. These networks operate in large measure serially, as an "assembly line" of specialized functions. Each network adds its own unique contribution to the processing of memories. Mathematical analysis of individual component circuits and the interactions among the circuits has resulted in the formulation of novel algorithms for temporal processing. Particular functions that emerge from the operation of specific circuits provide clues about their potential role in the processing of memories. The somewhat nonintuitive division of labor represented by this breakdown of processing into different components results in functions with considerable computational and psychological utility.

SYNAPTIC LEARNING IN THE BRAIN

Natural Learning Rules

The conjecture that specific changes in the connections between brain cells would suffice for the encoding of long-term memory dates back at least to the turn of the century (Tanzi, 1893; Cajal, 1909). Hebb's refinement of this idea, in the form of a brief "postulate" in his 1949 monograph "The Organization of Behavior" (Hebb, 1949), introduced the requirement of contingent pre- and post-synaptic activity for synapse-specific increases in excitability or synaptic "weight".

Not until the 1970s did scientists first identify a candidate biological phenomenon and show that it conformed to the demanding constraints of specific permanent change (Bliss and Lømo, 1973; Bliss and Gardner-Medwin, 1973). It is only within the past decade that the biochemical characteristics of this phenomenon have begun to be well understood (Lynch and Baudry, 1984). The phenomenon, synaptic long-term potentiation or LTP, is now widely believed to be the substrate of much of human memory. LTP is a unique biological mechanism, and poses a difficult challenge in view of the ongoing homeostatic breakdown and replacement processes that occur in the brain. Thus, the mechanism for LTP might be expected to exhibit specialized properties solely to address these biological constraints.

By definition, LTP provides only an increase in synaptic weights. However, other mechanisms such as LTP reversal and long-term depression (LTD) cause decreases in weights via other mechanisms, of as-yet uncertain duration. LTP increments proceed by relatively fixed-size steps, and only up to a fixed maximum strength or ceiling. LTP has complex temporal requirements for its induction. It is possible that this extra complexity is an epiphenomenon which occurs for reasons pertaining only to biological needs of the system. These complex properties might even be an impediment to the computations performed in neural circuitry, akin to the heat emitted by a light bulb. Alternatively, these specialized physiological properties may yield novel learning rules that confer useful computational abilities to the networks that use them, and indeed this is what has been found in many reported instances, e.g., (Granger, Ambros-Ingerson, and Lynch, 1989; Ambros-Ingerson, Granger and Lynch, 1990; Granger, et al., 1994).

We here describe a series of examples of networks with potential computational utility arising directly from novel learning rules based on the properties of LTP.

THE OLFACTORY-HIPPOCAMPAL PATHWAY

Figure 1 schematically illustrates the components of the circuitry formed by the hippocampus and its interfaces with the cortex. Various types of synaptic change have been demonstrated at every link in this system, and we take it as axiomatic that learning and memory in these structures is mediated by such synaptic change. The specific characteristics of LTP vary from circuit to circuit, and different functional properties arise from these different learning rules expressed in the different network architectures.

The superficial cell layers of entorhinal cortex are innervated by neocortical associational regions and by olfactory paleocortex. The superficial cells of entorhinal cortex in turn monosynaptically contact all three hippocampal divisions. These include the dentate gyrus, field CA3, and field CA1. In the dentate gyrus, granule cells contact local polymorph cells which in turn project back to the granule cells, forming a recurrent local circuit. The granule cells of the dentate gyrus also project directly to field CA3 via the mossy fibers, contacting the primary excitatory cells of CA3, the pyramidal cells. CA3 pyramidal cells densely contact neighboring CA3 cells, thereby making another, quite different, recurrent local circuit. The CA3 axons also have collaterals that project to field CA1. Field CA1 projects to the subiculum, which in turn projects to the deep layers of entorhinal cortex, thus completing a large loop spanning five distinct networks.

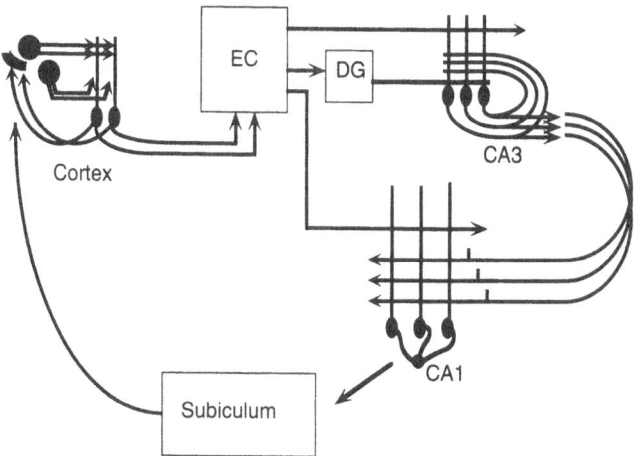

Figure 1. Circuitry of the hippocampus and interfaces with cortex. Cortical regions, including olfactory paleocortex and polysensory association cortices, project to entorhinal cortex (EC), which in turn innervates all three divisions of hippocampus: dentate gyrus (DG), field CA3, and field CA1. DG in turn generates the mossy fiber projection to CA3, and CA3 provides the Shaffer collateral projection to CA1. CA1 projects out, to deep layers of entorhinal cortex, to subiculum, and to pericallosal cortex, all of which in turn project back to association cortex.

      Potentiation of some form has been reported in each of these pathways (Jung, Larson, and Lynch, 1990; Bliss and Lømo, 1973; Green, McNaughton, and Barnes, 1990; Berger, 1984; Larson and Lynch, 1986). All but one of these instances of potentiation show evidence of being in the category usually designated LTP: (1) the changes are dependent on the presence of postsynaptic calcium via NMDA receptor channels, (2) they require postsynaptic depolarization coincident with presynaptic activation, (3) they do not interact with known presynaptic characteristics such as frequency facilitation, and, most importantly, (4) the changes are long-lasting, i.e., they persist over many days without significant decrement. The single exception is the central link in the cortico-hippocampal loop, the mossy fiber pathway from the dentate gyrus to field CA3. The potentiation that occurs in this pathway is quite distinct from LTP in that it is not NMDA-dependent, it does not require postsynaptic depolarization, it does interact with presynaptic frequency facilitation, and it lasts only hours.

      The following sections describe in detail the relevant anatomical and physiological characteristics of the piriform/entorhinal cortex and of hippocampal field CA1, and describe some hypotheses about their emergent functionality.

Piriform Cortex

Anatomy. Figure 2 schematically illustrates the major anatomical features of the primary olfactory (piriform) cortex and its primary subcortical input structure, the olfactory bulb. A key architectural feature is the presence of a loop consisting of (1) excitatory feedforward fibers from mitral cells in the olfactory bulb to cortical layer II/III cells, and (2) an inhibitory feedback projection from the cortex to the olfactory bulb. The primary excitatory cells of the olfactory bulb, the mitral/tufted cells, project monosynaptically to layer I of piriform cortex, where they synapse sparsely and nontopographically onto apical dendrites of cortical neurons whose cell bodies are located in layers II and III (Price, 1973). Layer II cells emit axon collaterals that project mainly in the caudal direction, contacting other cells in layers II and III, and forming the primary input to lateral entorhinal cortex (Zimmer, 1971; Wyss, 1981). The axons of layer III cells in caudal piriform cortex project mainly in the opposite direction, passing via layer III back to the olfactory bulb where they synapse not on the excitatory mitral/tufted cells, but rather on the granule cells, which in turn inhibit the mitral/tufted cells (Mori, 1987). The other major architectural detail that is important for modeling is the existence of excitatory-inhibitory interactions in local cortical circuits. Excitatory cortical cells outnumber inhibitory cells by roughly two orders of mangnitude. Inhibitory axons contact excitatory cortical cells relatively densely but within a small radius. Unlike excitatory neurons, inhibitory cells have no long axons. Thus the inhibitory cells roughly "tile" the cortical surface, each inhibitory cell having a region of influence that covers dozens to hundreds of excitatory cells. The resulting cortical "patches" can act as relatively coherent local circuits. Physiologically, inhibition is long, lasting hundreds of milliseconds, whereas excitation is brief (~ 10 ms). In simulations of such local-circuit patches, the excitatory neurons jointly innervate and receive feedback from a common inhibitory interneuron. The excitatory cell receiving the most input activation during a given activity cycle is the first to reach its spike threshold. Spiking excites the inhibitory cell which in turn prevents other excitatory cells from responding. The result is the natural generation of a simple competitive or "winner-take-all" mechanism, allowing only the most strongly activated cell in a patch to respond with spiking activity. Analysis has shown that this simple mechanism closely approximates an optimal winner-take-all mechanism (Coultrip, Granger, and Lynch 1992).

Synchronous Cyclic Activity. Normal unrestrained animals actively exploring an environment exhibit synchronous EEG activity at roughly 4-8 Hz, which is the "theta rhythm" of cyclic neural activity recorded in hippocampus (Hill, 1978). During an olfactory learning task, the entire olfactory-hippocampal pathway becomes synchronized to the theta rhythm (Komisaruk, 1970; Macrides, 1975; Eichenbaum, et al., 1987), and the latency of the peaks of the theta wave in each of the constituent circuits is roughly what would be expected due to synaptic transmission delays between the circuits (Macrides, 1975).

It has been demonstrated that this rhythm of 4-8 Hz is the optimal stimulation pattern for the induction of LTP. When afferents to a target are activated in brief (4-pulse) high-frequency (100 Hz) bursts separated by 200 ms (i.e., the 5 Hz theta rhythm), LTP is robustly induced, *in vitro* or *in vivo*, without the need for GABA blockers, voltage clamps, or physiologically implausible, seconds-long high-frequency stimulation (Larson and Lynch, 1986; Larson, Wong, and Lynch, 1986). The biophysical mechanism underlying the optimality of the 200 ms interval is known. The initial afferent burst causes both excitatory and inhibitory postsynaptic potentials in the target cell. These excitatory and inhibitory postsynaptic potentials (EPSPs and IPSPs) have different characteristic time courses, EPSPs lasting 10-20 ms and IPSPs lasting 100-300 ms. At 200 ms after the first burst, the initial IPSP has roughly returned to baseline, but the inhibitory synapse has become refractory (Mott and Lewis, 1991). Thus the second burst activates predominantly excitatory currents. These excitatory currents summate during the brief burst, enabling them to surpass the threshold of the voltage-sensitive NMDA receptor channel (Collingridge, Kehl, and McLennan, 1983). This allows the NMDA channel to open and enables Ca^{++} to enter the cell. Ca^{++} is the chemical messenger that triggers the cascade of events that induce LTP (Lynch, et al., 1983).

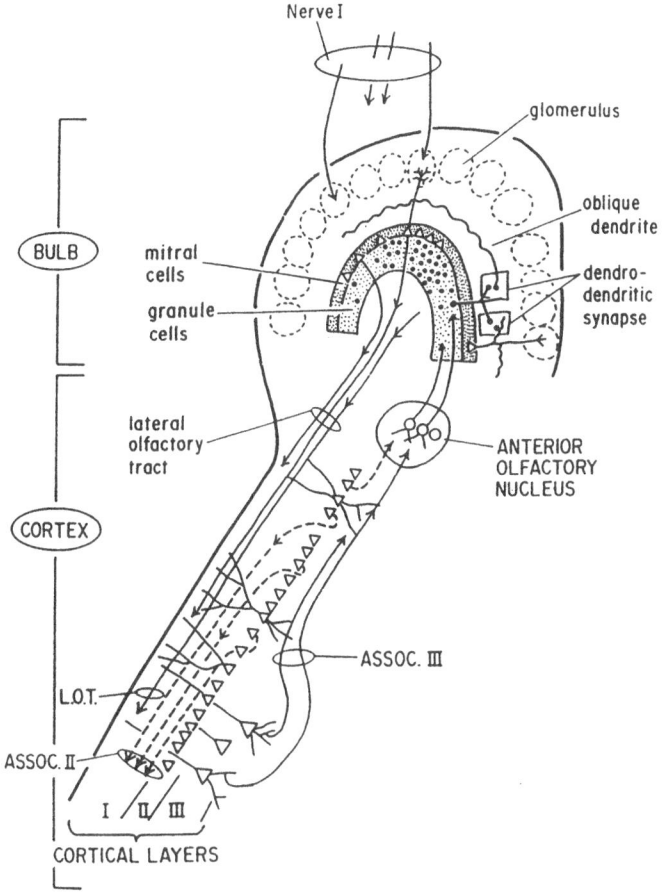

Figure 2. In the olfactory system, receptor cells in the nasal epithelium project topographically via the first cranial nerve to the excitatory mitral cells in the olfactory bulb. These cells in turn project sparsely and nontopographically to layer I of olfactory cortex, contacting apical dendrites of cells in layers II and III. Cortical cells project back, both directly and via the anterior olfactory nucleus, to the inhibitory granule cells of the olfactory bulb, which inhibit mitral cells via dendrodendritic synapses.

Emergent Computation. LTP in olfactory cortex (Roman, Staubli, and Lynch, 1987; Kanter and Haberly, 1990; Jung, Larson and Lynch, 1990) is induced via the same stimulation patterns as those found during learning in behaving animals, i.e., brief bursts at the theta rhythm (Komisaruk, 1970; Otto, et al., 1991). This suggests that a model of cortical memory function should incorporate this synchronized operating mode as well as the parameters for induction and expression of LTP. The formulation of hypotheses and the computational analysis of the olfactory system incorporating these features of LTP induction and expression have led to findings that suggest that both the encoding of sensory cues and the organization of the resulting memories occur via mechanisms not typically seen in neural network models. Implementation of a repetitive sampling feature meant to represent the cyclic sniffing behavior of mammals (Komisaruk, 1970) produced a system that exhibited successively finer-grained encodings of learned cues over sampling cycles. Each sampling cycle includes feedforward activity from the olfactory bulb to the cortex followed by feedback from the cortex to the olfactory bulb. Because this feedback activates long-lasting inhibition in the olfactory bulb, the next sample of the input arrives against an inhibitory background in the olfactory bulb, effectively masking part of the input, and thereby resulting in different activity patterns in the olfactory bulb and the cortex. Thus, re-sampling a fixed cue generates different responses in the olfactory bulb and cortex with each new sampling cycle.

In the model, cortical cells whose inputs become potentiated respond selectively to those classes of inputs on which they have been previously exposed, or "trained". Because of synaptic strengthening, variants of a trained input all come to elicit the same pattern of cortical cell firing. The result of learning in the model, then, is that the cortical response only coarsely reflects its inputs, not distinguishing among slightly different inputs. This corresponds to the common statistical operation of clustering, which arises readily from correlational rules in networks. This type of operation has been described by many researchers (Malsburg, 1973; Grossberg, 1976; Rumelhart and Zipser, 1985). However, it raises the question of how fine distinctions can be made given the lost acuity.

The feedback from cortex to the olfactory bulb granule cell layer selectively inhibits those portions of the olfactory bulb that give rise to the cortical firing pattern. Inhibition in the olfactory bulb lasts for hundreds of milliseconds (Nicoll, 1969). Re-sampling then causes new activity in the olfactory bulb and cortex, against the background of this long-lasting inhibition. The resulting cortical response corresponds to odor components not shared across members of the broad category. For instance, the first sample response to any flower will all be identical, signifying that floral odors are all members of a single category. Subsequent samples of the odor correspond to differences among different flowers, thereby effectively distinguishing among subcategories of floral odors. In this model, learning via LTP in the model generates a multilevel hierarchical memory that uncovers statistical relationships inherent in collections of learned cues, and during retrieval sequentially traverses this hierarchical recognition memory.

Having derived an algorithm from the operation of these networks, that algorithm can be analyzed for its computational properties. The standard analysis computes the number of sites required to store a given memory, i.e., the space cost of the algorithm. It also computes the number of processing steps required for storing each memory, i.e., the time cost. The algorithm described above turns out to exhibit time and space costs rivaling those of the plethora of standard hierarchical clustering algorithms in the literature. The space cost for storage is linearly proportional to both the number of items to be stored and to their dimensionality. It corresponds to the number of fibers in the input tract. The three time costs of the algorithm for each input are (1) the time required for summation of inputs to the target cell, (2) the time required to compute "winners" of the lateral inhibitory competition in local-circuit patches, and (3) the time required for weight modifications or learning to occur. Because of the inherent parallelism of the algorithm, (1) is proportional to the log of the dimensionality of the input, (2) is proportional to the log of the number of inputs, and (3) is constant. The network rapidly converges during training, and to process a collection of n inputs of dimensionality N takes (n (log n) (log nN)) steps in parallel (Ambros-Ingerson, Granger, and Lynch, 1990).

Temporal Processing in Field CA3

Hippocampal field CA3 is a unique exception to the extremely sparse connectivity characterizing most telencephalic networks. CA3 exhibits relatively dense recurrent associational fibers, which together produce a rich pattern of dynamical activity over time, suggesting that CA3 may be part of a reverberating short-term memory system.

Simulations of field CA3 based on anatomical and physiological data have shown that brief afferent activation can generate recurrent activity lasting for hundreds of milliseconds, and that the pattern of activity is determined to some extent by the initial input. These findings led to the suggestion that field CA3 might function as a kind of "holding memory," that bridges the temporal gap between transient neural signals and behavioral time (Taketani, et al., 1992). To serve as a holding memory, however, a given activity patterns would have to be repeated regularly over time, so that two cues separated by time could be integrated.

Physiological rules for the induction and expression of LTP in response to time-varying afferent signalling (Larson and Lynch, 1989) were incorporated into simulations of CA3. Repeated afferent stimulation at the theta rhythm leads to the potentiation of a small number of synapses in the model, strengthening a sequential chain of target cells. After such repeated stimulation or "training" with a pattern, subsequent brief (10 ms) stimulation with the input pattern sets up a sequence of recurring responses in the model, specific to the input, and repeating regularly with every repetition of the "carrier" theta wave.

Learning of recurring patterns in hippocampal field CA1, the primary target of projections from CA3, enables CA1 to repeat a pattern so as to bridge the time between the occurrence of two cortical (perforant path) afferents A and B, even when they are separated by many seconds. Thus, the coordinated activity of fields CA3 and CA1 could theoretically recognize the constituents of a coherent scene even when viewed from different vantage points, and at different rates of scanning. It has been suggested (Taketani, et al., 1992) that the coordinated functional capability of fields CA3 and CA1 which arises from their unique anatomical designs and their different requirements for induction and expression of LTP, constitutes the unique contribution of the hippocampus to the processing of spatial information, as well as to non-spatial tasks requiring the integration of disparate sensory cues over time.

Hippocampal Field CA1

Activity Over Time. Field CA1 of the hippocampus receives afferent inputs from entorhinal cortex and from the adjacent hippocampal field CA3. Both input systems make sparse, nontopographic contact with CA1 cells. Excitatory-inhibitory local circuit interaction in field CA1 resembles that in cortex, such that lateral inhibition effectively causes excitatory cells to compete for the opportunity to respond to inputs. The temporal characteristics of inputs to CA1 from cortex and CA3 raise the question of how the behavior of the local circuits in CA1 is distributed over time. Thus, in the model, the instantaneous competitive system has been extended to incorporate lateral inhibitory activity distributed over time. The earliest inputs produce a subset of facilitated cells, subsequent inputs produce a subset consisting of only the best responders from the first set, and so on, thus "honing" the subsets of responding target cells down to the few units that respond best to the particular sequence of input (Granger et al., 1994).

Non-Hebbian LTP. Synchronous bursts of activity that arrive during sequentially adjacent peaks of the theta-rhythm carrier envelope are optimal for inducing LTP in synapses on pyramidal cells in field CA1 (Larson, Wong, and Lynch, 1986). The mechanisms through which such sequences of inputs induce LTP have become increasingly well-understood (Mott and Lewis, 1991). The low probability that there will be complete synchrony of afferents during behaviorally relevant physiological activity raises the issue of how to characterize LTP induced by asynchronously arriving stimulus-evoked afferent activity within the envelope of single peaks of the theta rhythm. A "Hebbian" co-activity rule would predict that as asynchronous afferents arrive, increased depolarization of the target neuron over the staggered arrival times will cause later inputs to be strengthened more than earlier inputs. However, *in vitro* experiments in hippocampal slices using three small electrical stimuli applied in a staggered sequence over 70 ms showed that the earliest inputs potentiated their synapses the most, while subsequently arriving inputs caused successively less potentiation (Larson and Lynch, 1989; see Figure 3a).

Computational evaluation of this non-Hebbian order-dependent LTP induction rule addresses the question of how the resultant learning might be expressed during subsequent performance. Biophysical simulations (Figure 3b) predict that sequentially arriving inputs with different synaptic potencies will evoke the greatest depolarization in a target cell when they are activated in the order of their strengths, with the most-potentiated input activated first. Given the order-dependence of the LTP rule, this means that a "trained" cell reacts most strongly to the same temporal sequence that was used to induce LTP in it. This selective response constitutes a form of recognition; cells responding to a previously-presented sequence "recognize" that sequence by virtue of their differential response to it. Thus potentiation of an input sequence "codes" a cell to recognize that sequence subsequently, causing the cell to act as a form of "sequence detector" or sequence recognizer. The predicted effect occurs robustly across a wide range of simulated conditions, one of which is illustrated in Figure 3c.

Emergent Computational Model of Field CA1. The physiological characteristics of LTP in field CA1, incorporated into a network of simplified cells derived from biophysical and local-circuit models described above, provide a means to investigate the computational properties of a specific set of non-Hebbian LTP-mediated mechanisms for learning and performance. The resulting learning algorithm processes an input sequence by performing order-dependent potentiation of the synapses of those cells whose responses survive lateral inhibition in the competitive patches. The synapses of responsive cells are potentiated to a level that depends on the order in which their inputs were activated. The contact with the first input is most strongly potentiated and the contact with the last input is least strongly potentiated.

Based on this LTP learning rule, a model performance algorithm has been developed to exploit the properties of potentiated synapses. "Recognition" of a previously-presented sequence is defined in terms of the selective response of target cells to those sequences on which they have been "trained", i.e., the sequences which were used to induce LTP. The performance algorithm preferentially activates only those cells whose synapses have been appropriately potentiated at each input step. A network consists of groups or patches of cells each of which responds simultaneously and in parallel. The set of responding cells in each group is successively "honed" over the duration of an incoming temporal sequence. "Recognition" is said to occur only if all groups or patches of cells in the model produce at least one active cell in response to an input sequence. This responsiveness indicates that the network has previously been "trained" on the input sequence. If each patch in the network contains at least one responsive cell, the input is said to be accepted or recognized; but if, at the end of an input sequence, any patch in the network contains no cells that survive the "honing" process, then the input is said to be rejected, i.e., not recognized.

During performance, summation within a neuron is based on potentiated synaptic weights. However, during the learning process, summation is performed using only naive synaptic strengths, regardless of intervening changes to these synaptic strengths. This is intended to model the distinction between changes to AMPA versus NMDA receptor conductance with LTP induction (Muller, Joly, and Lynch, 1988; Kauer, Malenka, and Nicoll, 1988). NMDA receptors themselves remain relatively unpotentiated, whereas AMPA receptors express significant potentiation (Muller and Lynch, 1988; Muller, Joly, and Lynch, 1988; Kauer, Malenka, and Nicoll, 1988). During episodes of learning or LTP induction, the NMDA receptor mediated conductance dominates cell responses, and therefore is the main factor that

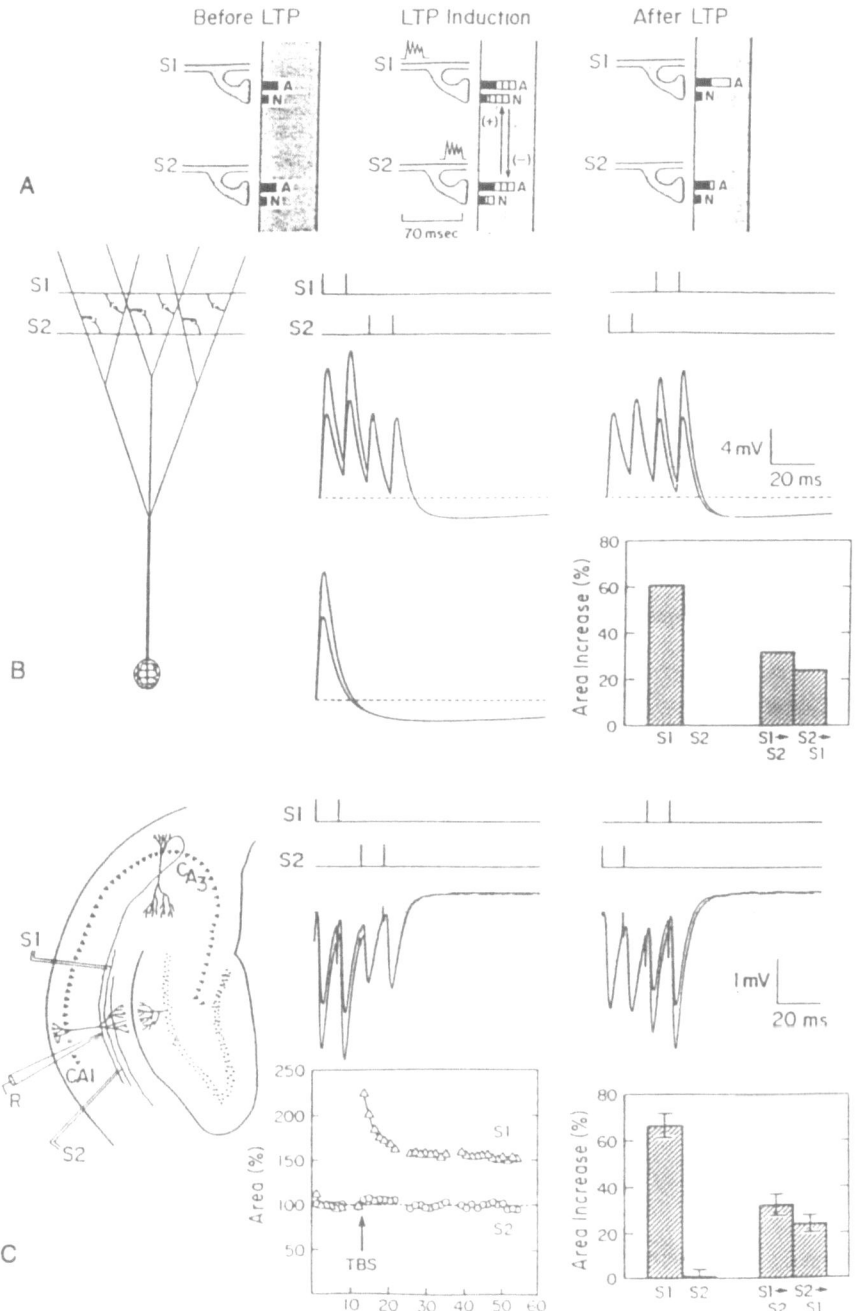

Figure 3. Order-dependent potentiation: **A.** excitatory synaptic afferents to a single target neuron activate the two primary types of excitatory glutamate receptors AMPA (A) and NMDA (N). Synapse 1 (S1) and synapse 2 (S2) carry brief bursts of high-frequency activity. Afferent S1 is activated 70 ms before S2. Initially, S1 activity opens only AMPA receptor channels; as this activity depolarizes the target, the

(Continued)

determines which target cells will "win" the competition and be potentiated. One computational consequence of this AMPA-NMDA distinction is notable. In contrast to the great majority of learning algorithms used in artificial neural networks, prior learning via LTP has little or no effect on subsequent LTP, preventing the formation of "attractor" cells, i.e., cells that tend to respond to any of a number of inputs similar to those previously trained.

The resulting extreme selectivity of cells, in combination with the order-dependency of the LTP-based learning and performance rules, confers unusually large capacity to the network. The capacity of a given network can be measured in terms of its errors as a function of the number of sequences stored. Two types of errors can be distinguished: errors of "collision", in which a particular set of cells responds to more than one temporal sequence during training, and errors of "commission", in which a target cell responds during performance to a string on which it has not been trained. Theoretical analysis of error rates for the LTP-based network indicates that these rates remain extremely low even with heavy "loading" of the network, i.e., training of a relatively small network with large numbers of sequences. It was found, for example, that a network of 1,000 cells could be trained to recognize 10,000 sequences of length ten, with an error rate of 0.0001. It has also been shown that the capacity of the network scales extremely well, i.e., larger networks learn comparably larger numbers of sequences, in contrast with many neural network approaches in which the network must be exponentially larger than the number of items it is to learn. In particular, given a network consisting of M competitive local circuits, each consisting of C cells, and each cell supplied with A synapses, the number (n) of sequences of length S that can be learned without exceeding an error rate E is:

$$n = \frac{C \, \log(1 - E^{1/SM})}{\log(1 - (1/A))} \tag{1}$$

Figure 4 is a contour graph of the relationship among the number of patches in the network, the number of cells in each patch, and the number of sequences that can be stored without exceeding a fixed recognition error of $p(\text{commission}) = 0.001$.

If cells are assumed to receive 10,000 synaptic contacts on their dendrites, then the point at the upper right corner of the contour graph corresponds to a 100,000-cell network (100 patches of 1,000 cells each) storing approximately 5×10^7 (50 million) random sequences of

---

Fig. 3 (cont.)

voltage-sensitive NMDA receptor channels become activated. Subsequently, S2 activity arrives, opening AMPA receptor channels. The resulting continued depolarization causes the NMDA channel at S1 to remain open, causing further potentiation at the S1 site (retrograde facilitation). The depolarization at S1 shunts some current from S2, somewhat lessening its potentiation (anterograde suppression). The result is that S1 becomes more potentiated than S2, due to the combined retrograde facilitation and anterograde suppression (Larson and Lynch, 1989). **B**. Biophysical simulations of cell depolarization via sequential activation of differentially potentiated afferents. S1 is strongly potentiated and S2 not potentiated. The first bar graph indicates the 60% potentiation of S1 and the lack of potentiation of S2. S1 and S2 are activated both in the order in which they were potentiated (S1-S2) and in the opposite order (S2-S1). The simulated depolarization in response to the sequence S1-S2 is 21% larger than that to the sequence S2-S1. **C**. Experimental test of the prediction from the biophysical simulation, using extracellular recording from field CA1 in a hippocampal slice. As in the simulation, S1 was strongly potentiated, and S2 left unpotentiated. Extracellular field potentials were measured in response to S1-S2 versus S2-S1. Shown at bottom left is a typical experiment lasting 50 minutes, with LTP induced in the S1 pathway at 12 minutes via theta-burst stimulation (Larson and Lynch, 1986), resulting in stable potentiation of about 66% ±5%. At top are extracellular EPSPs in response to the sequences S1-S1-S2-S2 and S2-S2-S1-S1, each measured 2 minutes before and 8 minutes after the potentiation episode (calibration = 1 mV, 20 ms). The bar graph (bottom right) shows the percent increases of the extracellular response area to S1 and S2 individually, and to the two sequences. In 18 experiments, the response to the sequence S1-S1-S2-S2 was significantly greater than the response to the reverse sequence (p < 0.01; one-tailed paired t(17) = 2.75).

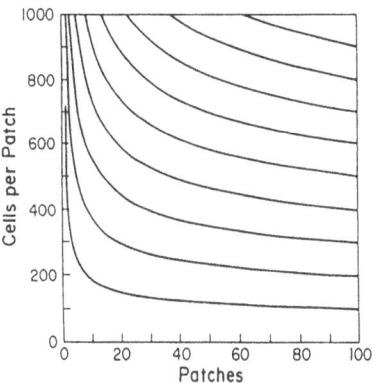

Figure 4. Capacity and scaling properties of the sequence-recognition network. Contour lines indicate the number of sequences of length 10 that can be learned without exceeding a fixed error probability of 0.001, for cells with 10,000 synapses each. The abscissa gives the number of patches in the network, and the ordinate denotes number of cells per patch. Each contour curve corresponds to a specific number of learned sequences. The lowest curve denotes 5 million learned sequences; following the curve shows that this many sequences can be learned by a network of 5 patches (x-axis) with about 250 cells per patch (y-axis), or by a network of 40 patches with 125 cells per patch, etc. The contour curves increase by 5 million per curve; the highest curve shown corresponds to 45 million learned sequences (accomplished via 60 patches of 1000 cells each ranging to 100 patches of 900 cells each). The largest network for which theoretical values are plotted corresponds to the point at the upper right corner of the contour graph: this network consists of 100,000 cells (100 patches of 1000 cells each) and has a capacity of 50 million sequences, with error rate $E < 0.001$.

length 10, with a recognition error rate of 0.001. It is worth noting that 100,000 cells with 10,000 synaptic contacts apiece constitutes a relatively small network in real terms (less than the size of one field CA1 in a rat), and yet 50 million learned sequences is equivalent to one novel sequence learned every ten seconds for eight hours each day for fifty years, with a 0.001 recognition error rate. This probably exceeds the extent of the rat's capacity by orders of magnitude; allowing for possible additional errors of transmission, contact and noise, it still provides a scheme capable of accounting for the dramatic capacity of our own memory systems. Standing (1973) found that subjects exposed for only seconds to each of 10,000 pictures nonetheless exhibited 90% recognition rates for those stimuli when tested weeks later. Neural network models do not exhibit the capability for this kind of rapid learning, long retention and large capacity; rather, they typically require extensive training and cannot store many items without using very large networks. Surprisingly, what might have been thought to be excessively low-level biological detail on the induction and expression of LTP gives rise to networks that directly address the problematic issue of how large memories can be implemented in brain circuitry.

The functional model of CA1 that emerges from this analysis is extremely unusual compared to typical "neural network" operation. In particular, artificial "neural networks" typically perform some type of generalization, i.e., respond similarly to similar inputs, even those on which they were not trained. Indeed, this capability is often taken as canonical for brain circuits, mapping intuitively onto the need for generalization in living organisms. The present analysis of LTP in field CA1 of hippocampus, however, yields an algorithm that learns temporal sequences without any significant generalization. This of course does not imply that generalization is not performed by organisms, or by different brain networks. Rather, it raises the question of functional interpretation of individual modular components (e.g., CA1) embedded in much larger systems (e.g., the corticohippocampal pathway). The primary inputs to field CA1 come from hippocampal field CA3, not a sensory field, and the primary outputs from CA1 project to subiculum and parahippocampal (entorhinal) cortex, not motor fields. The relationship between the function of CA1 and the overall function of the organism is thus far from straightforward. We hypothesize that a high-capacity, sequence-dependent, non-generalizing "accept-reject" or "match-mismatch" function such as that achieved by the model circuit described here may, in combination with the functions of the other constituent circuits of the cortico-hippocampal pathway, be of considerable utility in internal memory functions such as recognition of recency, expectation, and changing views with movement (Lynch and Granger, 1992).

CONCLUSION

In summary, the constituent networks comprising the olfactory cortico-hippocampal pathway each display strikingly different cell types, anatomical local circuit architectures, density of connectivity, and mechanisms for synaptic plasticity. The result of these differences is a set of systems with fundamentally different functional activity in response to behaviorally relevant physiological stimulation. Due in part to the different longevity of potentiation in the different anatomical systems, processing via these serial steps not only adds separate qualities to the resultant memory but may also mediate memory for duration of events in the recent past or "knowing when". Taken together, the elements of the cortico-hippocampal pathway can be viewed as serial processing steps in an "assembly line", each making a unique contribution to the formation of memories, including recognition memory in cortical networks, transient memory via the dentate gyrus mossy fiber projection to field CA3, very brief or "bridge" memory from dynamical activity intrinsic to CA3, and sequential expectation memory from CA1 capable of linking events across time. It is reasonable to conclude that hippocampal output signals, resulting from its stepwise componential analysis of potential memories, can trigger cortical activity of a type needed to render cortical synapses plastic. Such a proposal, although quite distinct from any of the many extant hypotheses of hippocampal function, is nonetheless consistent with a broad range of data suggesting that cortical memory relies on intact cortico-hippocampal processing, and provides an alternate interpretation of a complicated body of findings.

REFERENCES

Ambros-Ingerson, J., Granger, R., and Lynch, G., 1990, Simulation of paleocortex performs hierarchical clustering, *Science*, 247:1344.

Berger, T., 1984, Long-term potentiation of hippocampal synaptic transmission affects rate of behavioral learning, *Science*, 224:627.

Bliss, T. V. P. and Gardner-Medwin, A., 1973, Long-lasting potentiation of synaptic transmission in the dentate area of the unanesthetized rabbit following stimulation of the perforant path, *J. Physiol. Lond.*, 232:357.

Bliss, T. V. P. and Lømo, T., 1973, Long-lasting potentiation of synaptic transmission in the dentate area of the anesthetized rabbit following stimulation of the perforant path, *J. Physiol. Lond.*, 232:334.

Collingridge, G. L., Kehl, S. L., and McLennan, H., 1983, Excitatory amino acids in synaptic transmission in the Schaffer collateral-commissural pathway of the rat hippocampus, *J. Physiol. Lond.*, 334:33.

Coultrip, R., Granger, R., and Lynch, G., 1992, A cortical model of winner-take-all competition via lateral inhibition, *Neural Networks*, 5:47.

Eichenbaum, H., Kuperstein, M., Fagan, M., and Nagode, J., 1987, Cue-sampling and goal approach correlates of hippocampal unit activity in rats performing an odor-discrimination task, *J. Neurosci.*, 7:716.

Granger, R., Ambros-Ingerson, J., and Lynch, G., 1989, Derivation of encoding characteristics of layer II cerebral cortex, *J. Cognitive Neurosci.*, 1:61.

Granger, R., Whitson, J., Larson, J., and Lynch, G., 1994, Non-Hebbian properties of LTP enable high-capacity encoding of temporal sequences, *Proc. Natl. Acad. Sci.*, 91:10104.

Green, E., McNaughton, B., and Barnes, C., 1990, Exploration-dependent modulation of evoked responses in fascia dentata, *J. Neurosci.*, 10:1455.

Grossberg, S., 1976, Adaptive pattern classification and universal recoding: I. parallel development and coding of neural feature detectors, *Biol. Cybern.*, 23:121.

Hebb, D. O., 1949, "The Organization of Behaviour", Wiley, New York, NY.

Hill, A., 1978, First occurrence of hippocampal spatial firing in a new environment, *Exp. Neurol.*, 62:282.

Jung, M., Larson, J., and Lynch, G., 1990, Long-term potentiation of monosynaptic EPSPs in rat piriform cortex in vitro, *Synapse*, 6:279.

Kanter, E. D. and Haberly, L. B., 1990, NMDA-dependent induction of long-term potentiation in afferent and association fiber systems of piriform cortex in vitro, *Brain Res.*, 525:175.

Kauer, J. A., Malenka, R., and Nicoll, R., 1988, A persistent postsynaptic modification mediates long-term potentiation in the hippocampus, *Neuron*, 1:911.

Komisaruk, B. R., 1970, Synchrony between limbic system theta activity and rhythmical behavior in rats, *J. Comp. Physiol. Psychol.*, 70:482.

Larson, J. and Lynch, G., 1986, Induction of synaptic potentiation in hippocampus by patterned stimulation involves two events, *Science*, 232:985.

Larson, J. and Lynch, G., 1989, Theta pattern stimulation and the induction of LTP: the sequence in which synapses are stimulated determines the degree to which they potentiate, *Brain Res.*, 489:49.

Larson, J., Wong, D., and Lynch, G., 1986, Patterned stimulation at the theta frequency is optimal for induction of long-term potentiation, *Brain Res.*, 368:347.

Lynch, G. and Baudry, M., 1984, The biochemistry of memory: a new and specific hypothesis, *Science*, 224:1057.

Lynch, G. and Granger, R., 1992, Variations in synaptic plasticity and types of memory in cortico-hippocampal networks, *J. Cognitive Neurosci.*, 4:189.

Lynch, G., Larson, J., Kelso, S., Barrionuevo, G., and Schottler, F., 1983, Intracellular injections of EGTA block induction of hippocampal long-term potentiation, *Nature*, 305:719.

Macrides, F., 1975, Temporal relationships between hippocampal slow waves and exploratory sniffing in hamsters, *Behav. Biol.*, 14:295.

Mori, K., 1987, Membrane and synaptic properties of identified neurons in the olfactory bulb, *Prog. Neurobiol.*, 29:275.

Mott, D. and Lewis, D., 1991, Facilitation of the induction of long-term potentiation by $GABA_B$ receptors, *Science*, 252:1718.

Muller, D. and Lynch, G., 1988, Long-term potentiation differentially affects two components of synaptic responses in hippocampus, *Proc. Natl. Acad. Sci.*, 85:9346.

Muller, D., Joly, M., and Lynch, G., 1988, Contributions of quisqualate and NMDA receptors to the induction and expression of LTP, *Science*, 242:1694.

Nicoll, R., 1969, Inhibitory mechanisms in the rabbit olfactory bulb: Dendrodendritic mechanisms, *Brain Res.*, 14:157.

Otto, T., Eichenbaum, H., Weiner, S., and Wible, C., 1991, Learning-related patterns of CA1 spike trains parallel stimulation parameters optimal for inducing hippocampal long-term potentiation, *Hippocampus*, 1:181.

Price, J. L., 1973, An autoradiographic study of complementary laminar patterns of termination of afferent fibers to the olfactory cortex, *J. Comp. Neurol.*, 150:87.

Ramon y Cajal, S., 1909, "Histologie du Systeme Nerveux de l'Homme et des Vertebres", Moloine, Paris.

Roman, F., Staubli, U., and Lynch, G., 1987, Evidence for synaptic potentiation in a cortical network during learning, *Brain Res.*, 418:221.

Rumelhart, D. E. and Zipser, D., 1985, Feature discovery by competitive learning, *Cognitive Sci.*, 9:75.

Standing, L., 1973, Learning 10,000 pictures, *Quart. J. Exp. Psychol.*, 25:207.

Taketani, M., Ambros-Ingerson, J., Myers, R., Granger, R., and Lynch, G., 1992, Is field CA3 a reverberating short term memory system? *Soc. Neurosci. Abstracts*, 18:1211.

Tanzi, E., 1893, I fatti e le induzioni nell'odeierna istologia del sistema nervoso, *Rev. Sperim. d. Frenatria et d. Medic. Legal.*, 19:419.

von der Malsburg, C., 1973, Self-organization of orientation sensitive cells in the striate cortex, *Kybernetik*, 14:85.

Wyss, J., 1981, Autoradiographic study of the efferent connections of entorhinal cortex in the rat, *J. Comp. Neurol.*, 199:495.

Zimmer, J., 1971, Ipsilateral afferents to the commissural zone of the fascia dentata demonstrated in decommissurated rats by silver impregnation, *J. Comp. Neurol.*, 23:393.

MODULARITY OF SEQUENCE LEARNING SYSTEMS IN HUMANS

Steven W. Keele and Tim Curran

Department of Psychology
College of Arts and Sciences
University of Oregon
Eugene, OR 97403
USA

INTRODUCTION

Humans excel at a variety of learned and highly skilled activities in which complex sequential behavior is distributed over time. The major theme of this chapter concerns the hypothesis that sequence learning and production of sequences of activities involves not a single function, but rather is made up of multiple components. For example, in playing a piano, pitch is mapped to key position and key position is mapped to the motor system for bringing the arms, hands, and fingers to the keys. In addition to this spatial mapping, the pianist must learn the sequence of notes or keys that correspond to a piece of music. The sequential representation must indicate not only which note or key is next in a series, but must also specify the intervals at which the keys should be hit and with what intensity. In other activities, dancing for example, trajectory through space, and not just the target of movement, must be specified. It is likely that some of these functions are independent of one another, both in the psychological sense that one function can be affected with minimal or no influence on another, and in a neurobiological sense in that they depend on different brain regions. This chapter will focus on a selected aspect of skill, the representation of learned sequences, and will consider only those representations that specify the succession of events. One of the issues to be addressed is the relationship between the representation of a sequence and the motor system that actually produces the sequence. Evidence will be presented that sequence representation is relatively abstract and independent of the implementation system. A second line of evidence to be presented suggests that the sequential representation itself has constituent parts or modules.

Without a theory to describe the components of sequential representation and performance, it is difficult to design a focused investigation of the neurobiological underpinnings of skill. A considerable amount of research on "procedural learning" has been based on an assumption, not always stated, that sequence learning is less advanced and less differentiated into functions than is verbal or "declarative" learning. Moreover, it is often assumed that sequence learning occurs in some putative motor system of the brain, such as the basal ganglia or the cerebellum. If sequential behavior involves a complex of modules, however, it seems more likely that different neural systems would provide different components, resulting in distributed representation. Evidence that sequence representation is independent of motor implementation systems suggests that sequence representation originates outside of brain regions that are devoted primarily to selection of particular motor effectors and to actual motor production. Evidence for complex sequential structures and for different modules of representation suggests that a number of different brain regions are involved in sequence representation. A long-term goal of psychological studies is to work out the complexities of sequence learning from a psychophysical point of view in the hope that it will facilitate the analysis of the neural systems that underlie it.

The fact that less research has been done on sequential learning and behavior than on other domains such as verbal memory does not mean that it is more primitive or less important. Speech and language are pinnacles of human achievement. Speech requires the sequencing of a small set of phonemes into a myriad of words, and the sequencing of these words to produce phrases and sentences. Clearly, sequential learning is a prominent aspect of human language.

Besides the sequential aspects of language – be they expressed in speech, writing, sign, or typing – humans also exhibit impressive sequential behavior in other domains. They express music in song, in instrument, in dance. They knit, build cabinets, and acquire the complicated skills of sports. Such widespread capabilities of learning new and exotic forms of sequential behavior indicate that specialized brain systems for sequential learning in humans may generalize beyond the domain of language. Although some theorists have suggested that a key in human evolution is the development of language-specific brain systems, our own investigations have been guided by the notion that humans have evolved mechanisms especially attuned to learning sequential constructions and that subserve both language and nonlanguage.

This idea was articulated in more general form some years ago by Rozin (1976). He suggested that in the course of evolution, particular computational mechanisms arise to solve particular animal problems. In humans, and to a lesser extent in other animals, the computational mechanisms often have evolved further, to the extent that they have become separable from the task of origin and generalizable to other tasks (c.f., Greenfield, 1991, regarding sequential representation in infants, adults, and chimpanzees). This accessibility of a computational module by a variety of inputs and outputs, Rozin argued, lies at the heart of human intelligence. One of Rozin's primary examples concerns phonetic representation. Part of human speech capability stems from decomposition of speech sounds into elementary phonemes that can be reordered to produce different words. In humans, the phonetic representation that subserves speech can also be tapped into by a uniquely human invention, visual symbols called graphemes that can map onto phonemes and serve as a basis for reading. Thus, a module involved in speech is accessible through vision, a different input than anticipated in the course of evolution. A similar view of modularity has been advanced from a neurobiological perspective by Mesulam (1985, 1990) who suggests that local neural networks underlie specific cognitive operations. These local networks participate in a variety of complex behaviors through their large-scale interaction with other computational networks.

Recently, other examples of common computational modules that underlie diverse human activities have been described. One such computation that has inspired much of our sequence work concerns timing. Ivry (e.g., Ivry and Keele, 1989; Ivry and Gopal, 1992; Keele and Ivry, 1991) has presented evidence that a timing mechanism, operating in the range of a few hundred milliseconds to a second or two and localized in the cerebellum, underlies a variety of motor and perceptual tasks. Evidence for this idea comes from the following observations: (1) Timing of intervals in repetitive motor tapping is disrupted by lesions of the lateral cerebellum; (2) Speech dysarthria resulting from cerebellar damage reflects disruption of precise temporal relationships between speech components, as in the voice onset time of stop consonants, but does not affect nontemporal speech properties such as vowel formant structure; (3) Perceptual judgments of time between auditory events are disrupted by cerebellar damage, but loudness judgments of the same events are not; (4) Judgments of the velocity of moving visual displays, which depend on temporal information, are impaired in patients with cerebellar damage, but positional judgments of the same displays are not.

It has been argued that the same lateral regions of the cerebellum are necessary for classical conditioning (e.g., Thompson, 1986; 1990). Many forms of conditioning involve very precise timing in which the interval between a conditioned stimulus and a conditioned response corresponds to the interval between the conditioned stimulus and an unconditioned stimulus. This timing relationship often has adaptive value. For example, a conditioned eye blink that temporally anticipates a noxious stimulus to the eye may prevent the noxious stimulus from having damaging effects. Lesions to the cerebellum impair or abolish precisely-timed forms of classical conditioning, but have little effect on other types of conditioning, such as emotional conditioning. For example, in experiments where a tone is followed by a small electrical shock near the eye, lesions of the cerebellum may affect the linkage of the tone to eyeblink, but they do not affect the linkage of the tone to the autonomic response of change in heart rate elicited by the same shock (Lavond et al., 1984). Moreover, there is evidence that different systems within the cerebellum play different roles. Lesions of the nucleus interpositus of the cerebellum abolish precisely timed conditioned responses, while lesions of cerebellar cortex leave conditioning intact but with responses occurring at inappropriate times (Perret, Ruiz, and Mauk, 1993).

These studies suggest that in humans a particular class of computation, timing in the millisecond range, is separable from the performance of an individual task, and that the cerebellar cortex plays an essential role in the timing computation. We call a system that performs a class of computations and that can be interfaced with different inputs or outputs a module.

In this chapter we examine other components that contribute to skill, concentrating on psychophysical studies of sequence learning. We provide evidence that sequence representation is modular in the sense that it is separable from the motor systems that actually implement movement. Thus, sequencing resembles timing in that an abstract relationship is transferrable among different input/output systems. Secondly, we provide evidence for different sequential learning systems that are in certain respects independent of one another. We review some network models of sequence learning that are beginning to provide insight into possible computational mechanisms of learning. In addition, we discuss ways in which the psychophysical studies could be applied to an analysis of neural mechanisms involved in sequencing.

A MODEL TASK FOR STUDYING SEQUENTIAL REPRESENTATION

In biology and psychology, it is common to study particular, species-specific behaviors. To study sequencing, one might examine behaviors as diverse as language and speech, locomotion, musical performance, birdsong, or mouse grooming. However, the fact that humans are adept at learning a variety of sequences, many of which probably depend on common computational systems, has motivated the design of model tasks that differ from most naturalistic tasks, but have certain experimental advantages. A model task should comprise critical features of important human sequential tasks but use simple procedures amenable to experimental manipulation. The model task should be learnable within a short time frame. Appropriate model tasks can also be employed with animal and infant subjects in order to relate psychologically defined components to neural substrates and their development.

Several different model tasks have been developed to study sequencing. In one pioneering effort, Restle and Burnside (1972) used a linear array of six lights that corresponded to six response buttons. The lights came on successively in patterns such as 1234666662323543, where the numbers refer to lights from left to right. A subject's task was to learn to press a key corresponding to the next anticipated light in a sequence. The lights were presented at a fast pace so that subjects frequently made late responses, responding to one light after a subsequent one had already appeared. Late responses predominated at particular places in the sequence – at the end of a regularly changing sequence in one direction (1234), a sequence in the reverse direction (543), a set of repetitions (66666), or a trill (2323). These break points, where a subject was slow in anticipating the next light, suggested a simple but powerful principle, namely that the internal representation of a sequence had a hierarchic structure. That is, a sequence is stored and retrieved as a series of chunks, each chunk having its own internal structure.

The idea of hierarchic representation was subsequently elaborated by others. Povel and Collard (1982), rather than presenting lights, simply showed subjects a series of numbers (e.g., 321234) that represented the order in which subjects were to press 4 keys in a repeating sequence. The lengths of intervals between successive responses suggested that different individuals parsed such sequences in different ways. For example, some subjects might parse the sequence as a backward run (321) followed by a forward run (234), exhibiting a rather large transition time between 1 and 2. Other subjects might exhibit a parsing such as (32) (1234). Yet others might represent within the sequence the trill 212 preceded by 3 and followed by 34. The important point is that one and the same sequence is subject to different internal and hierarchic representations that can be deduced from the temporal output structure. Very similar ideas have been suggested by Rosenbaum (1987).

Not all investigations have used key pressing as a model task for sequencing. Gordon and Meyer (1989) taught subjects short sequences of 4 nonsense syllables. Sometimes they asked subjects to prepare to produce one sequence but then unexpectedly signaled them to perform a different one composed of the same nonsense syllables but differently ordered. Examining the time to reprogram a sequence led them to the now familiar conclusion that the internal representation of an event sequence, rather than being a linear string of associations, actually was hierarchic. In these experiments, a string of four elements was coded as two concatenated strings of two elements each.

These studies all involved explicit learning, meaning that subjects were either told or otherwise became aware of the exact order in which events occurred. In other situations, where events occur and are responded to in some particular order, the sequential structure is not apparent to the learner. This type of learning is referred to as implicit learning. Although a performance criterion may indicate that the sequence has been learned, the subject is not aware that any learning has taken place. For example, when children learn language prior to beginning school, they typically are not told the rules of word ordering that constitute the grammar of their language. Nevertheless, they are capable of producing correct sequences. It is not uncommon even for adults to be unable to describe the rules that govern their choice of word order, even though their grammar is invariably correct. For review and discussion of the distinction between implicit and explicit learning see Berry (1994), Reber (1989), and Shanks and St. John (1994).

For investigating questions involving explicit versus implicit sequence learning, a paradigm originally developed by Nissen and Bullemer (1987) has proven useful. Subjects view a screen with 3, 4, or 5 designated positions in a horizontal line. On each trial a visual signal, such as an X-mark, can appear at any position. Beneath the screen are corresponding response keys. The subject's task is to press the key that corresponds to the position of the visual signal – key 1 for signal position 1, etc. Reaction time is measured from signal onset to key press. Following a key press, and usually after a fixed interval (e.g. 200 ms), the next signal occurs. Typically blocks of about 100 signals are presented, after which there is a short rest period.

Within a block of trials, the signals can occur in either random or sequentially structured orders. Random signals are usually presented with the constraint that the same signal is not presented twice in succession. Sequential signals occur in specific orders. In a sequence designated 13232......, signals occur at three positions, numbered 1 through 3 from left to right. A set of 5 signals occurs in the order designated, after which, without any discernible break, the sequence recycles. The first signal on a block of trials can start at any particular position within the sequence. The subjects' task is simply to respond to each signal as it occurs, trying to respond as rapidly as possible. Subjects learn the sequence structure, whether they report awareness or not. Sequence learning can be quantified by comparing subjects' reaction times when events occur in sequence with their reaction times when events occur at random. Such an index provides a measure of performance learning that does not require awareness of the learning.

Nissen and Bullemer (1987) introduced another manipulation to examine the role of attention in sequence learning. In the typical experiment, there is a 200 ms interval between one response on the primary task and the presentation of the next visual signal. During that interval a high or low pitched tone can be inserted, and subjects are asked to count the high-pitched tones. Usually performance on this distraction task itself is not of great interest; rather the distraction is used to interfere with subjects' attending to the relationship between successive events of the primary task.

The important contribution of Nissen and Bullemer's paradigm is that it allows assessment of sequential learning for sequences of a variety of types under conditions where explicit instruction is provided, or where no information about the sequence of stimulus is given. In addition, these experiments can be performed under conditions of distraction or full attention.

INDEPENDENCE OF SEQUENTIAL REPRESENTATION FROM THE MOTOR SYSTEM OF EXECUTION

The modular theory of sequence processing suggests that the same internal representation of a sequence of events can be interfaced with diverse motor systems for executing the sequence. Rozin (1976) had developed this general argument based on evidence that phonetic representations underlie not only speech production and perception but also the reading of visually presented words. A number of lines of evidence are consistent with the view. At the informal level, it is often noted that writing style is remarkably similar for the same person, whether it is performed on a small scale by the hand or on a large scale by the arm. Even exotic effectors such as head movements or elbow movements produce similar writing styles (Bernstein, (1947) as reported in Keele, Cohen and Ivry, 1990; Raibert, 1977). Wright (1990) averaged multiple writing samples to eliminate sample-by-sample variation and noted even more remarkable similarity between hand writing and arm writing, though some effector differences emerged as well.

These informal observations suggest that the same internal description of space guides different effectors in the production of figures, consistent with a view put forward by Berkenblit and Feldman (1988): "There is a neuronal level that creates an abstract image (verbal or graphic) of the forthcoming movement (a circle, line, etc.). Then a combination of effectors and a coordinative structure is specified...."

The informal observations of letter similarity across effector systems together with the notion of an abstract image suggest that at least certain aspects of the representation of a motor act involved in drawing a single graphic figure, such as a letter or a geometric shape, are accessible by different effectors. However, these observations raise the question of how low in a hierarchy of motor acts such modularity descends, and at what level motor representation becomes specifically designed for the responding effector.

In a study reported in preliminary form, Wright and Lindemann (1993) had subjects practice writing particular letters with the *nondominant* hand. As would be expected, fluency in producing the letters improved with practice, but the important question was how the improved fluency transferred to nonpracticed letters when writing with the nondominant hand. Nonpracticed letters that shared the same strokes as the practiced letters, even though the strokes were arranged differently, were produced as well as the letters that had been practiced. However, letters composed of strokes not contained in the practiced letters were not executed as well. These results suggest that in handwriting, practice with a specific effector is confined to the level of strokes, i.e., skill improvement is effector dependent. In contrast, the mechanisms involved in arranging strokes into letters must be effector independent because only the dominant hand would have had extensive experience with the stroke arrangements in the nonpracticed letters. Thus, practice using a specific effector improves stroke production but such practice is not necessary for stroke assembly, implying that some basis for the assembly already exists.

Given that specifications of motor action above the stroke level for single letters can be shared at least across hands, one would expect that even higher levels of description such as the specification of a series of letters in a written word can also be shared among different motor systems. Hillis and Caramazza (1988) examined two patients with posterior cortical damage both of whom suffered partial unilateral neglect. The patient with right hemisphere damage tended to make handwritten spelling errors on the left edge of words, often misordering letters, as in writing "rpiest" instead of "priest". The patient with left hemisphere damage tended to make errors on the right edge of the words, again often misordering letters. Of particular interest is the observation that when the patients were asked to spell words orally, both patients made a similar proportion and type of spelling errors as in handwriting. Oral spelling occurs over time, not space, suggesting that the common "locations" of spelling errors for oral and written spelling both made use of a common internal specification of letter order.

In summary, the specification of serial order appears to be abstract in the sense that different effector systems can draw upon the same internal description of movement through space.

The view that serial specification of letters or phonemes is completely independent of the effector system that will execute the movements is not universally shared. Although most studies of the relationship between sequential specification and the motor system imply some degree of independence, they do not necessarily indicate that there is complete independence. A prominent theory to describe sequential motor behavior developed by Jordan (1986; 1993) builds sequential representation into a network that contains effector-specific components, rather than separating sequential specification from the motor systems of action. His model can perhaps best be appreciated by considering the task of ordering phonemes to produce words in speech.

In Jordan's model network, two types of input units are combined via hidden units to jointly specify the features of the next phonemic output (this model is more fully described in a later section, c.f. Figure 8). One set of inputs is a so-called plan, which can be thought of as a global representation of the word. The other input is a set of state units that essentially maintain decaying memories of the phonemes already emitted in a sequence. At the beginning of a phonemic sequence, the state units are all initialized to zero, so the first phoneme is determined exclusively by the plan units. After the first phonemic output, some of the state units change, reflecting the speech features that were involved in the immediately preceding phoneme. As a result the next output is a product of the original plan, plus the new state. The output units in the model are not abstract representations of phonemes that could be fed to motor speech apparatus. Rather the outputs are features of speech, such as lip rounding, tongue position, etc., and the state units that preserve information about prior outputs do not represent memories of abstract phonemes but instead represent the speech features that made up the phoneme. As a result of learning, the network develops the ability to produce the next appropriate set of speech features upon receipt of feedback that the speech features of the preceding phoneme have been emitted. This network, therefore, produces sequential behavior, but it does so within a system that has intrinsic speech-related outputs.

What would be the advantage of using speech features as output rather than abstract representations of phonemes that could then be shipped off to another modular network that would produce appropriate motor activity? The reason is that with practice, motor action becomes fluent so that one aspect of movement melds smoothly into another. One goal in constructing a series of movements is to minimize the amount of change in a motor effector during the transition from one movement to another. The mutual interaction of nearby movements is called co-articulation. Co-articulation maximizes the efficiency of movement with the sole constraint that the individual components still remain interpretable. To produce a smooth series of movements presumably requires specification of the movement apparatus. Thus, in typing, a particular kind of combined finger movement would be required to produce two keystrokes in quick succession. In speech, the motor apparatus might need a very different kind of motor control to make the transition between adjacent phonemes fluent. In actual fact, research in both speech and typing shows a tremendous amount of co-articulation between successive motor actions. Although some co-articulation might be explained in other ways, it was this feature that motivated Jordan's decision to incorporate the executing motor organ into the actual sequential representation.

Thus, some evidence suggests that sequential specification of motor acts is independent of the selection of motor effectors that will produce the actions. Nevertheless, there is also at least a theoretical basis for questioning complete modular separation.

Our own studies have used quantitative assessments of the transfer of learning to distinguish between a theory of modular representation and one in which sequential specification is intrinsic to the effector system. All of our studies have used a variant of the design developed by Nissen and Bullemer (1987).

To determine whether different motor effectors share a common sequence representation (Keele et al., 1995), subjects were trained to respond to sequential stimuli using one of two motor systems. The first involved using three fingers (index, middle, ring) to depress three keys and the second involved moving the arm back and forth to strike the keys with the index finger only.

Subjects began with two blocks of visual signals presented at random. The visual signals were an X-mark that appeared at various locations. The first experiment was conducted without a distraction task. Subjects were told to respond as rapidly as possible to each visual signal by pressing a corresponding key. Although they were told nothing about the presence of a sequence, many became aware of its presence. Different subjects received different sequences, but all were five elements in length, involving 3 positions, 2 of which were repeated within the sequence. An example is 13232...., where the numbers refer to order of positions on a screen at which a visual X-mark appeared. A preliminary *practice* period with successive events presented at random familiarized the subjects with the mapping from stimuli to key-press responses, but did not allow any learning of sequential order. The random practice period also obscured the fact that later events might be presented in sequence. Following practice, subjects typically received 6 to 8 *learning* blocks in which signals were presented in sequence. Some subjects were not told about the sequence. Following learning, subjects entered a *transfer* phase in which some were required to switch to a different effector system. At transfer, half the subjects continued to use the same effector system as in initial learning; the other half changed to the previously unpracticed motor system. During transfer, subjects typically received one block of random trials to familiarize them with the new response arrangements, followed by random, then sequenced, then random blocks. The difference in reaction time between random and sequenced blocks during this final phase was taken as a measure of the amount of sequential information acquired during learning that transferred to new conditions with a different response mode. If the sequential representation were independent of the motor system of execution, there should be complete transfer of knowledge.

Figure 1 shows the results. For the group that used the same effector during the transfer phase as during initial learning, the reaction time for events that occurred in sequence was about 130 ms less than for random series of stimuli. For subjects who switched to a previously unpracticed effector, the reduction in reaction time was approximately the same and did not differ statistically from the former, suggesting a common sequence representation for the two effector systems.

However, it is possible that the random blocks preceding the sequence test extinguished learning, and the 130 ms sequence advantage actually represents new learning on the single sequence block of the transfer phase. To examine that possibility, a control group of subjects received random events throughout the entire "learning" period and changed effector system during the transfer period, so that the sequence was introduced on a single test block. Those subjects also showed faster reaction times on the sequence block than on the surrounding random blocks, suggesting that some learning occurred on that block alone. Nevertheless, the sequence advantage was substantially and reliably less than that for subjects with previous practice on the sequence.

These results suggest that *all* sequence knowledge acquired in the context of responding with one effector system transfers to a different effector system, because amount of transfer did not depend on whether or not the effector changed. This finding further suggests that sequential representation resides in a separate module from implementation systems.

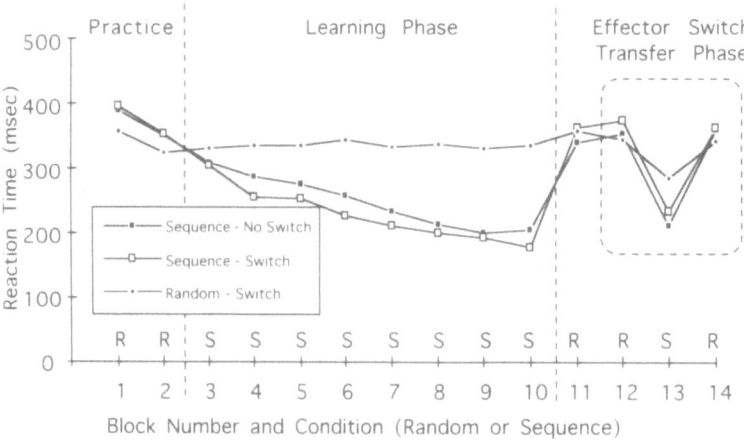

Figure 1. Performance of two groups of subjects who were presented with sequences during the learning phase. During transfer, one group changed the responding effector from fingers to arm or vice versa. The other retained the same effector. Sequence learning is indicated by the difference between sequence block 13 and random blocks 12 and 14, as shown in the inset. The letter R on the abscissa indicates a random block and the letter S a sequence block except for a control group that received random events on all of blocks 1-12 and 14. All blocks are under single-task conditions.

A hallmark of explicit memory is the flexibility with which it can be expressed (e.g., Cohen and Eichenbaum, 1993). To test whether transfer of sequential knowledge between effectors is due to explicit memory, or whether transfer also occurs when subjects are unaware of the sequence, we used a distraction procedure that has been shown to abolish awareness of a sequence for almost all subjects (Cohen, Ivry, and Keele, 1990; Nissen and Bullemer, 1987). A tone was inserted in the short 200 ms interval between each keypress response and the next visual stimulus of the primary task. At the end of a block of trials, subjects reported the number of high-pitched tones.

Figure 2 shows the results of this experiment. The addition of the secondary task caused reaction times on the primary task to increase, and it also reduced the total amount of sequence learning as assessed by the difference between random and sequence conditions during the transfer period. Nonetheless, sequence learning was reliable, and the amount of learning did not depend on the effector that was used during transfer. A control group that had not experienced the sequence until the critical test phase showed no evidence of learning. Thus, under conditions of distraction, the shorter reaction time for the sequence is the exclusive result of learning prior to transfer rather than new learning during the transfer phase. This experiment suggests that sequential knowledge is independent of the motor system that expresses it, even under distracting conditions that minimize explicit learning.

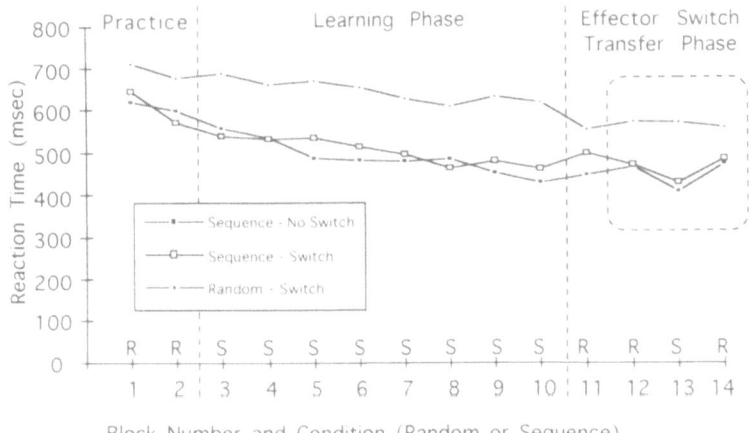

Figure 2. Identical experiments as in Figure 1, except that all blocks are under dual-task conditions, in which tones were presented between visual stimuli and subjects were instructed to report the number of high-frequency tones.

The two experiments just described suggest that sequential representation is not in the motor system because the knowledge is completely transferable from one motor effector to another. However, they do not distinguish between two other possibilities having to do with the nature of the representation. The form of the sequence representation could be either the order in which the stimuli occur, in this case a visual/spatial representation, or the order in which keys are pressed. Presumably, key press order is not the same as a motor code, because the same response key can be pressed with different motor effectors.

To differentiate between these two types of sequence representation, a third experiment was conducted. It introduced verbal responses in addition to finger responses so that transfer could occur with changes in the nature of the response. Like the second experiment, this one employed a distraction task throughout. During learning, one group of subjects responded verbally to signal position with the words left, middle, and right; the other group responded manually with key presses. During transfer, both groups responded verbally. Verbal reaction times were measured from voice-onset times.

The results shown in Figure 3 in some ways resemble those of the first two experiments, but differ in other ways. Some sequence knowledge acquired manually did transfer to verbal responding. That sequence knowledge, though small in magnitude, was statistically greater than that exhibited by a control group making manual responses that had only been presented with random events during training. These results suggest that, at least some of the sequence knowledge acquired in responding to a series of visual events describes the order of the events, not the order of responding. It is also the case, however, that the group who practiced with verbal responses throughout learning exhibited greater sequential knowledge during the transfer phase than the group that had practiced manually and transferred to verbal responding.

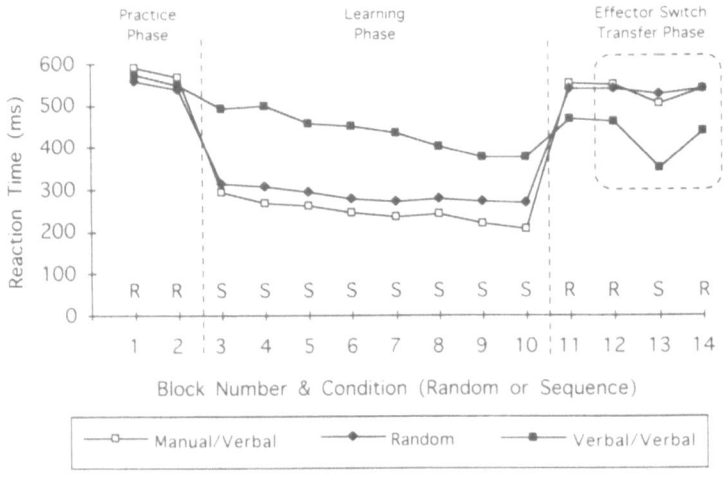

Figure 3. Performance of two groups of subjects, one group that practiced with manual responses, and a second that practiced with verbal responses. A control group also practiced with manual responses, but events were random. During transfer, all groups used verbal responses. All blocks are under dual-task conditions.

There are at least two possible explanations for incomplete transfer of sequence learning between manual and verbal responding. Less sequence learning might occur with manual responses than with verbal responses, so there would be less information to transfer from the manual to the verbal mode. Second, although some sequence information resides in stimulus order, some may also reside in response order. Further experiments will be necessary to distinguish among these possibilities.

Other researchers have also provided evidence that sequential information resides in a code independent of the motor system. In a design somewhat similar to our own, Stadler (1989) found that sequential information about perceptual events transferred from key pressing with one set of fingers to responding with a different set of fingers. Mayr (1995) also demonstrated that some sequential representation is tied to stimulus representation, completely independent of response requirements. Subjects in Mayr's study pressed keys that corresponded to the identity of geometric shapes. The shape on a given presentation appeared in one of four different locations, but location was irrelevant to response selection. Unknown to the subjects, the geometric shapes and hence the order of responses occurred in one particular sequential order. The locations of the shapes occurred in a different and uncorrelated order. By occasionally reverting to random order either in shapes or in positions and observing declines in reaction time, Mayr was able to show that subjects had acquired sequential knowledge not only of the upcoming shape, which determined the response, but also of its position, despite the fact that position was not the response determinant subjects had been instructed to use. When learning took place without distraction several subjects became aware of the sequences, but similar results occurred in a tone-distracted study that virtually blocked awareness.

A study by MacKay (1982), using German-English bilinguals to examine sequence transfer, provides additional insight into the nature of the sequence representation. He presented subjects with sentences in one language or the other and observed their improvement in speaking speed over 12 repetitions of the same sentence. When the sentence was then read aloud in the other language, preserving the same ordering of concepts, MacKay found 100 percent transfer of the previous sequence learning. While most people would intuitively predict some transfer, the important observation is that transfer was complete. This outcome indicates that the speed improvements were localized to a sequential representation not only more abstract than particular movements of speech apparatus but also more abstract than a specific language or word order. In another of MacKay's experiments, subjects spoke a randomly ordered series of words, repeating them in the same random order each time. When switched to the alternate language, no transfer of learning occurred.

How might MacKay's apparently discrepant findings for the two situations be explained? He proposed a theory of hierarchic representation of sequences (see MacKay, 1987 for a much expanded treatment of his theory and many other sequential phenomena). His model divides the representational system into three modules – conceptual, language-specific, and motor. High-level sequential representation resides in the conceptual module. A sentence is represented first at the level of abstract object-action concepts. At successively lower levels of a hierarchy, the complex concepts are differentiated until they correspond to individual concepts that can be denoted by different words in a different language. It is important to note, however, that even at this level, concepts are not words; the same concept can underlie both an English and a German word. For accomplished bilinguals, who already have fluent articulatory abilities, the novelty of a new sentence is restricted primarily to the highest levels. Thus, when a novel sentence is practiced, learning is restricted primarily to new conceptual structure. It is this information that transfers from one language to the other. The reason why transfer fails for randomly ordered words as opposed to those words embedded in a sentence, is that conceptual structure helps convey the meaning of individual words, and such structure is lacking for random words. Consider a word such as "right" that has numerous meanings, referring variously to spatial position, lack of error, political orientation, etc. Sentence structure helps restrict the conceptual meaning. Strictly serial order representation with no hierarchical structure fails to specify precise meaning and hence limits transfer to words of another language.

Once the conceptual representation reaches its terminal level, individual concepts are translated into language-specific words through a second module. This module too contains hierarchic representation. A word is successively broken down into components, first into syllables, and ultimately into a phonetic representation. The phonetic representation is then interfaced with a third module that specifies articulatory components.

MacKay's theory and his own empirical observations are consistent with the present evidence that sequential representation of action is separable from the motor system that implements the action. In MacKay's theory, articulatory control systems can produce individual articulations, but do not contain any information about what subsequent articulations will be. Any anticipation of articulation comes not from prior articulation per se but from higher levels of hierarchical representation that are functionally separable from articulation.

In summary, a number of studies, all using a transfer methodology, have suggested that sequential representations that guide a sequence of actions reside in a module that is separable from the effector system itself. Although these studies suggest that separability of sequential representation from implementation applies to both conditions of attended and unattended learning, there is evidence that sequential memories acquired under conditions of distraction involve at least partially separate modules from those requiring attention.

INDEPENDENT ATTENTIONAL AND NONATTENTIONAL SEQUENCE REPRESENTATIONS

A number of experiments have sought to specify sub-components of the sequence learning system (Nissen and Bullemer, 1987; Nissen, Willingham, and Hartman, 1989; Nissen, Knopman, and Schacter, 1987). These studies have used relatively complex sequences of the type, in which visual signals are presented at successive positions, with key-pressing reaction time used as a measure of sequence learning. Nissen and colleagues found that a distracting task prevented subjects from learning complex sequences that were easily learned without distraction. As long as there was no distraction, even patients with Korsakoff's syndrome, who suffer severe amnesia as a result of chronic alcoholism, were able to learn the sequence, although they were unable to express awareness of it (Nissen and Bullemer, 1987; Nissen, Willingham, and Hartman, 1989). Administration of scopolamine, an anticholinergic drug, reduced sequence awareness in normal subjects but did not prevent sequence learning (Nissen, Knopman, and Schacter, 1987). Thus, the work of Nissen's group suggests that attention, in terms of freedom from distraction, is needed for learning a complex sequence, but that attentional learning does not necessarily lead to awareness (see also Willingham, Nissen, and Bullemer, 1989).

In our laboratory, we (Cohen, Ivry, and Keele, 1990) found that under certain circumstances, even with a concurrent secondary task, certain types of sequences were learned. Sequences of the sort 13232.... where numbers refer to spatial positions, were learned, as were slightly longer sequences involving more elements, of the sort 132314..... However, when sequences of the type 132312.... were used, a distraction task blocked sequence learning. The latter two sequences are extremely similar, involving a different event only at one sequence position. Why is one learnable under distraction and the other less so?

We noted that a sequence such as 132312.... has a certain ambiguity. Each possible event occurred twice, but on each occasion, it was followed by a different event. Such a sequence is impossible to learn based only on direct, pairwise associations. The sequence used in all of Nissen's studies has this ambiguous character. We hypothesized that such ambiguity is solvable by a coding mechanism that parses a sequence into chunks, allowing the learning of order within a chunk. This mechanism is essentially one of hierarchic coding. Sequences of the sort 13232... and 132314... have, in contrast, one uniquely occurring event, and more than one unambiguous ordering within a sequence. Thus, in 132314..., event 4 is followed only by event 1 and event 2 is followed only by event 3. Such unique associations, we hypothesized, allowed learning of the entire sequence by a non-hierarchic mechanism. In short, our initial studies led us to postulate two distinct forms of sequence learning, one hierarchic and the other non-hierarchic.

The finding by Cohen et al. (1990) of an interaction between attention and sequence structure led to two extended lines of investigation. The first line (Curran and Keele, 1993) followed up the suggestion that there might be two independent forms of sequence learning, one of which requires attention but is capable of learning sequences with ambiguous associations, presumably by mechanisms of hierarchic representation. The second line (Keele and Jennings, 1992) involved computational investigations of possible mechanisms of hierarchic and nonhierarchic forms of sequence learning.

We (Curran and Keele, 1993) performed four experiments to examine the hypothesis that there are two independent forms of sequence learning, one requiring freedom from distraction and the other not. For convenience, the two forms of learning will be called attentional and nonattentional, to distinguish between their relative susceptibility to distraction. These two forms of learning, it was hypothesized, do not communicate their contents to one another. When attention is available, the attentional system acquires information in parallel with the nonattentional system. It was hypothesized that the attentional system needs attention not only for acquisition of sequence knowledge, but also for the conversion of that knowledge, once acquired, into either performance or awareness. Finally, we hypothesized that the two learning forms differ in their capability. Specifically, only the attentional system is capable of learning sequences thought to require hierarchic coding; both systems can learn sequences that do not have ambiguity of association.

A first experiment varied the amount of learning that occurs in a nondistracted state by either explicitly telling subjects of the presence of a sequence or not telling them. This manipulation influences the amount of attentional learning and allows a test of the hypothesis that there will be no change in the amount of knowledge in the nonattentional system, because of the assumption that attentional learning is not available to the nonattentional system. Knowledge in the nonattentional system can be assessed by adding a distraction task following initial learning.

There were two basic conditions during training, and two groups of subjects. Under both conditions, the first two blocks of trials were random. Then, the informed learning group was told that signals would occur in a particular order. The order was described, and subjects were given a minute to study it. Different subjects received different signal orders, but all involved four events, two of which occurred twice and two that occurred once in a repeating 6-event cycle (e.g., 143132...). These sequences are learnable with distraction (Cohen et al, 1990).

These same orderings were given to a second set of subjects, but they were not informed that a sequence was present. A questionnaire administered prior to the transfer phase indicated that some of the uninformed subjects had become aware of the sequence on their own and were able to describe parts of it; other subjects did not express awareness. Although the specific awareness criterion used affects the number of subjects placed in one group or the other, the exact dividing point is not critical. The main point is that a group with explicit knowledge, a group that discovered the sequence themselves (more aware), and a group that expressed little or no awareness (less aware) differed in the amount of sequence knowledge exhibited when there was no distraction task. The results shown in Figure 4 were in accord with expectations. When there was no secondary task, variations in degree of explicit knowledge had a clear effect on performance. The informed learning group had very fast reaction times when the sequence was present, and slowed considerably when events returned to random order. Subjects who became aware of the sequence on their own performed as well as informed subjects after some practice. Subjects who expressed little or no awareness of the sequence, still learned under single-task conditions, but showed a somewhat reduced sequence effect.

After a training period without distraction, all groups were transferred to a situation where the tone-counting distraction task was added. Sequence learning was reassessed by comparing performance on blocks of random events and blocks of sequenced events. Despite variations among groups in single-task learning, once the distraction task was added, the three groups performed comparably. All showed significant evidence of residual sequence knowledge, but statistically speaking, the residual knowledge was equivalent for all three groups. Results of our earlier reported experiments, illustrated in Figures 1-3, indicate that the small sequence effect that occurs under the dual-task conditions was not due to new learning on the single block of sequenced trials during the transfer phase.

The results in Figure 4 are consistent with the hypothesis that variations in attention-based learning are not transferable to the nonattentional system. Such results are rather remarkable. One group of subjects had been told precisely the nature of the sequence, and they could parlay that knowledge into extremely fast performance when there was no distraction. Indeed, mean reaction times of about 200 ms after some practice suggest that at least some of the responses actually anticipated the next stimulus, because in the absence of anticipation, reactions times are seldom that fast. Nonetheless, that knowledge was of no use when a distraction task was added, because performance dropped to that of a group that expressed no awareness of the sequence. A related, and even more powerful point is made in the next experiment.

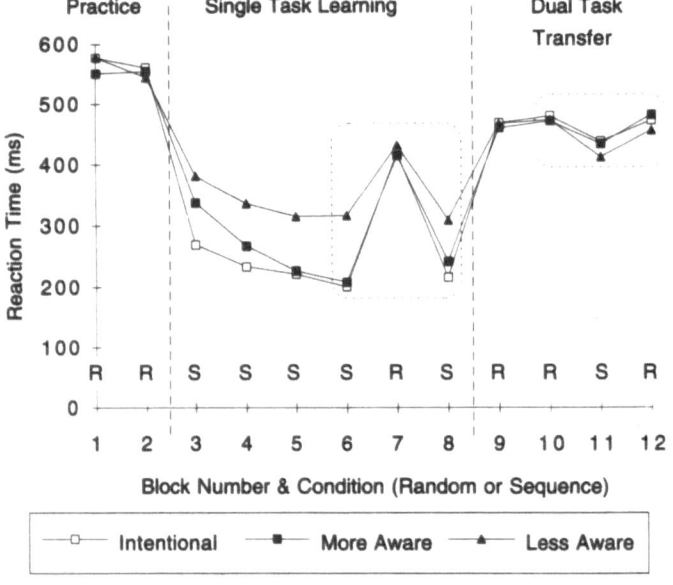

Figure 4. During single-task learning, one group of subjects was informed of the sequence, another became aware of the sequence on its own, and a third group did not become aware. The figure also indicates performance when a secondary task was added. Random blocks are designated by R and Sequence blocks by S.

In the experiment just described, sequence knowledge was acquired under a condition with no distraction task. A stronger prediction is that knowledge acquired by the non-attentional system under conditions of distraction is equivalent to knowledge acquired when free from distraction. To test this, two groups of subjects were run much as before, except that no diagnostic of sequence knowledge was given during the learning phase. Following two initial dual-task practice blocks with random events, the distraction task was removed for one group, and that group was explicitly told the nature of the sequence. For the other group, not only were they not told about the sequence, but all training occurred under dual-task conditions. Cohen, Ivry, and Keele (1990) had shown that under such distraction conditions, few if any subjects became aware of the presence of a sequence.

As seen in Figure 5, performance during the transfer phase was poorer in general for the group that had practiced under single-task conditions. Undoubtedly this was because they were less proficient at interweaving the two tasks, given that they had less practice in doing so. Nevertheless, while both groups showed a reliable difference between the sequence test block and the two surrounding random blocks, indicating that they had learned the sequence, that measure was not reliably different between the two groups. The group that had practiced under dual-task conditions and the group that had practiced under single-task conditions showed equivalent transfer of knowledge to the test phase. As before, it seems quite remarkable that a group with explicit knowledge of the sequence showed no better sequence performance than a group without such knowledge, once a secondary task was added. The results suggest that none of the explicit knowledge had transferred to the nonattentional system. In that sense, the two systems are independent.

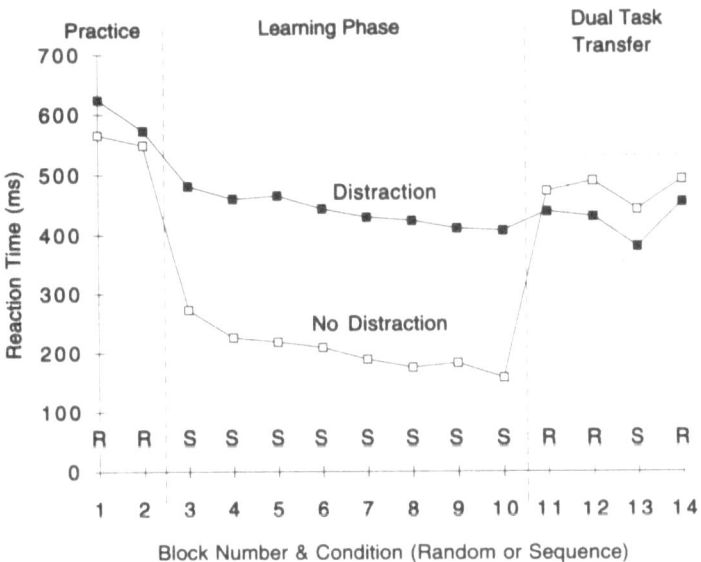

Figure 5. One group of subjects learned the sequence with no distraction following explicit instruction on the nature of the sequence. The other group learned implicitly with distraction , i.e., was never told about the sequence. The distraction prevented awareness of the sequence.

One potential criticism of these two experiments is that the amount of sequence knowledge during the critical test phase may in fact differ under different conditions, but something about the procedure itself prevents its manifestation. Perhaps the dual-task setting puts some kind of ceiling on the amount of learning exhibited. Thus an experiment was designed to address this concern. If the initial sequence learning occurs under dual-task conditions, presumably learning occurs only in the nonattentional system. If the secondary task were removed following dual-task learning, reaction times should improve. Because there would initially be no knowledge in the attentional system, the benefit of sequential conditions over random conditions should remain unaltered. It is useful to re-examine Figure 4 in which transfer was in the opposite direction, from single- to dual-task. There, even the unaware group showed a larger sequence effect during single-task conditions than under dual. We take this to mean that as long as there is no distraction, subjects learn more in single-task conditions than in dual. Lack of distraction allows some learning of the type we call "attentional" despite being unaware. This idea is supported by the finding that "attention-based" sequence learning can occur under administration of scopolamine and in patients with Korsakoff's syndrome, both groups with reduced awareness (Nissen, et. al., 1987, 1989). The prediction is, however, that when conditions are reversed, going from dual- to single-task conditions, the sequence effect will be equivalent under both dual- and single-task settings.

To test this prediction, a single group was examined. This group was initially trained under dual-task conditions with one random block inserted to allow assessment of the amount of learning. Then in the transfer phase, the distraction task was removed and sequential knowledge again assessed. The sequence block in the last phase was the sole occasion on which a sequence had been experienced under single-task conditions, and the prediction is that single-task performance would be no better than dual-task performance. The results are shown in Figure 6.

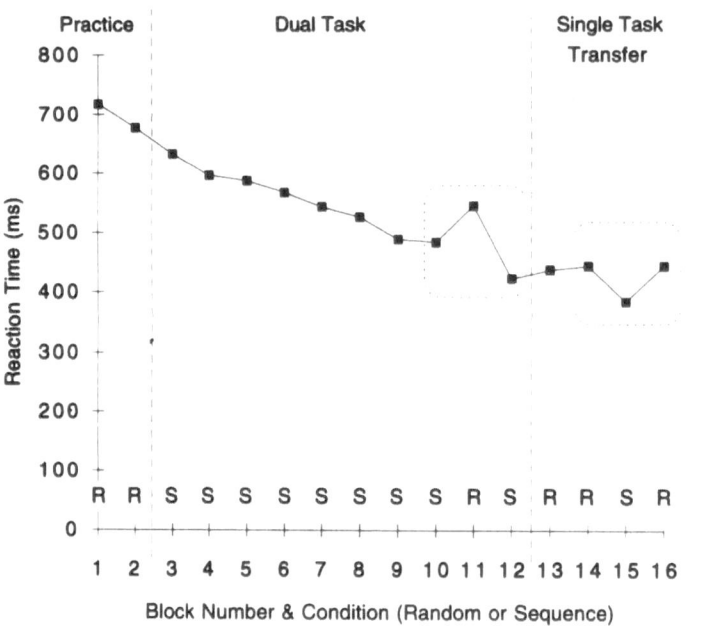

Figure 6. Performance of subjects trained under dual-task conditions and then transferred to single-task conditions.

Reaction times on the first two blocks of single-task conditions were no faster than the preceding dual-task conditions because the last block under dual-task conditions had been with a sequence, while the first two single-task blocks were random. These factors counterbalance each other. The critical point is that there was no statistical difference in performance on the sequence between single-task and dual-task conditions. It appears that during the original dual-task learning, only the hypothesized nonattentional system was available for learning, and only that knowledge source had any useful information about the sequence during single-task transfer.

The experiments described so far show that a nonattentional system can learn sequences that contain some points at which one event uniquely predicts the next event in a sequence. The work of Cohen, Ivry, and Keele (1990) had shown that ambiguous sequences of the sort 132312.... are difficult to learn with distraction. What remains to be demonstrated is that knowledge of sequences of this latter sort, when learned under distraction-free conditions, presumably by the attentional system alone, is blocked when distraction is subsequently added. Such a demonstration would argue that attention is needed not only for learning ambiguous sequences, but also for performance once learning has occurred.

To test this an experiment was designed in which, a single group of subjects initially learned an *ambiguous* sequence under single-task conditions. Because earlier experiments (e.g., Keele and Jennings, 1992) had suggested that ambiguous sequences (e.g., 132312...) sometimes could be learned to a marginal extent even under dual-task conditions, we made the sequence more complicated. We continued to use 3 events but embedded them in a 9-element cyclic sequence such as 132312123.... Following initial single-task training, including the diagnostic test of sequence knowledge, subjects transferred to the dual-task condition and sequence knowledge was again assessed. Subjects showed clear evidence of sequence learning of the ambiguous sequences in the absence of distraction, but subsequent addition of distraction abolished signs of learning. These results suggest that attention is needed not only to code events by place in a sequence, but also to keep track of place in the sequence during performance.

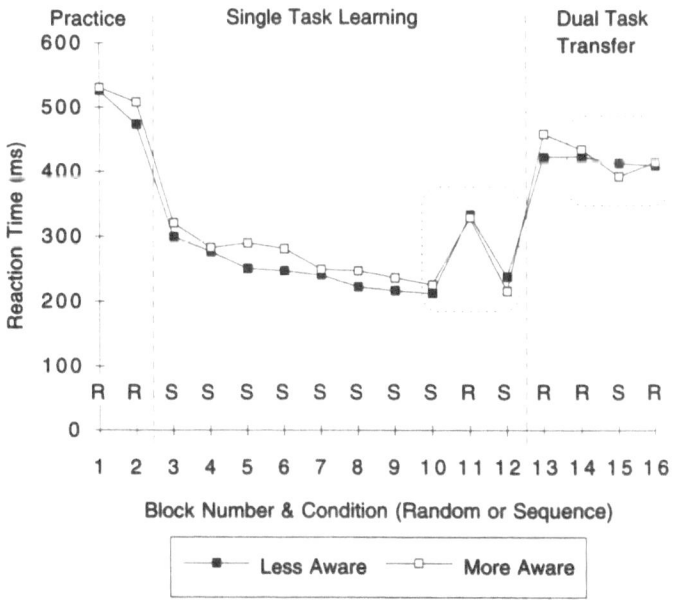

Figure 7. Performance of subjects trained under single-task condition with *ambiguous* sequences, in which each sequence event is followed by a different event depending on the place in the sequence. The figure also shows performance of the same subjects when a secondary task is added.

An observation related to this last experiment was described by Nissen and Bullemer (1987). They note that when subjects were presented with an ambiguous sequence under dual-task conditions, no learning was manifest. Moreover, when the distraction was subsequently removed, not only was there no immediate evidence of sequence knowledge, but subsequent single-task learning showed no acceleration. These observations, when coupled with our own, suggest that sequences entirely composed of ambiguous associations are not stored in a nonattentional memory system.

Recent unpublished work (Goschke, personal communication), has largely supported this hypothesis. Goschke examined sequences of 12 elements, constructed from 6 different signal locations. Each location occurred twice in a sequence, but was in each case followed by a different event. In this paradigm, if place in the sequence is ignored, each event is followed by one of two other events each with a probability of 0.5. A random control condition was included where each event could be followed by any of the other 5, yielding transition probabilities of 0.2. When pairwise associations are probabilistically predictive, learning of transitional probabilities can speed reaction times (see also Jackson and Jackson, 1992; Stadler, 1992). Goschke found significant learning of the ambiguous sequences under distraction and argues that this result reflects learning of transition probabilities without the learning of context that would definitively specify the element at a particular place in the sequence. Nevertheless, he considers freedom from distraction to be necessary for building a representation that uses context to specify place in the sequence.

We have presented evidence that the nonattentional learning mechanism cannot learn sequences that are entirely composed of ambiguous associations. Nevertheless, significant learning of such sequences has been shown to occur under distraction (Keele and Jennings, 1992; Reed and Johnson, 1994). The results are inconclusive because distraction is unlikely to completely eliminate attention, and some nonattentional learning of ambiguous associations might also occur under greatly extended training. Furthermore, the difficulty of the secondary task can vary between experiments. Cohen et al (1990) showed that the difficulty of tone counting increased with the number of targets. Thus, it may be critical that Reed and Johnson found ambiguous sequence learning when 30% - 70% of the tones were targets, but Cohen et al. failed to find learning with 50% - 75% targets. Despite the contradictory nature of some of the results, there is consistent evidence that the ambiguous components of sequences containing both ambiguous and unique associations are learned under distraction (Curran and Keele, 1993; Frensch, Buchner, and Lin, 1994; Keele and Jennings, 1992). Thus, the hypothesized nonattentional mechanism must learn more than simple, pairwise associations. In our section on computational models we will consider candidate mechanisms.

Overall, the preceding experiments make a strong case for independent attentional and nonattentional learning systems. This again raises the question of whether these two systems differ in their learning capabilities. We have suggested that the attentional system has the capability of parsing sequences into chunks to build a hierarchic representation. Although such a conclusion is speculative, evidence to support it comes from a variety of explicit sequence-learning tasks cited in the introduction to this chapter (Restle and Burnside, 1972; Povel and Collard, 1982; and Gordon and Meyer, 1989), which showed that explicitly described sequences are coded in hierarchic form. Hierarchic coding in implicit memory has recently been investigated by examining the effects of exogenously parsing the sequence into sub-chunks with the insertion of temporal pauses at regular intervals. Stadler (1993) found that learning a completely ambiguous sequence was much better when the sequence was chunked in this way, even when subjects were not informed that a sequence was present.

There are at least two reasons why a hierarchic representation might depend on attention. First, hierarchic coding may require some kind of short-term memory process that preserves earlier portions of a sequence in order to chunk them with later portions. A distraction task might interfere with short-term memory, thereby preventing chunk formation (a similar idea has been expressed by Frensch, Buchner, and Lin, 1994; Frensch and Miner,

1994). A second possibility is that an attentional mechanism might serve a kind of place-keeping function. For a sequence with ambiguous associations, in order to know what event follows a current event, it may be necessary to keep track of the current position within the sequence, and attention may be necessary for that function. In another implicit learning paradigm, artificial grammar learning, distraction impaired the ability to learn positional information, lending some credence to such an idea (Dienes, Broadbent, and Berry, 1991).

Stadler (1995) has proposed an alternative explanation for the effects of distraction that does not assign an important role of "attention". Stadler suggests sequences are learned as unique runs (or chunks) of stimuli, and that the boundaries of these chunks are influenced by extraneous cues. For example, chunking patterns can be shaped by the insertion of temporal gaps at consistent places within the sequence so that insertion of random gaps disrupts learning (Stadler, 1993). According to this theory, consistent grouping allows the same chunks to be consistently encoded. Conversely, random grouping leads to encoding a large number of inconsistent chunks that are more poorly learned due to fewer repetitions. Because insertion of random gaps has effects that are very similar to the effects of tone counting, tone counting may merely interfere with the normal organization of the sequence (Stadler, 1995). Thus, Stadler suggests that transferring to and from conditions of distraction has deleterious effects because the organization of the sequence is changed. Further research is necessary to test the implications of these various theories of the effects of attention and distraction on sequence learning.

The empirical work on sequence learning reviewed thus far has provided a number of useful insights for our modular theory of sequence learning. First, although sequence learning clearly benefits from explicit knowledge (Cohen and Curran, 1993; Curran and Keele, 1993; Perruchet and Amorim, 1992; Willingham, et al., 1989), other work suggests that sequence learning does not require explicit knowledge (Nissen and Bullemer, 1987; Nissen, Knopman, and Schacter, 1987; Nissen, et al., 1989; Stadler, 1989, Reed and Johnson, 1994; Willingham, Greenley, and Bardona, 1993; Willingham, et al., 1989). Therefore, further work is necessary to distinguish between the mechanisms for implicit or unaware learning and those for explicit sequence learning. Work from our own lab suggests that, even when learning occurs implicitly, the learning that occurs under distraction is qualitatively different from learning that occurs when attention is fully available (Cohen, Ivry, and Keele, 1990; Curran and Keele, 1993). Finally, we must differentiate between the mechanisms for learning and representing sequences and mechanisms for activating the motor systems that are controlled by these representations (Keele et al., in press; Stadler, 1989). One way to investigate these questions is through the development of computational models.

COMPUTATIONAL EXPLORATIONS OF SEQUENCE LEARNING

Two important and related network models of sequence learning have been developed by Jordan (1986; 1995) and Elman (1990). We (Keele and Jennings, 1992) have explored Jordan's model to see whether it can account for our empirical results, and Cleeremans (1993) has similarly examined Elman's model. These particular models have provided two benefits. First, they have allowed us to formulate more precise ideas about possible meanings of terms like parsing, hierarchy, and chunking, and the role such factors play in sequence learning. Second, the models have suggested gaps in the existing data that must be filled before further computational progress can be made.

Jordan's model is a network of connections between input units, hidden units, and what we call prediction units (see Figure 8). The input units can be viewed as stimulus patterns that when processed through hidden units, produce output patterns on units that "predict" what the next response should be. We assume that the activation patterns on prediction units reflect the extent to which a particular response is primed.

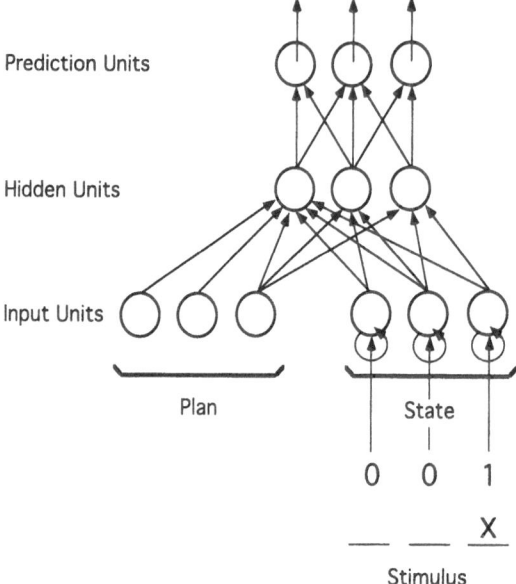

Prediction Units

Hidden Units

Input Units

Plan

State

0   0   1

___ ___ X

Stimulus

Figure 8. Jordan's (1986; 1995) network model of sequential behavior. Not all connections are shown. Each plan and state unit connects to each hidden unit and each hidden unit connects to each prediction unit. Prediction errors, the difference between predicted and actual responses, are used to adjust weights by back propagation. The actual response is determined by a presented stimulus. The presented stimulus also feeds a single state unit with a weight of 1.0. Each state unit feeds back on itself with a fixed weight, μ.

The input layer consists of two segregated sets of units. One set, called plan units, retains an unchanging activation pattern over a sequence of stimuli. One might think of the plan units as representing a higher-order representation of a sequence to be performed, much like a concept of a word. The "word" representation as embodied in the plan units would remain activated until all the constituent phonemes are produced and a different "word" is activated. Thus, plans act as a high level node in a hierarchic arrangement.

The second set of input units are called state units. Activation of the state units is a function of the stimulus on the current trial, t, as well as the state units' activation on previous trials $(t\text{-}n)$. Stimulus positions are coded locally with a value of 1 passed to the current stimulus/state position, and 0 to all others. The state units also feed back upon themselves with a recurrent connection of weight $\mu < 1$. The result of these two inputs to each state unit is a representation of the stimulus that is influenced by the locations of prior stimuli. One state unit will be strongly activated by the current stimulus. That and other state units may have residual activation from past states. The rate of decay of past states depends upon the parameter setting of the recurrent connection with weight μ. If μ is low, less than about 0.2, the current stimulus tends to dominate, and there is little residual memory of events further back. If μ is high, greater than about 0.8, there is little loss of past states so that events of the distant past retain too great an influence. For intermediate values of μ, the state units provide a kind of moving window which represents the context in which stimuli occur. It is the maintenance of context that is critical for building associations between nonadjacent events.

The hidden units combine input from both plan and state units. The particular weightings assigned to the connections from state and plan units to the hidden units change as a function of learning. The hidden units provide input to the prediction units, and those weightings also change as a function of learning. The discrepancy between the actual response as determined by the stimulus (the "teacher") and the predicted response which is set to be the desired pattern of output is the source of error that propagates backward through the network resulting in gradual changes of weights until the network reliably produces the desired outputs.

The operation of the network occurs as follows: A pattern of activation appears on the plan units representing a sequence to be produced. Different sequences would have different plans. The state units, in the absence of a prior output, are all set at zero. The combination of plan input and zero input from the state units feeds through the network to produce the prediction of the first stimulus event. At that point in time, an external signal, such as a stimulus in a particular position, triggers the first response output in the series. Any prediction unit that does not match the response defines an error, and the error is then used by a back-propagation algorithm to adjust connection weights from state and plan units to hidden units and from hidden units to prediction units. The result of such weight adjustment is such that if exactly the same patterns of state and plan activity were to be fed through the network again, error would be reduced. At the next time step, the combination of plan units, which have remained fixed, and state units, one of which has been altered, is different than on the preceding time step. This new input pattern will lead to a different prediction than before as the information flows through the network.

The separate contributions of plan units and state units can be appreciated by considering a series of outputs that are identical to a particular place in a sequence. For example, in speech production of the word elegant vs. elephant, the first three phonemes are identical, and through that point, state units go through identical settings. What is it, therefore, that allows the system to correctly branch to the "g" in one word and the "ph" in the other? Outputs depend not on state units alone but on the conjunction of the plan and state units as mediated by connections to hidden units. Since the plan inputs are different for an intent to pronounce two different words, the conjunctions of plan and state are different, once the critical fourth phoneme is reached, and it is the difference in conjunction that results in different flow patterns.

Consider also a case in which identical stimuli occur at two different places in a sequence but are followed by different stimuli at those two places (e.g., 1213). How is the network able to accommodate that? Although two stimuli may themselves be identical, and therefore send identical input to the state units, the residual activity in the state units would typically differ, reflecting differing prior contexts for the two identical events. Thus, the total state unit pattern would differ for the two places in the sequence having identical stimuli, resulting in different predictions for what comes next.

The role of the plan units is somewhat different from the typical function of a node in a hierarchy. The plan units do not directly cause events to occur. Rather, plan units that remain constant over a series of events act concurrently with state units to jointly determine the next element in a series. Still, it is appropriate to think of the plan units as a level of representation higher in hierarchy than state units, because the plan units change less frequently and allow identical states to elicit plan-dependent predictions.

Jordan's model captures in computational form several of the concepts that we have invoked to explain our empirical results on sequence learning. Plan units can be viewed as implementing the concept of hierarchic representation or chunking. Parsing can be viewed as a process that resets state units to zero at the end or before the beginning of a sequence, thus marking its beginning and end. In computational form, one could imagine the last element of a sequence always to be a "null" element that causes resetting. One might suppose that attentional distraction blocks parsing processes that discover starting and ending points and disables the plan systems that assign representations to chunks.

To determine whether these concepts are adequate to explain features of our data, we (Keele and Jennings, 1992) ran a series of simulations in which the model learned the same sequences that we presented to human subjects. We examined sequences of the type 132312..., in which all pairwise associations were ambiguous, as well as sequences with some uniquely occurring elements (e.g., events 2 and 4 in the sequence 132314...). The simulations involved blocks of 120 successive signals, i.e., 20 cycles through a sequence, and 10 blocks of trials were given. The first block of a trial could start at any position in the sequence. On each step in a series, the pattern of activation of prediction units was compared to the actual next stimulus and the discrepancies were used to modify connection weights in the network.

We ran simulations with three different versions of the Jordan model to assess the roles of parsing and plan-dependent representation. In one simulation, although plan units were present, they never changed. In that sense, the plan units represent nothing about a particular sequence. Moreover, although state units started at zero on the first trial of a block of 120 trials, after that they were free running so there was no demarcation of the end of a cycle through a sequence. These manipulations were intended to simulate nonattentional learning, where we supposed parsing and hierarchic representation are unavailable. In this mode, the system works as a relatively sophisticated associational system. We say associational, because the recurrent loops in the state units provide a decaying memory of past events, allowing such memories to participate in associational learning that spans intervening items.

The Jordan system stripped of representation and parsing was able to readily learn the sequences with uniquely occurring elements. Although the ambiguous sequences also were eventually learned, such learning was substantially slower. When parsing was added such that state units were reset to zero whenever a cycle of the sequence ended, there was little change in learning rate of the sequences that contained unique events, but learning of ambiguous sequences improved dramatically, becoming as rapid as the former. Similarly, if instead of parsing, plan-dependent representation was implemented, learning of ambiguous sequences improved to equal that of sequences with uniqueness. We implemented hierarchic representation simply by assigning one pattern of activation on the plan units to part of a sequence (e.g., for the 132 of the sequence 132312) and a different plan pattern for the other part of the sequence (i.e., 312).

Some insight into the mechanism by which the network learned the sequences was provided by an additional simulation. Again we stripped the system of parsing by not resetting state units at the end of a cycle, and eliminated hierarchic representation by not changing plan units for different parts of a sequence. In this case, we complicated an ambiguous sequence (e.g., 132312...) by adding a unique element, making the sequence longer and seemingly more difficult (e.g., 1323124.....). The strictly associational system learned the longer sequence with a unique element more readily than the shorter but completely ambiguous sequence. To see whether this outcome would also occur with *human* subjects under nonattentional conditions, we compared two groups of subjects. One group received longer sequences with one unique event and another group received shorter ambiguous sequences. Both groups were tested with the secondary distraction task of tone counting, which would presumably block parsing and hierarchic representations. The group receiving the longer sequence learned more readily than the group receiving the shorter sequence, confirming the prediction of the model. Indeed, in one replication of the experiment, the group receiving the shorter, ambiguous sequence did not learn at all.

Why is it that adding a unique element to an otherwise all-ambiguous sequence is beneficial? The explanation can be found in the nature of recurrent feedback to state units. Whenever a uniquely occurring stimulus appears, it activates a state unit on the next iteration. Although the activation of the state unit is partially renewed after each successive event, it gradually decays. The unique state event serves as a kind of marker that helps distinguish events from one part of a sequence from otherwise identical events in another part of the sequence. Exactly the same function is supplied by altering the pattern on plan units at different parts of the sequence. That is, the different plans that accompany different parts of a sequence endow those parts with unique features that help disambiguate associations. Resetting of the state units, or what we call parsing, serves a similar function of disambiguating otherwise similar events.

A general lesson emerges from these simulations. One reason why hierarchic coding like that in the Jordan model is so beneficial is that it provides an auxiliary cue to disambiguate the same items in different sequential contexts. Consider once again the speech example of "elegant" versus "elephant". Despite the fact that in early portions, identical series of phonemes occur, co-occurrence of a plan embodied in plan units provides a disambiguation, allowing an associational machine to branch in appropriate directions.

An outstanding problem with our particular simulations is that we have not endowed the system with an ability to discover chunks and assign representations on its own. We have simply shown that if chunks are preassigned, or sequences are pre-parsed, then the general associational system has a much easier time learning the events within a chunk. Some other network simulations, in particular one by Cleeremans (1993a; Cleeremans and McClelland, 1991), do not have this limitation. Cleeremans' network has some similarities to a Jordan net, though it is based on a slightly different architecture, the serial recurrent network (SRN) developed by Elman (1990). Instead of having recurrent feedback of a state unit on itself, the SRN has recurrent feedback within a hidden layer. Hidden unit activation is determined by the current stimulus as well as the hidden unit activation on the previous trial. Thus, the hidden unit representation of a given stimulus is a graded function of the representation of previous stimuli. This system is able to learn sequences that have partial to complete ambiguity.

The explanation for why the Elman network can learn ambiguous sequences is similar to why the Jordan net learns, especially when the plan units are functioning. First of all, the hidden units capture some of the recent past history of a string of events. Such prior context helps disambiguate a sequence. Thus, in the above sequence, the event to follow position 3 can be disambiguated if a memory is retained of the item preceding 3. However, since the recurrent connections are themselves plastic, unlike Jordan's state units (see Cleeremans 1993a) this recurrent influence changes with learning. An analysis of the information being "learned" at the hidden layer reveals that some of the units come to represent not just the preceding item, but small clusters of preceding items. Thus, hidden units that represent chunks act much like plan units in the Jordan system to help disambiguate otherwise ambiguous pairwise associations. Importantly, the "chunks" in the hidden cells of the Elman system are self-discovered in the process of learning a sequence.

One assumption often made about secondary tasks is that they prevent the focusing of attention on a primary task and result in an increased signal to noise ratio in network connections. Cleeremans and McClelland therefore simulated the effects of distraction by adding noise to the hidden-unit input. Given that some of the hidden units eventually represent small subseries of events, adding noise to the hidden-unit input impairs the construction of chunks. They found that such added noise greatly impairs the learning of ambiguous sequences, but had a much reduced effect on sequences containing at least some unique associations. Such results were qualitatively similar to the empirical data of Cohen, Ivry and Keele (1990).

Despite the fact that the Cleeremans and McClelland model captures some aspect of the process by which chunks are discovered, it does not predict basic features of studies (Curran and Keele, 1993) involving transfer between attentional and nonattentional states. In his more recent work, Cleeremans (1993b) developed a "dual SRN" model that is able to simulate the empirical results. The model employs both a serial recurrent net, as described earlier, and a short-term memory buffer that has independent knowledge about a sequence when no distraction is present. This short-term buffer interacts with the basic network by explicitly predicting each sequential element and allowing those predictions to modify the network's hidden-unit representation of the sequence. Put another way, the new model offered by Cleeremans still has two different knowledge systems, one for explicit knowledge and one for implicit knowledge. Rather than being strictly independent, however, the explicit knowledge can be an input source to the implicit system, though not vice versa.

Despite the architectural differences between Keele and Jennings' (1992) adaptation of the Jordan network and Cleereman's (1993b) dual network, both models provide similar insights relevant to our modular concept of sequence learning. Both models have a basic associative learning mechanism at the core. The representational capabilities of this associative mechanism can be enhanced when allowed to interact with higher-level processes. In the Jordan network these higher level processes include parsing via state-resetting at the end of a sequence, and hierarchic organization via plan units. In Cleereman's model, predictions generated by the short-term buffer constrain the hidden-unit representations of the associative system. Furthermore, the associative mechanisms of both models represent more than pairwise associations. Both allow for a kind of contextual representation of inputs such that the representation of an event is influenced by prior events. Such features allow ready learning of sequences that have a mixture of unique and ambiguous associations. In both models, however, the manner of learning higher level representations that allow acquisition of more complicated sequences – plans in the Jordan model and the explicit knowledge system in the Cleeremans model – is unspecified. All that can be said is that availability of such information facilitates sequence acquisition.

To this point there has been fruitful interaction between empirical discoveries about sequence learning and computational analysis. This interaction has indicated a need for additional empirical analysis that would test the underlying assumptions of different models. In particular, it appears that two future developments would be useful in guiding computational models. First, at the empirical level, we have insufficient evidence about the order in which different subparts of a sequence are learned. Secondly, the different computational models make different assumptions about the architecture of a sequence learning system. That is, each sequence learning system has a number of subparts, configured in different ways. Empirical analysis in sequence learning needs to be oriented toward further decomposition of the processes involved so that these processes can be incorporated into computational models.

POSSIBLE NEURAL SUBSTRATES FOR SEQUENCE LEARNING

Despite these needs for further work, the joint empirical and computational work has suggested brain systems or structures that might be involved in sequence learning. The motor-independence of sequential representation is consistent with the idea that there are distinct neural locales for sequential representation and conversion to motor activity. Similar suggestions have been advanced in studies of patients having ideomotor apraxia, a form of apraxia in which movement is intact and fluent, but inaccurate in the patterns produced (Heilman, Rothi, and Valenstein, 1982; Gonzales and Heilman, 1985). In making a salute, for example, movements may approximate salutes in some aspects but miss in others. Heilman and colleagues found such apraxic syndromes to occur following lesions either of posterior parietal cortex or of frontal cortex, the latter presumably involving areas of premotor cortex. However, if these same patients are asked to observe pairs of gestures one correctly performed and one poorly performed and indicate the correct gesture, patients with frontal damage perform well. Patients with parietal damage perform poorly. Thus, the patients with parietal lesions do poorly not only on motor production but on perceptual recognition; the patients with frontal lesions do poorly only on the production.

It seems plausible that the brain areas controlling sequential learning and performance may show a distribution of function that is similar to that seen in Heilman's apraxic patients. For visual-spatial actions, a parietal mechanism could subserve sequential learning and memory. In this respect it is useful to recall the patients with lesions in posterior parietal cortex described by Hillis and Caramazza (1988). These patients produced sequencing errors in either written spelling or oral spelling. The same mechanism that subserves sequential learning might also specify the location of future responses and interact with frontal mechanisms that specify articulators. Such fronto-parietal interactions have been hypothesized to subserve learning-dependent control of action, especially involving spatial tasks, in a number of domains (e.g., Fuster, 1993; Goldman-Rakic, 1990; Goodale, 1993; Passingham, 1993).

Our conclusion that sequential representation is effector independent leads us to speculate that at least visual-spatial sequences are represented in parietal cortex, whereas representations that control sequential performance are represented in frontal cortex. Frontal cortex may also contribute to sequence learning by functioning like plan units in Keele and Jennings' (1992) simulations. Rizzolatti and Gentilucci (1988) review single-cell work from their laboratory demonstrating that inferior regions of frontal cortex have properties suggestive of plans. Cells in inferior premotor cortex (inferior area 6) of the monkey become active when the monkey makes particular kinds of arm or mouth movements. Some cells become active during precision grasps involving finger and thumb; others become active for whole-hand grasps involving all fingers. Some cells are active if the monkey grasps with either the hand or the mouth. The cells involved in grasps do not become active if the hand is configured in similar ways but for purposes other than grasping. Many of the cells continue to fire throughout a series of actions that comprise a behavior. Thus, a "precision grasp cell" might start to fire as the monkey's arm is reaching toward a target, and continue to fire as the hand opens and then closes on the object. Such properties are reminiscent of the way plan units behave in the Jordan model, in which a plan represents a particular action sequence rather than specific components of the action and therefore remains constant throughout the execution of all of the components specified by the plan.

It is especially interesting in the work of Rizzolatti and Gentilucci that some neurons in inferior area 6 represent a similar grasp whether accomplished by the mouth or by the hand. Such results are in accord with our suggestions that a sequential representation is independent of the effector system of execution. Some cells in inferior area 6 are active not only during a monkey's grasp, but become active when the monkey observes a similar grasp performed by the experimenter (Rizzolatti, personal communication). Again, this suggests that the cells describe sequential events independently of the particular effector of execution.

Inferior area 6 of the monkey may be homologous to Broca's region in human cortex, long thought to be involved in human speech and language. Recently, Greenfield (1991) has suggested that Broca's area is specialized not for speech and language per se, but for hierarchic organization of event sequences. She points out, for example, that development of hierarchical control of various action sequences in infants occurs simultaneously with the development of hierarchical control of phonemes in speech. Moreover, hierarchical control of action and language reach the same relatively undeveloped stage in chimpanzees, compared to humans.

The observations by Rizzolatti and Gentilucci and arguments by Greenfield are consistent with a view that inferior portions of premotor cortex play a role in the chunking of events into sequences. In turn, one might speculate that such premotor regions interact with parietal regions to specify the particular events that make up a visuospatial sequence. In the context of Jordan's model, this would place plan units in premotor cortex, and state and possibly hidden units in parietal cortex. On the surface this scheme appears inconsistent with the earlier suggestion that apraxia due to parietal lesions results from a representational deficit while apraxia due to frontal lesions results from translation to particular motor effectors.

However, it is quite likely that different frontal regions or distinct distributed neural assemblies underlie these separate functions, especially given the variability of lesion sites and symptoms in patients with apraxia.

These speculations about the cortical loci of sequence representation differ from suggestions sometimes seen in the literature that sequence representation is largely subcortical. Previous research has shown that patients with basal ganglia dysfunction due to Parkinson's disease or Huntington's disease show impaired sequence learning (e.g., Ferraro, Balota and Conner, 1993; Jackson et al., in press; Knopman and Nissen, 1991; Willingham and Koroshetz, 1993). Our own suggestion is that the basal ganglia, rather than being a storage locus for sequence representation, are involved in sequence production. Although not a focus of this review, the hypothesis that the basal ganglia are involved in production has been explored in preliminary work in our laboratory (Hayes et al., 1995). When subjects were explicitly taught short sequences composed of two parts, patients with Parkinson's disease exhibited a substantial deficit at the transition point from one part to another whenever the identity of the second part differed from that of the first part. Such results suggest that the basal ganglia are part of a system that implements a shift from one sub-sequence representation to another, but that they might not be the site where sequence representation itself occurs. Such a hypothesis is in line with a broader hypothesis that the basal ganglia provide set-shifting functions across a range of domains.

Our long-term hope is that psychological analysis of the modules that make up sequential representation and production will form a reasonable basis for a neurological analysis. Our current thoughts are that the representation is distributed across posterior and frontal cortical regions and that additional frontal regions are involved in effector specification. The basal ganglia are part of an implementation system that allows progression through the representation in real time.

SUMMARY

Sequence learning may be comprised of a number of dissociable modules which subserve particular functions. Experimental evidence suggests that sequential representations are not tied to any particular effectors involved in executing responses, but exist at a more abstract level that specifies sequences of stimuli and/or responses rather than specifying specific actions.

The presence or absence of distraction, while not affecting independence of the representation from the effector system, does influence the kind of sequences that can be learned. Full attention allows the learning of more complex sequences that contain repeated events but in different orders in different portions of the sequence. Computational modeling suggests that attention enables mechanisms that parse a sequence to operate so that order within parts of the sequence can be represented. That is, attention allows hierarchic coding to take place.

In short, sequence learning and performance appears to be comprised of a complex of representational and control processes. Successful linking of brain mechanisms to sequence learning and production will require both an elaborated theory of cognitive processes and consideration of a diverse array of neural systems.

FOOTNOTE

Tim Curran is now on the faculty of Psychology at Case Western Reserve University in Cleveland, Ohio.

REFERENCES

Alexander, G. E., Crutcher, M. D., and DeLong, M. R., 1990, Basal ganglia thalamo-cortical circuits: Parallel substrates for motor control, oculomotor, "prefrontal" and "limbic" functions, *Prog. Brain Res.*, 85:119.

Berkenblit, M. B., and Feldman, A. G., 1988, Some problems of motor control, *J. Motor Behav.*, 20:369.

Berry, D.,1994, Implicit Learning: Twenty five years on. A tutorial, *in*: "Attention and Performance XV: Conscious and Nonconscious Information Processing", C. Umiltë and M. Moscovitch, eds., MIT Press, Cambridge, MA.

Cleeremans, A., 1993a, "Mechanisms of Implicit Learning: Connectionist Models of Sequence Processing", MIT Press, Cambridge, MA.

Cleeremans, A., 1993b, Attention and awareness in sequence learning, *in*: "Proceedings of the 15th Annual Conference of the Cognitive Science Society", Erlbaum , Hillsdale, NJ.

Cleeremans, A. and McClelland, J. L., 1991, Learning the structure of event sequences, *J. Exp. Psychol.: Gen.*, 120:235.

Cohen, A., and Curran, T., 1993, On tasks, knowledge, correlations, and dissociations: Comment on Perruchet and Amorim. *J. Exp. Psychol.: Learning, Memory, and Cognition*, 19:1431.

Cohen, A., Ivry, R. I., and Keele, S. W., 1990, Attention and structure in sequence learning, *J. Exp. Psychol.: Learning, Memory, and, Cognition*, 16:17.

Curran, T., and Keele, S. W., 1993, Attentional and nonattentional forms of sequence learning, *J. Exp. Psychol.: Learning, Memory, and Cognition*, 19:189.

Dienes, Z., Broadbent, D., and Berry, D., 1991, Implicit and explicit knowledge bases in artificial grammar learning, *J. Exp. Psychol: Learning, Memory, and Cognition*, 17:875.

Elman, J. L., 1990, Finding structure in time, *Cognit. Sci.*, 14:179.

Ferraro, F. R., Balota, D. A., and Connor, L. T., 1993, Implicit memory and the formation of new associations in nondemented Parkinson's disease individuals and individuals with senile dementia of the Alzheimer type: A serial reaction time (SRT) investigation, *Brain and Cognition*, 21:163.

Frensch, P. A., Buchner, A., and Lin, J., 1994, Implicit learning of unique and ambiguous serial transactions in the presence and absence of a distractor task, *J. Exp. Psychol.: Learning, Memory, and Cognition*, 20:567.

Frensch, P. A., and Miner, C. S., 1994,Effects of presentation rate and individual differences in short-term memory capacity on an indirect measure of serial learning, *Memory and Cognition*, 5:95.

Fuster, J. M., 1993, Frontal lobes, *Curr. Opin. Neurobiol.*, 3:160.

Goldman-Rakic, P. S., 1990, Cellular and circuit basis of working memory in prefrontal cortex of nonhuman primates, *Prog. Brain Res.*, 5:325.

Gonzalez, L. J., and Heilman, K. M., 1985, Ideomotor apraxia: Gestural discrimination, comprehension and memory, *in*: "Neuropsychological Studies of Aapraxia", E. A. Roy, ed., North Holland Publishers, New York, NY.

Goodale, M. A., 1993, Visual pathways supporting perception and action in the primate cerebral cortex, *Curr. Opin. Neurobiol.*, 3:578.

Gordon, P. C., and Meyer, D. E. , 1987, Control of serial order in rapidly spoken syllable sequences, *J. Memory Language*, 26:300.

Grafton, S. T., Hazeltine, E., and Ivry, R., 1995, Functional mapping of sequence learning in normal humans, *J. Cognit. Neurosci.*, in press.

Greenfield, P. M., 1991, Language, tools and brain: The ontogeny and phylogeny of hierarchically organized behavior, *Behav. Brain Sci.*, 14:531.

Heilman, K. M., Rothi, L. J., and Valenstein, E., 1982, Two forms of ideomotor apraxia, *Neurology*, 32:342.

Hillis, A. E., and Caramazza, A., 1988, The graphemic buffer and attentional mechanisms, Report no. 30, Cognitive Neuropsychology Laboratory, Johns Hopkins University, Baltimore, MD.

Ivry, R. I. and Gopal, H. S., 1992, Speech production and perception in patients with cerebellar lesions, *in*: "Attention and Performance XIII", D. Meyer and S. Kornblum, eds., MIT Press, Cambridge, MA.

Ivry, R. I. and Keele, S. W., 1989, Timing functions of the cerebellum, *Cognit. Neurosci.*, 1:134.

Jackson, G., and Jackson, S., 1992, Sequence structure and sequential learning: The evidence from aging reconsidered, Technical Report No. 92-9, Institute of Cognitive and Decision Sciences, University of Oregon, Eugene, OR.

Jackson, S. R., Jackson, G. M., Harrison, J., Henderson, L., and Kennard, C., 1995, Serial reaction time learning and Parkinson's disease: Evidence for a procedural learning deficit, *Neuropsychologia*, 33:577.

Jordan, M. I., 1986, Serial order: A parallel distributed processing approach, ICS Report 8604, Institute for Cognitive Science, University of California, San Diego, La Jolla, CA.

Jordan, M. I., 1995, The organization of action sequences: Evidence from a relearning task. *J. Motor Behav.* 27, 179-211.

Keele, S. W., Cohen, A., and Ivry, R. I., 1990, Motor programs: Concepts and issues, *in*: "Attention and Performance XIII", M. Jeannerod, ed., Lawrence Erlbaum Associates, Hillsdale, NJ.

Keele, S. W., and Ivry, R., 1990, Does the cerebellum provide a common computation for diverse tasks: A timing hypothesis, *in*: "The Development and Neural Bases of Higher Cognitive Function", A. Diamond, ed., *Ann. NY Acad. Sci.*, 608:179.

Keele, S. W., and Jennings, P., 1992, Attention in the representation of sequence: Experiment and theory, *Human Movement Science*, 11:125.

Keele, S. W., Jennings, P., Jones, S., Caulton, D., and Cohen, A., 1995, On the modularity of sequence representations, *J. Motor Behav.* 27:17.

Knopman, D., and Nissen, M. J., 1991, Procedural learning is impaired in Huntington's disease: Evidence from the serial reaction time task, *Neuropsychologia*, 29:245.

Lavond, D. G., Lincoln, J. S., McCormick, D. A., and Thompson, R. F., 1984, Effects of bilateral lesions of the dentate and interpositus nuclei on conditioning of heart-rate and nictitating membrane/eyelid responses in the rabbit., *Brain Res.*, 305:323.

MacKay, D. G., 1982, The problem of flexibility and fluency in skilled behavior, *Psychol. Rev.*, 89:483.

MacKay, D. G., 1987, "The Organization of Perception and Action", Springer-Verlag, New York, NY.

Mayr, U., 1994, "Spatial attention and implicit sequence learning: Evidence for independent learning of spatial and nonspatial sequences, Technical Report. 94-13, Institute of Cognitive and Decision Sciences, University of Oregon, Eugene, OR.

Mesulam, M.-M., 1985, Patterns in behavioral neuroanatomy: Association areas, the limbic system, and hemispheric specialization, *in*: "Principles of Behavioral Neurology", M.-M. Mesulam, ed., F. A. Davis Company, Philadelphia, PA.

Mesulam, M.-M., 1990, Large-scale neurocognitive networks and distributed processing for attention, language, and memory, *Ann. Neurol.*, 28:597.

Nissen, M. J., and Bullemer, P., 1987, Attentional requirements of learning: Evidence from performance measures, *Cognit. Psychol.*, 19:1.

Nissen, M. J., Knopman, D. S., and Schacter, D. L., 1987, Neurochemical dissociation of memory systems, *Neurology*, 37:789.

Nissen, M. J., Willingham, D., and Hartman, M., 1989, Explicit and implicit remembering: When is learning preserved in amnesia, *Neuropsychologia*, 27:341.

Pascual-Leone, A., Grafman, J., Clark, K., Stewart, M., Massaquoi, S., Lou, J.-S., and Hallett, M., 1993, Procedural learning in Parkinson's disease and cerebellar degeneration, *Ann. Neurol.*, 34:594.

Passingham, R. E., 1993, "The Frontal Lobes and Voluntary Action", Oxford University Press, Oxford, UK

Perrett, S. P., Ruiz, B. P., and Mauk, M. D., 1993, Cerebellar cortex lesions disrupt learning- dependent timing of conditioned eyelid responses, *J. Neurosci.*, 13:1708.

Perruchet, P., and Amorim, M., 1992, Conscious knowledge and changes in performance in sequence learning: Evidence against dissociation, *J. Exp. Psychol.: Learning, Memory, and Cognition*, 18:785.

Povel, D. J., and Collard, R., 1982, Structural factors in patterned finger tapping, *Acta Psychol.*, 52:107.

Raibert, M. H., 1977, Motor control and learning by the state space model, Technical Report AI-M-351, NTIS AD-A026-960, Massachusetts Institute of Technology, Cambridge, MA.

Reber, A. S., 1989, Implicit learning and tacit knowledge, *J. Exp. Psychol.: Gen.*, 118:219.

Restle, F., and Burnside, B. L., 1972, Tracking of serial patterns, *J. Exp. Psychol.*, 95:299.

Rizzolatti, G., Gentilucci, M., 1988, Motor and visual-motor functions of the premotor cortex, *in*: "Neurobiology of Neocortex", P. Rakic and W. Singer, eds., Wiley, New York, NY.

Rosenbaum, D. A., 1987, Successive approximations to a model of human motor programming, *in*: "The Psychology of Learning and Motivation", G. Bower, ed., Academic Press, New York, NY.

Rozin, P., 1976, The evolution of intelligence and access to the cognitive unconscious, *in*: "Progress in Psychobiology and Physiological Psychology", J. M. Sprague and A. N. Epstein, eds., Academic Press, New York, NY.

Shanks, D. R., and St. John, M. F., 1994, Characteristics of dissociable human learning systems, *Behav. Brain Sci.*, 17:367.

Stadler, M. A., 1989, On the learning of complex procedural knowledge, *J. Exp. Psychol.: Learning, Memory, and Cognition*, 15:1061.

Stadler, M. A., 1992, Statistical structure and implicit serial learning, *J. Exp. Psychol.: Learning, Memory, and Cognition*, 18:318.

Stadler, M. A., 1993, Implicit serial learning: Questions inspired by Hebb (1961), *Memory and Cognition*, 21:819.

Stadler, M. A., 1995, The role of attention in implicit learning, *J. Exp. Psychol.: Learning, Memory, and Cognition*, 21:674.

Thompson, R. F., 1986, The neurobiology of learning and memory, *Science*, 233:941.

Thompson, R. F., 1990, Neural mechanisms of classical conditioning in mammals, *Philos. Trans. R. Soc. Lond.*, B329:161.

Willingham, D. B., 1992, Systems of motor skill, *in*: "Neuropsychology of Memory", L. R. Squire and N. Butters, eds., The Guilford Press, New York, NY.

Willingham, D. B., Greenley, D. B., and Bardona, A. M., 1993, Dissociation in a serial response time task using a recognition measure: Comment on Perruchet and Amorim (1992), *J. Exp. Psychol.: Learning, Memory, and Cognition*, 19:1424.

Willingham, D. B., and Koroshetz, W. J., 1993, Evidence for dissociable motor skills in Huntington's disease patients, *Psychobiol.*, 21:173.

Willingham, D. B., Nissen, M. J., and Bullemer, P., 1989, On the development of procedural knowledge, *J. Exp.Psychol.: Learning, Memory, and Cognition*, 15:1047.

Wright, C. E., 1990, Generalized motor programs: Reexamining claims of effector independence in writing, *in*: "Attention and Performance XIII", M. Jeannerod, ed., Lawrence Erlbaum Associates, Hillsdale, NJ.

Wright, C. E. and Lindemann, P., 1993, Effector independence in hierarchically structured motor programs for handwriting. Presented at 34th Annual Meeting of the Psychonomic Society, Washington, DC.

CLASSIFICATION OF SPATIOTEMPORAL PATTERNS WITH APPLICATIONS TO RECOGNITION OF SONAR SEQUENCES

Joydeep Ghosh[1] and Larry Deuser[2]

[1]Department of Electrical and Computer Engineering
The University of Texas
Austin, TX 78712-1084
[2]Tracor Applied Sciences
6500 Tracor Lane
Austin, TX 78725-2050
USA

INTRODUCTION

Many tasks performed by humans and animals involve decision-making and behavioral responses to spatiotemporally patterned stimuli. Thus the recognition and processing of time-varying signals is fundamental to a wide range of cognitive processes. Classification of such signals is also basic to many engineering applications such as speech recognition, seismic event detection, sonar classification and real-time control (Lippmann, 1989; Maren, 1990).

A central issue in the processing of time-varying signals is how past inputs or "history" is represented or stored, and how this history affects the response to the current inputs. Past information can be used explicitly by creating a spatial, or static, representation of a temporal pattern. This is achieved by storing inputs in the recent past and presenting them for processing along with the current input. Alternatively, the past events can be indirectly represented by a suitable memory device such as a series of time-delays, feedback or recurrent connections, or changes in the internal states of the processing *cells* or "neurons" (Maren, 1990).

This chapter reviews possible artificial neural network mechanisms for the classification of signals with time-varying spectral characteristics. Such signals can also be of varying lengths or time durations. Signals obtained from passive sonar are used as a case study to evaluate some of these mechanisms.

Before classification can be attempted, a signal has to be separated from background noise and clutter, and represented in a suitable form. Since most artificial neural network approaches to classifying such signals are based on *static* pattern recognition techniques, each signal must first be described by a vector of numerical attributes, called a *feature vector*. These attributes or features typically include one or more of the following: (1) signal energy in different frequency ranges or bands, (2) time duration of the signal, (3) location and magnitude of the highest peaks in the time and/or frequency domain, and (4) a sequence of numbers such as autoregressive or wavelet coefficients, that are obtained from a particular mathematical model of the signal (Ghosh, Deuser, and Beck, 1992). The selection of an appropriate set of attributes is indeed critical to the overall system performance, since it fundamentally limits the performance of any classification technique that uses those attributes (Ghosh, Deuser, and Beck, 1990).

If the set of features are measured or extracted over the *entire* signal duration, then the signal is represented by a single feature vector. This feature vector can then be fed to a static artificial neural network classifier that has been trained to identify similar feature vectors. Popular static classifiers include the multilayer perceptron and radial basis function networks (Ghosh, Deuser, and Beck, 1992).

Such a system is not ideal because representing each signal by a single feature vector results in a blurring of the spatiotemporal pattern characteristics and may lead to a loss of information that would be useful for discrimination. Another, more detailed way to view a time-varying nonstationary signal is to treat it as a concatenation of quasi-stationary segments (Hermand and Nicolas, 1989). Each segment can be represented by a feature vector obtained from that region. Thus the entire signal is represented by a sequence of feature vectors that constitute a spatiotemporal pattern. A popular way of representing such signals is by a two-dimensional feature-time plot called a spectrogram which shows the sequence of feature vectors obtained by extracting attributes from consecutive segments that are equally spaced and possibly overlapping. Thus the number of segments representing a signal is not necessarily pre-determined, and consequently the number of segments in a spectrogram may vary in length.

Any neural network applied to the signal detection task must address the following issues:

(1) *Cueing*: Generally, any network that performs a recognition task must be informed of the start and end of the pattern to be recognized within the input stream. If this is so, the classifier is called a cued classifier. An uncued classifier, on the other hand, is fed a continuous stream of data from which it must itself figure out when a pattern of interest begins and when it ends. This process of locating and extracting relevant information from a larger pattern is called segmentation.

(2) *Spatiotemporal Warping*: Often the original pattern gets distorted in temporal and spatial domains as it travels through the environment. A robust classifier must be insensitive to these distortions. If the classifier is not sophisticated in this aspect, then the distorted signal received needs to be preprocessed to annul the effects of the environment. This is called equalization and is widely used in communication systems. The characteristics of the communication or signal transmission medium are "equalized" by the preprocessor, so that the output of this processor is similar to the original signal. Alternatively, mechanisms such as the dynamic time warping algorithm (Kung, 1993) can be built into the classifier to make it less susceptible to the distortions.

ARTIFICIAL NEURAL NETWORK TECHNIQUES FOR SPATIOTEMPORAL PATTERN RECOGNITION

In this section we summarize the principal artificial neural network based approaches to processing of spatiotemporal patterns. These approaches are primarily categorized by how they account for past history while processing the current input at the same time. Details of several networks sketched below can be found in textbooks (Hertz, Krogh, and Palmer, 1991; Haykin, 1994), and research papers (Maren, 1990; Ghosh, 1993). For a description of some possible biological mechanisms for spatiotemporal processing, the reader is referred to (Shamma, 1989).

What are the desirable properties of any sequence processing network? From a cognitive standpoint, such networks should represent the past as *context* or in terms of hypotheses. Moreover, the context information or hypotheses should be dynamically maintained and updated. Hypotheses lead to expectations specific to a given environment and past history, and the violation of such expectations prompts the update or modification of the hypotheses. Thus, an ideal network should be capable of self-organization or training using local learning rules. In biological systems, hypotheses are arranged in hierarchies for reasons of economy and improved representational power (Ambros-Ingerson, Granger, and Lynch, 1990; Braitenberg, 1990). In artificial systems, the ability to generalize, predict and explain the results, is a much sought-after property.

Time Delay Neural Networks

At present, the most widely applied artificial neural networks are multilayered feed-forward ones such as the multilayered perceptron and the radial basis function network. Such networks are called "static" since, for a given architecture and set of weight values, the network output is solely a function of the current input, and not of the past dynamics. It is not surprising that at present, the most popular approach to classifying temporal sequences is a rather brute force method wherein the current input as well as some successive past inputs are presented at the same time to a static feedforward network. Thus the effective input to the network is in the form of a matrix in which the columns are the successive values of the input feature vector, and the rows are the different components of the vector. This "spatialization" of a temporal sequence can be implemented by a tapped-delay line, a clocked structure which passes a signal through a series of unit delays and makes the signal value available after each delay. This matrix then forms the effective input to a layered feedforward network with at least one hidden layer of nonlinear cells.

The spatialization described above can also be obtained by using only a single feature vector as the input, but replacing each weight from an input to a hidden unit by a finite impulse response filter (Wan, 1990; Haykin, 1994). Generalizing this idea, even the weights from a hidden node to an output node can be replaced by tapped-delay lines. The resulting network is known in the artificial neural network community as the time delay neural network (Waibel, 1989) or finite impulse response neural network (Wan, 1990).

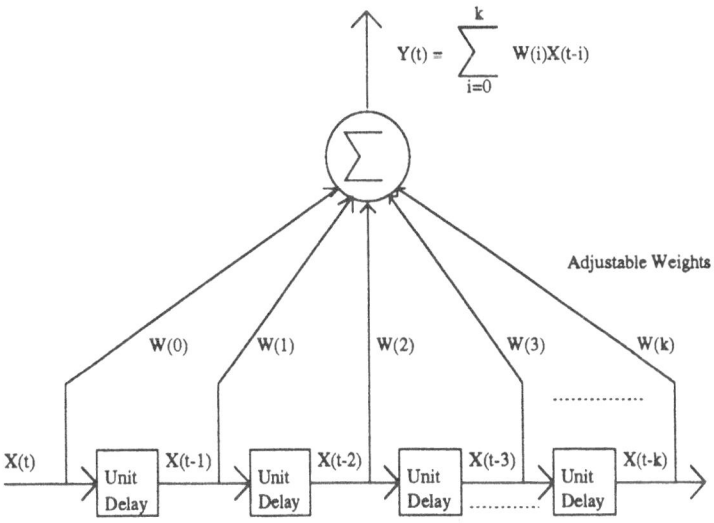

Figure 1. A Finite Impulse Response Filter.

Figure 1 shows a finite impulse response filter with a single input. Figure 2 depicts a time delay neural network with a single hidden layer and with time delays in the input layer. Each hidden node in the network has an output obtained by applying a nonlinear function, usually the sigmoid function, to the weighted sum of its inputs. Clearly, the network of Figure 2 can be visualized as a layered feedforward network in which individual weights in the first layer have been substituted by finite impulse response filters with the structure of Fig. 1. Observe that for any of these layers, the input values for a column of cells at a particular time are the same as that of the preceding column at the preceding time instant. This allows one to unfold the network in time (Haykin, 1994), replicating each unit at each time instant. The weights of a time delay neural network can be updated by moving down the gradient of the error surface of this unfolded network, but the resulting equations are cumbersome. Fortunately, simpler update equations can be obtained by assuming that weights change slowly compared to the input clocking time period (Wan, 1990). Notable successes using time delay neural networks include phoneme recognition using a network with two hidden layers (Lang, Waibel, and Hinton, 1990; Waibel and Hampshire, 1989) in which successive layers were not fully connected, but only had localized connections. This served to reduce the number of parameters or weights, and led to better generalization. The finite impulse response neural network with temporal back-propagation learning algorithm has received recent publicity because it performed very well in a recent contest on the prediction of time series (Weigend and Gershenfeld, 1993).

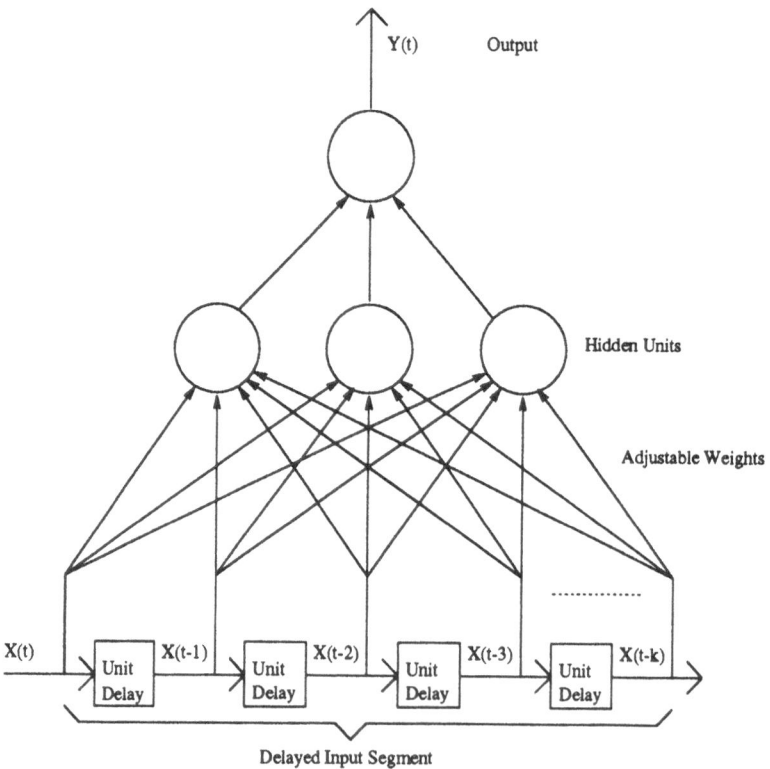

Figure 2. A Time Delay Neural Network with one hidden layer. If the inputs X are vectors, then the input plane will be a two-dimensional array.

The basic time delay neural network has several drawbacks such as the use of a fixed, predetermined window size and time delay, and slow learning due to large network size. However, the familiar feedforward dynamics on which it is based help it retain its popularity. Several researchers have used different strategies to overcome the limitations of the basic time delay neural network. It is well known that variable signal delays along axonal pathways are omnipresent in the mammalian brain and play an important role in neurobiological information processing (Braitenberg, 1990). Day and Davenport (1993) have designed a feedforward network in which the fixed time delays are replaced by variable time delays which are adjusted during training. Such time delay neural networks with adaptive weight delays have been used for predicting chaotic time series (Lin, Dayhoff, and Ligomenides, 1992) and realizing trajectories in 2-D space (Lin, Ligomenides, and Dayhoff, 1993).

To counter the restrictions of predetermined sampling rate and window size, Hopfield and Tank (Tank and Hopfield, 1987; Hopfield and Tank, 1989) proposed the use of broadening and smoothing filters on the inputs to obtain a "concentration-in-time" network. In the discrete time domain, this idea corresponds to obtaining each delay line memory by convolving with a suitable kernel function, such as the exponential trace memory or gamma memory (Mozer, 1993; de Vries and Principe, 1990). The gamma memory model (de Vries and Principe, 1990; de Vries and Principe, 1991; Principe, Kuo, and Celebi, 1994) which uses the gamma function as the convolving kernel, in fact is a generalization of both time delay neural network and the concentration-in-time model. This model is notable because it obviates *a priori* signal segmentation and allows network resolution to be traded off with memory depth.

Function Space Neural Networks

Our interest in time delay neural networks is partly due to the fact that they are a particular realization of the function space neural network family, whose powerful approximation capabilities have been formalized recently (Sandberg, 1991; Sandberg, 1992a; Sandberg, 1992b). As opposed to static networks that attempt to realize some unknown function, function space neural networks generate a map from one function space to another. In particular, they can map certain time-varying signals onto distinct real numbers (Sandberg, 1994), and can thus be used for classifying such signals.

Function space neural networks have a conceptually simple structure, namely, a preprocessing linear operation stage such as convolution with suitable kernel function, followed by a nonlinear feedforward network with adequate mapping capabilities. In particular, the selection of a series of delta functions as the convolving kernels and the multilayered perceptron as the nonlinear network results in a time delay neural network with delays only in the input layer.

Function space neural networks have powerful approximation capabilities. For causal, time-invariant discrete-time nonlinear maps that satisfy certain continuity and finite memory conditions, a network in which the signal and its time-delayed values are fed into a multi-layered perceptron or radial basis function network, is sufficiently powerful for uniform approximation (Sandberg, 1991). Moreover, a broad class of continuous real valued functionals can be approximated as closely as desired with a network using an integration/ preprocessing stage wherein the input is convolved with suitable kernel functions, followed by a trainable non-linear feedforward network (Sandberg, 1992b). At present, key issues pertaining to the design and use of function space neural networks, including determination of suitable kernel functions, network size, the form of nonlinearity for different problem classes and convergence rates for alternative learning algorithms, are being actively researched.

Recurrent Networks

The underlying principle behind a wide range of recurrent networks is that past history can be captured through feedback connections among the processing cells. Partially recurrent networks using context units start with a layered feedforward network and add feedback with time delays to selected cells within the network in order to obtain a "context" or representation of past history. Figure 3 shows four architectures that use context units (Hertz, Krogh, and Palmer, 1991). Examples of recurrent networks for temporal pattern processing are (1) Elman's sequential recurrent network (Elman, 1990) of Fig. 3.a, that specifies a context from hidden units, and (2) a Jordan network (Jordan, 1989), that uses feedback from output units to the input layer (Fig. 3.b).

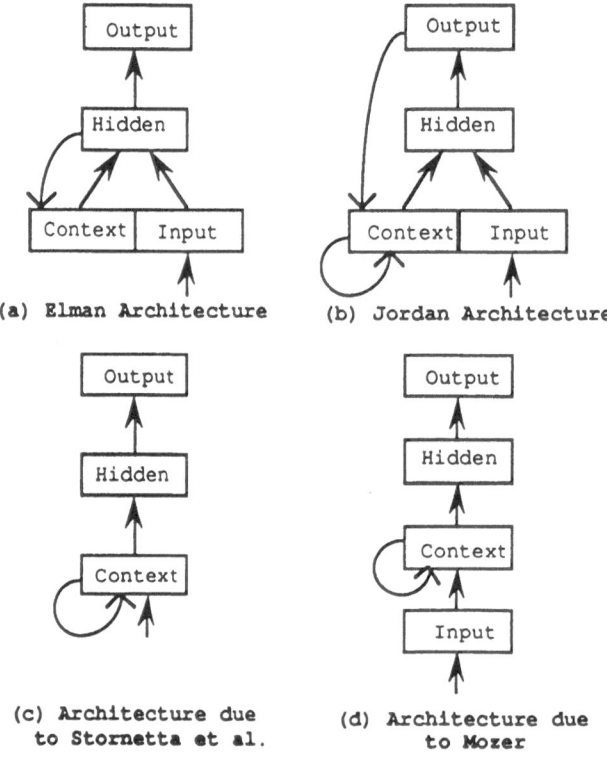

Figure 3. Partially recurrent networks with context units. The feedforward arrows represent fully connected adjacent layers while the feedback arrows represent connections only from the ith unit in the source layer to the ith unit in the destination layer.

Partially recurrent networks with context units assume that an ordered sequence of feature vectors is input, one at a time, to the network. By focusing the back-propagated error signal, they reduce the difficulty of temporal credit assignment. Temporal credit assignment is the problem of determining how past values of network weights affect the current output of the network. Because their recurrent connections act as temporal memory, partially recurrent networks also eliminate the need for a buffer to hold the input sequence and/or intermediate activity levels (Hertz, Krogh, and Palmer, 1991). The delayed feedback signals should provide a means to classify waveforms of arbitrary duration using a network of fixed size. Such networks have been used for speech recognition (Lippmann, 1989) and also to model finite state automata (Ghosh and Karamcheti, 1992). A finite state automaton is a theoretical device which can take on only a finite number of states. A set of rules governs the transitions between states, so that finite state automata can be used for example, to represent rules of grammar that govern the generation of sentence-like sequences. A two layer network with higher-order inputs and feedback from selected outputs has been recently used to differentiate among three trajectories, each of which consists of an ordered sequence of points in a Cartesian coordinate plane (Sun, et al., 1992).

To analyze continuous time inputs, several real-time recurrent back-propagation networks (Pearlmutter, 1989; Pineda, 1989; Werbos, 1988) have been proposed. These networks are described by systems of coupled first-order differential equations. The stable points of these systems provide the stored states or outputs. These networks are capable of on-line modification of weights, and have been used for learning trajectories and for some robotic manipulator control applications (Hertz, Krogh, and Palmer, 1991). Sequential recurrent networks require clocked inputs and are susceptible to spatiotemporal warping (Hecht-Nielsen, 1990). Real-time recurrent back-propagation and time-dependent recurrent back-propagation (Sato, 1990), on the other hand, do not require clocked inputs.

Unfortunately, recurrent networks are typically computationally intensive. More significantly, extensive simulations by several researchers have shown that both partially and fully recurrent networks are extremely sensitive to initial conditions, learning rates and strengths of feedback connections. They are also prone to unstable behavior. Thus, much more work needs to be done in order to understand the full range of their limitations and capabilities.

Attractor Transition Networks

Several recurrent networks based on transitions from one stable state to another have been investigated, particularly in the physics community. These approaches start with a Hopfield network, which contains symmetric connections between each unit and every other unit. It can be shown that the pattern of activity in a Hopfield network with symmetric connections is governed by an *energy surface*, analogous to a landscape of hills and valleys. The network will evolve so as to go downhill with every change in state. When it is not possible to further locally descend down the energy landscape, the network stops evolving. Thus the stable points of the network, which represent stored information, correspond to minima of the energy surface. In order to force transitions between stable states, asymmetric connections and/or connections with variable delays are added. These connections create energy surfaces whose minimum points change with time. A good survey of several such techniques is given in (Bell, 1988). Unfortunately, such networks can only store a small number of sequences, or trajectories, and the trajectories are very susceptible to following unintentional paths (Hertz, Krogh, and Palmer, 1991).

Biologically Inspired Approaches

Several researchers have studied spatiotemporal sequence recognition mechanisms based on neurophysiological evidence, especially from the olfactory, auditory and visual cortex. Representative of this body of work is the use of non-Hebbian learning proposed by Granger and Lynch, which, when used in networks with inhibitory as well as excitatory connections, can be used to learn very large temporal sequences (Granger, Ambros-Ingerson, and Lynch, 1991). Similar networks have been used to act as adaptive filters for speech recognition (Kurogi, 1987), or provide competitive-cooperative mechanisms for sequence selection (Changeux, Dehaene, and Nadal, 1987). At the neuronal level, irregularly firing cells have been proposed as a basic computational mechanism for transmitting temporal information through variable pulse rates (Dayhoff, 1990).

A study of pattern generation in olfactory bulb (Freeman, Yao, and Burke, 1988) has prompted investigation of sequence generation through the coupling of oscillating units (Wang, Buhmann, and von der Malsburg, 1990). Also, data from the visual cortex have been used to study temporal pattern association in a large collection of neurons, treating a large number of discrete units in the limit as a continuous field of activity (Chang and Ghosh, 1993). The results yield a neural mechanism for invariant pattern recognition, but cannot be readily implemented as a working system using current technology because of the large number of cells required.

All these efforts are concentrated on neurophysiological plausibility rather than being geared toward algorithmic classification of man-made signals. Issues such as computational efficiency and ease of implementation on computers, are secondary. Thus they are not treated further in this chapter.

Approach Based on Evidence Building

For classification of sequences, the simple idea of matching an incoming sequence with stored sequences of templates, one per class, has been present in the signal processing literature for a long time. The total match score is given by a (possibly weighted) average of the match scores in each time segment. A more sophisticated version of this idea is embodied in the dynamic time warping algorithm (Kung, 1993), wherein a dynamic programming framework is used to contract or dilate stored template sequences in a search for the best match to an input sequence. When the best match is found, the input sequence is recognized as belonging to the class of the matching template sequence.

Dynamic time warping can be extended to form a *dynamic-in-time network* that maintains, for each stored class, a level of confidence that a signal belonging to that class has been encountered. As feature vectors are presented in sequence to the network at each step in time, these confidence levels are updated. The network makes a decision if the confidence level exceeds a threshold for some class. In the neural network literature, this philosophy is apparent in the Avalanche Filter (Grossberg, 1970; Hecht-Nielsen, 1987), which models a flow of confidence level forward in time, increasing as each event in a sequence is received in the order matching a stored sequence. This philosophy is also used in the time-into-intensity map of Banzhaf (Banzhaf and Kyuma, 1991) that converts a temporal sequence into a spatial representation. A somewhat related method of temporal differences has been used for making sequence prediction more tractable (Sutton, 1988). Differences between subsequent predictions of a final outcome are used to adapt connections in a network.

The dynamic-in-time network that we have designed for the classification of sonar signals is also based on the evidence accumulation approach, with added features to make it more robust against time warping, noise and signal variability. We now examine the characteristics of underwater transients in the next section to get a better appreciation of the design goals for a sonar sequence classifier that is detailed in Section 4.

CHARACTERISTICS OF SHORT-DURATION SONAR SIGNALS

Short-duration underwater sounds contain valuable clues for source identification in noisy and dissipative environments (Deuser and Middleton, 1979; Chen, 1985). Biological sources such as sperm whale clicks, porpoise whistles, and snapping shrimp, as well as non-biological phenomena such as ice crackles and mechanical sounds, produce characteristic sounds of a very short duration, typically 5 to 250 ms. For example, porpoises radiate echolocation pulse trains. Each elementary pulse lasts from 15 to 20 ms, with an average interval of 40 ms between pulses. Such distinctive patterns can be identified by trained humans either by listening to the sounds or by looking at spectrograms of the processed sonar signals. However, attempts at automated classification of real-life acoustic signals based on purely spectral characteristics or on autoregressive modeling have met with very limited success over the past 25 years (J. Ocean Eng., Special Issue, 1987; Urick ,1975).

A careful study of short-duration underwater signals shows several reasons why algorithmic characterization and classification of oceanic acoustics is so difficult:

- The signals have rapidly varying statistics.
- The signals show significant variations in spectral characteristics and signal-to-noise ratio (SNR) due to differing sources or propagation paths, and due to multi-path propagation.
- Signals may overlap with one another.
- Signals must often be associated with other nearby signals for proper identification.

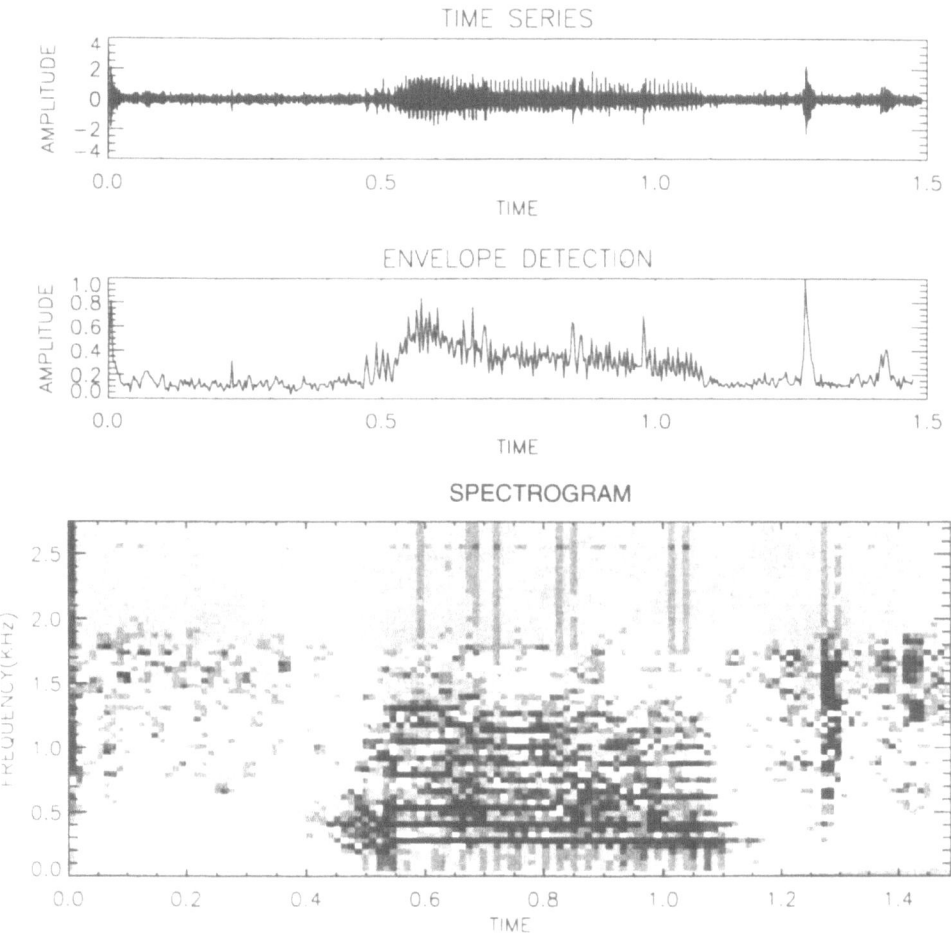

Figure 4. Underwater acoustic signal sequence caused by a toadfish represented as a time waveform (above), a time envelope waveform (center), and a time-frequency image called a spectrogram (below). The occurrence of the short duration signal at t=1.3 seconds following a longer duration primary burst of lower frequency (shown from t = 0.5 to 1.1 s), is characteristic of toadfishes. The association of these two events along with some other signal features, enable discrimination between toadfishes and other marine biological acoustic sources.

Some of these features can be observed in Fig. 4, which shows short-duration underwater signals from a toadfish. The occurrence of a shorter duration signal after the first signal of longer duration and lower frequency, is a distinguishing feature for toadfishes. If spectral information is computed over the duration of the event and used as a single feature vector, this event association is missed. Such characteristics further motivate the use of sequence based classifiers rather than static ones for sonar signal detection and classification, even though static networks are better understood and simpler to implement.

Signal Preprocessing and Feature Extraction

Figure 5 (from Simpson, 1990) shows four generic stages of a sonar signal processing system. Each stage contributes to the quality of the overall system. Since a general purpose cued classifier must detect the presence of a signal as well as classify it, measures of quality include not only classification accuracy, but false alarm rate, the number of times the system indicates the presence of a signal when there is none, and missed detection rate, the number of real signals that are present but not detected. Also important are the computational power required and confidence in classification decisions when they are made (Ghosh, Deuser, and Beck, 1992).

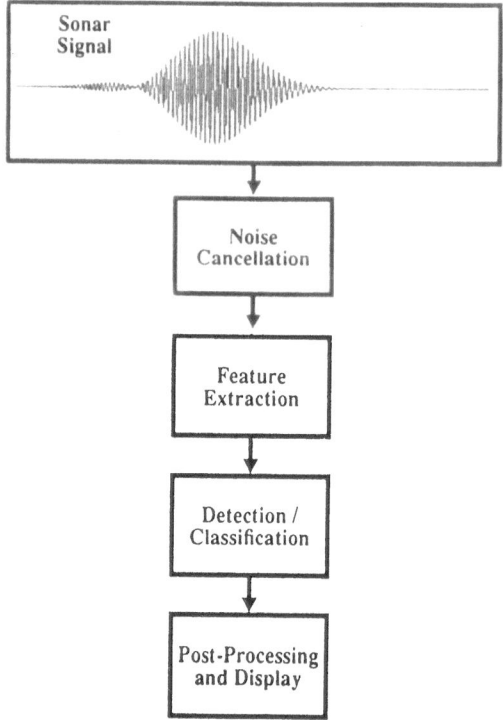

Figure 5. Four stages of a sonar signal processing system.

In order to classify events such as short-duration signals from an underwater acoustic sensor, it is necessary to eliminate or account for all effects that are not a result of the events of interest. In addition, to perform feature extraction, the presence of each type of signal of interest should result in a measurable difference in the observable features. These needs motivate background normalization.

For background normalization, an adaptive digital filter is used to remove statistical correlations from the long-term time sequence. The adaptation of the filter is relatively slow so that background effects such as ambient noise, interference, and sensor characteristics which show long-term correlation are removed but correlation of short-duration signals of interest remains. The adaptation rate of the filter is chosen based on the expected duration of signals of interest. This process is called whitening the data (Widrow and Stearns, 1985). After a signal, X(t), is extracted and whitened, it is represented by a series of feature vectors, V_1, V_2, . . . , sequenced in time (Pao, Hemminger, Adams, and Clary, 1987). Each feature vector is a mathematical descriptor of the signal observed within a particular time window. A set of partially overlapping time windows is typically used to describe the sequence.

Besides the issue of segmentation and the associated lengths of the time windows, a key design decision is the choice of the features used to represent the signal in each time window (Ghosh, Deuser, and Beck, 1992). For example, a vector sequence could consist of measurements of the energy at different frequency ranges extracted from successive time windows (spectral coefficients). Another approach is to assume a mathematical model for generation of the signal and use parameters of the model as elements of the feature vector. A popular model represents the value of a signal at a given time as a weighted average of its own past values. In order to compute the weights, or coefficients, of this *autoregressive* model, the statistics of the signal are assumed to be constant over small ranges of time (Hermand and Nicolas, 1989). The set of features extracted from a signal, represented as a two-dimensional array, can serve as an input to a time-delay neural network used for speech recognition (Lippmann, 1989).

For calculating feature vectors, the selection of window length depends both on the length and the nature of the signal. Segmentation of nonstationary signals has been extensively studied (Djuric, Kay, and Boudreaux-Bartels, 1992). Ideally, each signal is assumed to be composed of a few independent and stationary segments, so that its characteristics can be captured by a short sequence of feature vectors. If all of the signals of interest are of comparable duration, a predetermined window length can be used. However, it is common for the durations of acoustic signatures to vary, sometimes by more than two orders of magnitude. For the case of a variable signal length, a two-stage network can be used (Lefebvre, Nicolas, and Degoul, 1990). The first stage extracts the signals and sorts them into different categories according to signal duration. The second stage then applies the classification algorithm to each category.

The key question now is, what features describing the analog oceanic signal will provide the most amount of discriminating information for the classification algorithm? Some common ways of describing a time series is through autoregressive modeling, or through spectral characteristic, as determined by Fourier transforms or the cepstrum. Extensive studies have shown that values of both autoregressive and cepstral coefficients are sensitive to the SNR and phase of realistic sonar signals. They are also sensitive to the number of parameters chosen for the model, so that choosing a mathematical model with too many parameters results in noisy coefficients. Many oceanic signals are embedded in significant noise, mostly broadband. Phase depends on the estimated starting point of the signal, which is difficult to determine even with a good detector. Setting a high SNR threshold for lower false alarm also results in a poorer phase estimate.

We have also found that features which capture both broadband and narrowband components of signals are important for classification. In general, high frequency signal components correspond to short duration events and a relatively wide frequency range, or high bandwidth. Low frequency signal components correspond to long duration events and a relatively narrow frequency range, or low bandwidth. This suggests the use of wavelet transforms, which are similar to Fourier transforms for determining spectral content of a signal, but yield an analysis of signals over multiple resolutions or scales simultaneously. The advantage of the wavelet transform is that it is responsive to broadband signals at high frequency and to narrowband signals at low frequency and therefore captures important information from a signal at different resolutions at the same time. A good description of wavelets can be found in (Rioul and Vetterli, 1991).

It was confirmed in a previous study (Ghosh, Deuser, and Beck, 1990; Beck, Deuser, Still, and Whiteley,1991) that parameters obtained using wavelet transforms of the input signal yield better performance than those using autoregressive modeling or spectral coefficients, for classifying oceanic signals. For the results reported in this article, the wavelet transform is used to generate sixteen coefficients that describe the spectral characteristics of the signal. These sixteen coefficients are augmented with temporal descriptors such as total signal duration, and spectral measurements such as peak frequency, to yield a 25 dimensional feature vector from each window in time.

Table 1. Description of the data set of short-duration oceanic signals.

Class	Description	Training	Testing
0	Porpoise Sound	116	284
1	Ice 1	116	175
2	Ice 2	78	39
3	Whale Sound 1	116	129
4	Whale Sound 2	148	251
5	Background Noise	116	127
Total		690	1005

The data set of oceanic signals used as examples in this paper is described in Table 1. This table shows the types of sounds used, and how many of each type of sound were available for training and for testing the neural classifier. Each type of sound corresponds to a class. The goal of the neural classifier is to determine the class to which a particular input signal belongs. Available data is split into a training set and a test set. The training set is used during training to adapt parameters of the classifier. The test set is used after parameters have been fixed, and serves to verify the performance of the classifier. Training is *supervised*, since in the training set, a known desired output is related to each input and the difference between desired and actual outputs is used to adapt the parameters, or weights.

THE ADAPTIVE SPATIOTEMPORAL RECOGNIZER (ASTER)

In this section we introduce ASTER, a network tailored for the detection and classification of spatiotemporal patterns. Though it has some similarities with the avalanche filter of Grossberg (Grossberg, 1970) described above and the similar spatiotemporal networks of Hecht-Nielson (Hecht-Nielsen, 1987; Hecht-Nielsen, 1990), ASTER differs from them in significant ways. It can be more conveniently viewed as a nonlinear, adaptive version of the dynamic time warping algorithm described above. Though it could be used for classifying any type of temporal pattern,we have used it to classify short-duration oceanic sounds.

Let $V(t)$ be the n-dimensional feature vector obtained using a window centered at time t. As mentioned in the previous section, we shall be using a 25-dimensional vector that includes sixteen wavelet coefficients, as the feature vector of choice. An alternative to the spectrogram is to represent a signal as a path connecting two points in the R^n space of n-dimensional feature vectors. This path is the trajectory of $V(t)$ observed over a time interval $[t_0, t_1]$. The nature of time t is assumed to be discrete in this paper, so that each path yields a sequence of feature vectors, $V(t_0), V(t_0 + \Delta t), \ldots, V(t_1)$, which form the input to a *sequence classifier*. Moreover, for classification purposes we assume that each spatiotemporal pattern belongs to one of N predetermined categories or classes.

The ASTER monitors the sequence of inputs $V(t_0)$, $V(t_0 + \Delta t)$, ..., $V(t_1)$, indexed by time, and dynamically maintains, for each stored signal category or *class*, a separate probability estimate (confidence level) that the current input matches the stored class. After a new feature vector is presented at each time step, the confidences are updated for all classes. If, at any stage the confidence for a class exceeds a threshold value, the signal is considered to be detected and classified. Note that, unlike time delay neural network models, only one feature vector is used as input at any time, and past history is instead recorded as "residual activation" in the form of confidence levels. ASTER is also clearly different from the gamma networks where history is stored using a dispersive delay line that implements a short term memory.

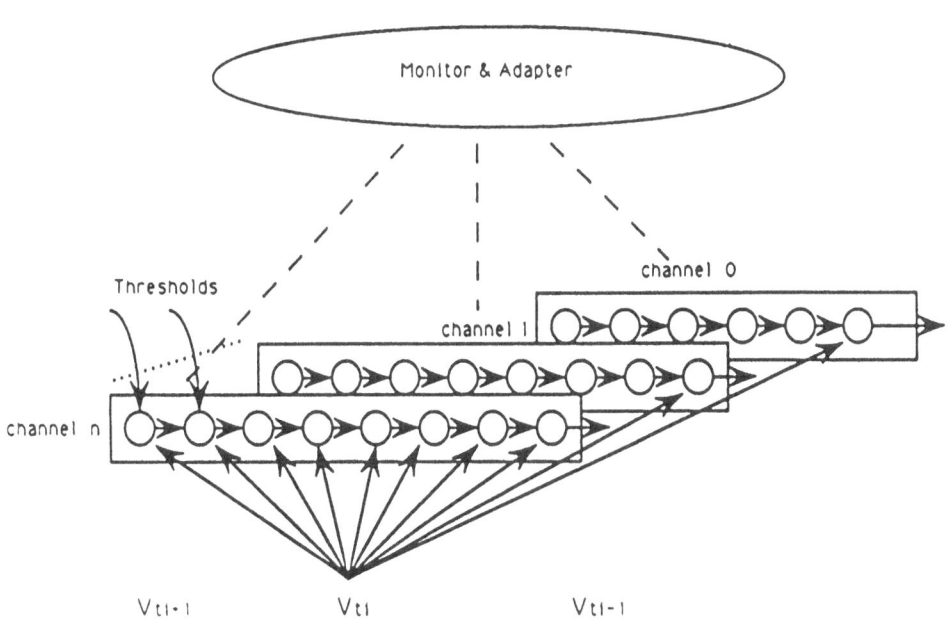

Figure 6. The Adaptive Spatio-Temporal Recognizer (ASTER).

As indicated in Fig. 6, the ASTER consists of a set of N *channels*, one for each of the N classes to which an input may belong. A channel consists of a sequential array of Similarity Estimation Units. SEU(i, j) denotes the *i*th similarity estimation unit in the channel corresponding to class C_j. The total number of similarity estimation units in the *j*th channel is w_j, which directly corresponds to the sequence length of interest for a typical signal of class *j*. A training procedure, described later, is used to assign at least one prototype feature vector per segment per class. The representative vectors are stored in the similarity estimation units such that the prototypes (indexed by k) corresponding to stage i of class C_j, are stored in SEU(i, j). These prototypes are denoted by $V_i^{j,k}$.

The testing phase is the process of passing previously unseen sequences through the network for classification. Feature vectors of the test signal are presented in temporal order to the network. For convenience of notation, let the time scale be normalized such that one feature vector is extracted and presented to the network per unit time. Then, at each time increment, t, and for all i and j, SEU(i, j) computes the confidence in the hypothesis that a signal belonging to class C_j started at time t-i, for $1 \leq i \leq w_j$. Note that all the similarity estimation units are updated at each time increment. The updated confidence SEU(i, j) is a monotonically increasing function of the similarity between its stored prototype(s) and the input feature vector, as well as of the confidence of SEU(i-1, j) evaluated at the previous time increment. Let the similarity between the input feature vector to the ith stage, $V_i(t)$, and the prototype feature vectors V_i ^{jk}, be denoted by $c_i{}^j(t)$. Several measures of similarity can be used:

Correlation, taken as the cosine of the angle between the input feature vector and the most similar prototype:

$$c_i^j = max_k \{ \frac{\overline{V}_i^{j,k} . V_i(t)}{\|\overline{V}_i^{j,k}\| . \|V_i(t)\|} \},$$

(1)

where $\|V\|$ represents the magnitude of vector V, and k indexes the different prototypes at the ith stage of the class j channel

Distance: This measure is a function of the Euclidean distance between the nearest prototype and the input vector, given by

$$c_i^j = max_k \{ exp(-\|V_i(t) - \overline{V}_i^{j,k}\|^2 / 2\sigma_{i,j}^2) \};$$

(2)

where V_i is the input vector and $V_i{}^{j,k}$ is the kth prototype for ith stage of class j channel.
The distance measure is used for ASTER in this article.
Let the confidence of SEU(i, j) at time t be denoted by $p_i{}^j(t)$. Then the confidence $p_i{}^j(t)$ is a function of the previous confidence $p_i{}^j_{-1}(t$-$1)$ and the current similarity $c_i{}^j(t)$ as follows:

$$p_0^j(t) = 0, \forall t,$$
$$p_i^j(t) = f(p_{i-1}^j(t-1) + c_i^j(t) - a_i^j(t)), \quad 1 \leq j \leq N.$$

(3)

where $a_i{}^j(t)$ is the local threshold parameter of SEU(i, j) at time t, and $f(.)$ is the hyperbolic tangent function $tanh(x) = \frac{e^x - e^{-x}}{e^x + e^{-x}}$.
If $f(.)$ was a linear function, then the confidence at stage i would essentially be a sum of the similarities obtained at previous stages, with a subtracted threshold. Choosing the hyperbolic tangent transfer function is key to the robustness of ASTER, as it builds up confidence quickly if the similarities are greater than the corresponding thresholds at the earlier stages, and this "running tally" is not greatly affected if $(c_i{}^j - a_i{}^j)$ is low for some intermediate stage because of noise or large signal variability in that stage. A classification decision is made when any of the confidences exceeds a suitable overall detection threshold value, typically 0.8-0.9. If each element of an input sequence has a very good match with the corresponding prototype, then the overall detection threshold may be reached at an intermediate similarity estimation unit even before the entire sequence is seen. The choice of the overall detection threshold should not be too low, else a few early matches can cause a false detection and/or wrong classification. Similarly, if the threshold is too high, slight deviations from the corresponding prototype sequence may be enough for a signal of interest to avoid detection.

Network Training

The performance of ASTER depends on the choice of prototypes $(V_i^{j,k})$ as well as the thresholds (a_i^j). Other design decisions include the number of stages or similarity estimation units used per channel, and the overall threshold used to make a decision. The thresholds strongly influence the noise suppression properties, false alarm rate, and probability of detection. Misclassification can be decreased by optimizing the choice of the prototypes. In the following description, the assignment of feature vectors to channels and selection of the values to be stored in each similarity estimation unit within the channels is performed during network initialization, while the thresholds are adapted later on. This separation simplifies and speeds up the training process.

Network Initialization: The average sequence size, w_j, for a signal of class j is first determined by applying a suitable segmentation algorithm (Djuric, Kay, and Boudreaux-Bartels, 1992) to all samples belonging to the training set for that class, or by determining the number of fixed duration windows that cover a typical signal. Once the number of windows required for storing the prototype sequence of vectors for a given class is determined, we need to determine the individual representative vectors for these sequences, i.e., the values of Vs for every sequence $V_1^j, ..., V_w^j$. These components are used to obtain the similarities, c_i's, at each similarity estimation unit. Also, the threshold function parameters (a_i^j)s need to be initialized. The most suitable value of the (a_i^j)s depend not only on the expected signal to noise ratios, but also the cost of misses or misclassification versus the cost of false alarms.

The prototype vectors, Vs, to be stored in the similarity estimation units, can be extracted by applying the Learning Vector Quantization algorithm (Kohonen, 1989) on the training sequences. Learning Vector Quantization is an adaptive version of the classical Vector Quantization algorithm whose aim is to represent a set of input vectors by a smaller set of codebook vectors or reference vectors so as to minimize a classification error. Typically the algorithm consists of the following: Step 1: The reference vectors are initialized by a random selection from the population or by performing K-means clustering on the training set. K-means clustering is a method of finding a set of reference vectors such that any vector in the population may be represented by a reference vector with a minimum overall representational error. Step 2: Following random sampling or K-means clustering, an adjustment procedure then assigns classes to the reference vectors. The reference vectors are adjusted iteratively by moving them closer to or further away from training inputs depending on whether the closest reference vector is of the same class as the input vector or not. The magnitude of the adjustment is proportional to the distance and to a learning rate parameter which decreases monotonically with time. Step 3: Unknown inputs are assigned the class of the nearest reference vector, using a suitable distance metric such as Euclidean distance. More details can be found in (Kohonen, 1989).

For each i, j, the prototypes, $V_i^{j,k}$ that are stored in SEU(i, j) are obtained by applying the Learning Vector Quantization algorithm on a set of sample feature vectors obtained from the ith segments of the training patterns from all classes. When the Learning Vector Quantization training is over, the reference vectors for each class are stored as prototypes in the corresponding similarity estimation units. Notice that this method assumes that we have at least one representative of each class beforehand. Sometimes, such prior knowledge may not be available for all the classes. By clustering the input samples, one may detect prominent clusters that do not seem to belong to any of the previously known classes. If this happens, such clusters can be labeled as new classes.

During the testing phase, the similarity measure, c_i, between the input vector $V_i(t)$ and the prototypes stored in SEU(i, j), is given by (2), where the widths $(\sigma_{i,j})$ are chosen to be comparable to the distance between a $V_i^{j,k}$ and its nearest prototype.

Threshold Adaptation: The similarity estimation unit threshold parameter $a_i^j(t)$ in (3) can be adapted by performing gradient descent on a suitable cost function. Performing gradient descent consists of computing or approximating the derivative of a function with respect to an adjustable parameter and using this derivative to adjust the parameter in a direction so as to minimize the function. One possible cost function is obtained by assuming that for a signal of a known class, the corresponding channel should reach a confidence close to 1 at the w_jth similarity estimation unit at the instant when the last segment of that signal is presented. At this instant, the confidences for the other channels should have much lower values.

The scheme for adapting thresholds as described above, assumes that all signals of the same class, j, are characterized by a fixed sequence length, w_j, though it does allow different lengths for different classes. To accommodate some variations in length within a class, a dynamic programming approach is used whereby similarity in each unit is computed by comparing an input with prototypes of a given stage as well as with prototypes at the preceding and succeeding stages as well. Note that because of the carrying over of confidence from one stage to the next, the output at stage w_j should still be higher than that for other channels at time t_m even if the sequence corresponding to a compressed or expanded signal of class j is slightly shorter or longer than w_j. The interested reader is referred to (Ghosh and Gangishetti, 1993) for a more mathematical and detailed description of ASTER.

THE REFERENCE NETWORK

The time delay neural network is used as a reference network for comparison with ASTER. This network was developed to explicitly handle the temporal aspect of incoming data. Typical feature vectors for sonar signals have between 16 and 32 components and are thus of rather high dimension, so these networks can get very large. For example, for the 25-dimensional vectors used in (Ghosh, Deuser, and Beck, 1992), if a signal of interest spans six time windows, then the corresponding time delay neural network has 150 inputs. To make the processing more tractable, dimensions are reduced using Kohonen's self-organizing feature map (Kohonen, 1990). This neural network maps the sequence of feature vectors onto a trajectory in lower dimensional (2-D) space. The resultant *reduced* feature vectors are then used as the actual inputs to a multilayered feedforward network.

Input Reduction

The usage of a self-organizing map network as the first stage of the reference network has been motivated by an earlier application of a similar map to the "phonetic typewriter" (Kohonen, 1988). In that application, individual phonemes of speech were used as input to a self-organizing map, and the output of the network organized itself into a map of the phonemes in which similar-sounding phonemes were generally close to each other. The map was roughly analogous to a keyboard with the property that phonetically similar letters are always found near each other. Each phoneme could be represented by the location of the map's winning output. In addition to reducing the dimensionality of the input vector space, the "phonetic typewriter" was also able to compensate for modest but consistent imperfections in the feature vector extraction stage, and for some time warping effects.

In our application, the self organizing feature map is used to generate a two-dimensional trajectory representative of an input sequence. It does this by relating vectors in a high dimensional input space to a lower dimensional output space. The output space or map chosen is a two-dimensional square array, all cells of which are fed simultaneously by the input vector $x(t)$ at time t. Each unit in the output map is characterized by its two-dimensional coordinate location in the map and by a weight vector. This weight vector has the same number of components as each of the vectors in the input space. Mapping an input vector from the input space to a two-dimensional vector in the output space consists of locating the unit in the output map whose weight vector is the closest to the input vector in some sense. The two-dimensional coordinate of this winning unit in the output map is then the output of the feature map. Training the map consists of initializing the weight vectors m_i of the array with random initial values $m_i(0)$. For discrete values of t:

(1) The output unit with the closest weight vector is determined as

$$\|x(t) - m_c(t)\| = min_i\|x(t) - m_i(t)\| \qquad (4)$$

(2) The weight vectors are updated by:

$$m_i(t+1) = \left\{ \begin{array}{ll} m_i(t) + \alpha(t)(x(t) - m_i(t)) & \text{for } i \in N_c(t), \\ m_i(t) & \text{otherwise.} \end{array} \right\}$$

where $N_c(t)$ denotes the size of a neighborhood around cell c. The parameters α and N_c are decreased with time. The asymptotic values of m_i define the vector quantization, or final dimension-reduced mapping. The input vectors used during this training process are obtained from a *typical* set of spatiotemporal patterns. It has been observed that the specific order of presentation of the input vectors is not significant while training.

　　To prepare data for the TDNN, each feature vector of a spatiotemporal sequence is presented to a trained self-organizing map. The output of the map is a reduced dimension representation of the input, and this is the feature vector that forms the actual input of the TDNN at each step in time.

Time Delay Neural Networks

　　Time delay neural networks were described in Sec. 2.1 of this chapter, where we also pointed out the potential power of their mapping capabilities. Referring to the schematic architecture of a typical time delay neural network as shown in Fig. 2, we see that the input layer is nothing but a tapped-delay line. Time delays are also used in the hidden layers, so that an upper layer receives activation from the units of the previous layer, as well as from delayed versions of the outputs of these units. Thus every hidden layer contains a two-dimensional array of time delay units, with the width of the array indicating the number of time delay steps employed. The number of delays may vary for each signal channel and for different layers. In Fig. 2 the number of delay steps for all units in the same layer is the same, although they may vary from layer to layer.

　　Let τ denote the unit of time delay. This unit is usually normalized to be one for convenience. The output, or activation value, of unit j in layer l, given the μth input pattern, is

$$a_{j,l}^\mu(t) = f(S_{j,l}^\mu(t)) \qquad (5)$$

where

$$S_{j,l}^\mu(t) = \sum_i \sum_{k=0}^{k_{l-1}-1} w_{j,i,k} \cdot a_{i,l-1}^\mu(t - k\tau). \qquad (6)$$

　　In these equations, i indexes the units at level $l - 1$, $f(x)$ is the sigmoid activation function, and k_l is the number of delay steps at layer l. The weight $w_{j,i,k}$ is the strength of the connection between the output of unit i in layer $l - 1$, delayed by k time steps, and unit j in layer l.

The learning rule and weight adjustment are derived from the back-error propagation paradigm. This method relates the errors in the output of a network to changes in weights through an approximation to the gradient of the output with respect to each individual weight. Errors are *backpropagated* from the output layer through each successive layer towards the input, at each stage multiplied by the weights of the appropriate connections. At each time step, previous activations of input (and possibly hidden) units are delayed along tapped delay lines. The output of the network is computed from (5), and (6). The desired output of the network, which is a representation of the class to which the input is known to belong, is used as a target value. The error of unit j in the output layer is

$$\delta^{\mu}_{j,l}(t) = (t^{\mu}_j - a^{\mu}_{j,l}(t))f'(S^{\mu}_{j,l}(t)), \tag{7}$$

where t^{μ}_j denotes the target value for unit j when the μth pattern is presented, $a_{j,l}{}^{\mu}(t)$ is the actual output, and $f'(\cdot)$ is the derivative of the sigmoid function.

If unit j is in the hidden layer, the error is computed as a weighted sum of the δ values of all units that receive output from unit j. We use only one hidden layer, denoted by $l = 1$, so that the error value of unit j is computed as:

$$\delta^{\mu}_{j,1}(t) = [\sum_{k_l} \sum_{m} \delta^{\mu}_{m,2}(t)w_{m,j,k}]f'(S^{\mu}_{j,1}(t)), \tag{8}$$

where m indexes the units of the output layer ($l = 2$).

An adjustment equal to the appropriate error multiplied by a learning rate parameter is added to each connection weight. The learning rate is a number between zero and one. High learning rates yield fast but noisy learning. Lower learning rates yield slow convergence but less sensitivity to variations between input patterns. A training *epoch* consists of presenting each pattern in the training set and updating weights after each presentation. The patterns are usually presented in a randomly shuffled order. The training process consists of several epochs, and at each epoch the mean squared error is computed as $\frac{1}{P}\sum^{P}_{\mu}\sum_{j}(t^{\mu}_j - a^{\mu}_j)^2$, where t^{μ}_j and a^{μ}_j are the target value and actual value for unit j when the μth pattern is presented, and P is the total number of training patterns. Weight adjustments are made at each time step during presentation of a training pattern. Learning is continued until the mean square error falls below an acceptable level.

TRAINING DATA AND SIMULATION RESULTS

The networks were studied using two types of signals, (1) Banzhaf plots (Banzhaf and Kyuma, 1991) used as artificial sonograms, and (2) 25-dimensional feature vectors extracted from 6 classes of underwater sounds of a biological origin, as described in Section 3.1 and Table 1.

The Banzhaf plots allow experimentation in a controlled environment. Each spatio-temporal signal is a superposition of two-dimensional gaussian patterns in time and in feature space. The plots specify intensities on 20 frequency channels, sampled over at most 30 time steps. Thus, a signal can be viewed as a spectrogram on a 20×30 array and interpreted as an artificially generated sonogram (Banzhaf and Kyuma, 1991).

Signals belonging to eight classes are generated as follows: a "template" signal is created for each class by assigning a certain set of values to parameters which specify the general shape and orientation of the Banzhaf plot. The rest of the signals in the class are generated by perturbing values of the template signal's generating parameters as well as by adding Gaussian noise. Figure 7 shows Banzhaf templates or prototypes for four different classes, and Figure 8 four examples of noise degraded versions of one of the prototypes. Note that there is some variation is signal duration both within a class and between classes.

The training set consists of half of the sample sequences from each class, chosen at random. The other half is used for testing. The input vectors are first normalized by scaling and adding an offset, so that signal values within each feature vector range from zero to one. A 5×5 self organizing feature map is trained using segments of the training set. To prevent undue emphasis on signal vectors with very low magnitudes, signal segments with energy below a chosen threshold are not normalized. Once trained, the self organizing feature map is used to generate a trajectory as explained above. The signals are not transmitted to the self organizing feature map separately, but are preceded and followed with arbitrary lengths of vectors obtained from background oceanic noise. Thus, the network is an uncued classifier; it must determine when a signal is present and then classify it. Generated trajectories are input

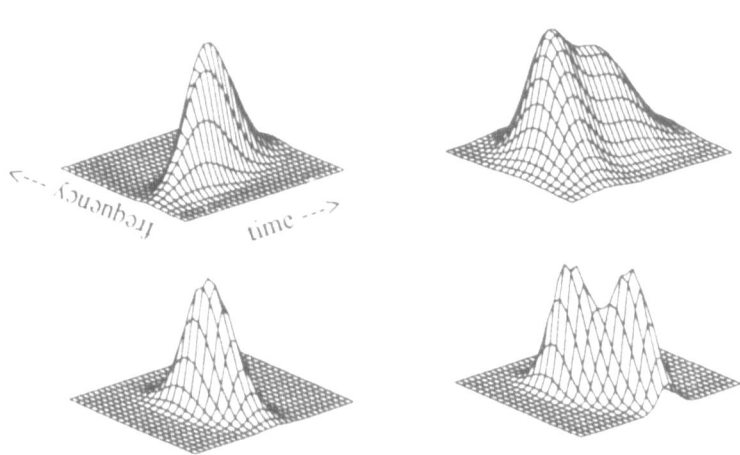

Figure 7. Example of Banzhaf prototype signals belonging to four different classes.

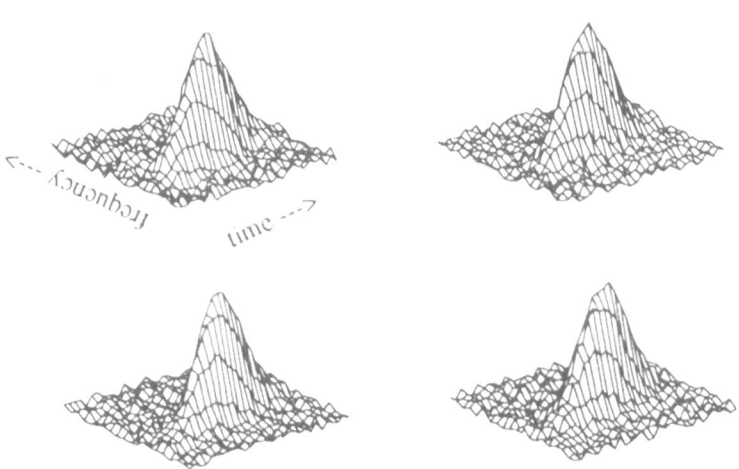

Figure 8. Four noise-degraded versions of the top-left prototype in Fig.7.

to the time delay neural network in real time. The time delay neural network chosen has three layers. The input layer has two units, each with ten delays, the hidden layer has six units, each with fifteen delays, and the output layer has nine units with no delays. Each of the output units correspond to one of the above classes, with the ninth class representing background noise.

The desired output, or target pattern of the time delay neural network is a function of time. When no signal is present, the desired output for each unit is equal to 1/9, so that no unit has a "winning" output. As a signal is presented, the desired output for the unit corresponding to its class increases linearly from 1/9 to one, representing an increasing confidence in presence of the signal as evidence accumulates in the tapped delay lines. While the desired output of the unit for the correct class increases to one, the desired outputs of the remaining units decrease linearly to zero, representing a decreasing probability that any of the other signals is present. The rate of increase or decrease of the desired outputs corresponds to the time delay length for the input layer (k_0), so that after k_0 time steps, the desired output has reached either one or zero. If the pattern being presented is longer than k_0, then the target output remains at one or zero until the end of the pattern.

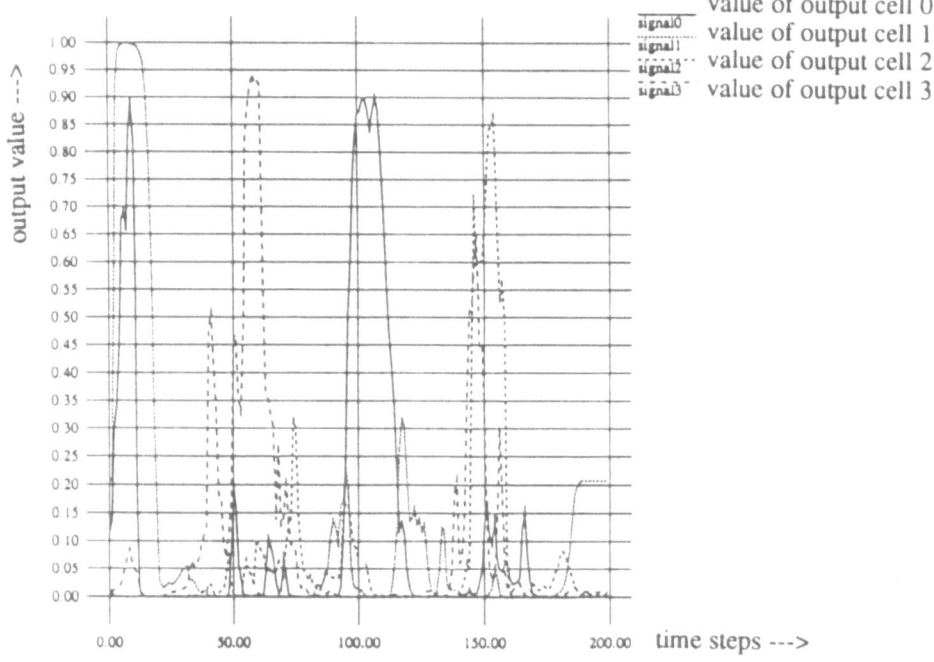

Figure 9. Response of the TDNN to four signals embedded in a sequence of 200 feature vectors. Sequences corresponding to single samples of classes 1, 3, 0 and 2 were input to the network at time instances 0, 40, 90 and 135 respectively.

Figure 9 shows a sample output of the network. Signals in classes one, three, zero and two were presented to the self organizing feature map beginning at time instances 0, 40, 90 and 135. For clarity only the outputs of units corresponding to classes zero through three are shown. Any acceptance threshold between 0.55 and 0.87 is able to correctly detect and classify the four occurrences of signals of interest. However, for signal one, the hypothesis that the signal could be of class zero is also quite strong. For this example sequence, the ASTER was able to correctly detect and classify the four occurrences. From our overall results obtained so far using both Banzhaf plots and bioacoustic signals, the time delay neural network performs better for fairly clean signals (signal-to-noise ratio above 5 dB), but the ASTER is superior for lower noise levels and is also more tolerant of time-warping. The thresholds in ASTER allow tuning to achieve the desired tradeoff between correct classification and false alarms. Moreover, our experience indicates that segmentation of non-stationary signals, feature selection and cross-channel coupling need to be studied in a unified framework in order to obtain further improvements in classification accuracy.

CONCLUDING REMARKS

Engineers and mathematicians have developed powerful tools over the years to analyze and process signals that evolve in time. These tools include time domain techniques such as autoregressive modeling, as well as spectral domain techniques such as Fourier transforms and filter theory for signal characterization and processing, respectively. These innovations were largely motivated by their analytical capabilities and ease of implementation using analog or digital hardware, rather than because of any biological plausibility. In contrast, neurobiologists, cognitive scientists and psychologists have been concerned for a long time with the representation and generation of temporal patterns in living beings, at various levels: from the neuronal to the behavioral. Both of these "macro" communities have much to contribute, and much to learn from each other, if they are to build real-time adaptive systems that effectively interact with dynamic environments.

This chapter highlighted some of the engineering-oriented approaches to the generation and recognition of pattern sequences using artificial neural networks. In particular, the utility of evidence accumulation as a technique for robust on-line processing was examined through the study of a network called ASTER, and applied to the detection and classification of underwater sonar signals. There is some evidence that dolphins and whales also use a type of evidence accumulation to differentiate among underwater signals. Adult dolphins often imitate "signature whistles" of interacting dolphins (Freitag and Tyack, 1993), and mother-offspring develop remarkably similar signature whistles (Sayigh, et al., 1990). Humpback whales vary their "song pattern" over time (Payne, Tyack, and Payne, 1983), and it appears that sperm whales vary their click rate depending on the number of whales present and the group behavior (Whitehead and Weilgart, 1990).

Recently, several researchers have been investigating artificial neural networks that are at least partially inspired by biological mechanisms for detecting or responding to event combinations, event ordering and delays between events. For example, short term memory is being modeled by a network of oscillatory dual-neuron units that can learn, recognize and reproduce complex temporal sequences (Wang and Arbib, 1990). Habituation is a means by which biological neural systems ignore repetitive, irrelevant stimuli. It has also been suggested to be a means of encoding temporal information. Habituation is caused by the reduction in certain synaptic strengths in response to repeated familiar stimuli, and has been mathematically modeled at the neuron level (Byrne and Gingrich, 1989). Recently, this model has been incorporated inside a time delay neural network. The resulting network performs significantly better than the simple time delay neural network for the sonar classification problem considered in this paper (Stiles and Ghosh, 1995). Such hybrid approaches to modeling and processing of time-varying signals or signal sequences provide several promising lines of research that warrant further exploration.

ACKNOWLEDGEMENTS

This research was funded in part by ONR contract N00014-92-C-0232 to Tracor Applied Sciences Inc., Austin, TX, with a subcontract to The University of Texas, and by NSF grant ECS-9307632. We thank Steven Beck, Srinivasa Chakravarthy and Richard Palmer for their contributions at various stages of this project.

REFERENCES

Ambros-Ingerson, J., Granger, R., and Lynch, G., 1990, Simulation of paleocortex performs hierarchical clustering, *Science*, 247:1344.

Banzhaf, W. and Kyuma, K., 1991, The time-into-intensity-mapping network, *Biol. Cybern.*, 66:115.

Beck, S., Deuser, L., Still, R., and Whiteley, J., 1991, A hybrid neural network classifier of short duration acoustic signals, *Proc. IJCNN*, 1:119.

Bell, T., 1988, Sequential processing using attractor transitions, *in*: "Proceedings of the 1988 Connectionist Models Summer School", Morgan Kaufmann Publishers, San Mateo, CA.

Braitenberg, V., 1990, Reading the structure of brains, *Network*, 1:1.

Byrne, J. H. and Gingrich, K. J., 1989, Mathematical model of cellular and molecular processes contributing to associative and nonassociative learning in *Aplysia*, *in*: "Neural Models of Plasticity", J. Byrne, and W. Berry, eds., Academic Press, San Diego, CA.

Chang, H. J. and Ghosh, J., 1993, Pattern association and retrieval in a continuous neural system, *Biol. Cybern.*, 69:77.

Changeux, J.-P., Dehaene, S., and Nadal, J.-P., 1987, Neural networks that learn temporal sequences by selection, *Proc. Natl. Acad. Sci. USA*, 84:2727.

Chen, C., 1985, Automatic recognition of underwater transient signals-a review, *Proc. ICASSP*, 1270.

Day, S. P. and Davenport, M. R., 1993, Continuous-time temporal back-propagation with adaptable time delays, *IEEE Trans Neural Networks*, 4:348.

Dayhoff, J., 1990, Regularity properties in pulse transmission networks, *Proceedings of the Third International Joint Conference on Neural Networks*, 3:621.

de Vries, B. and Principe, J. C., 1990, The gamma model - a new neural net model for temporal processing, *Neural Networks*, 5:565.

de Vries, B. and Principe, J. C., 1991, A theory for neural networks with time delays, *in*: "Advances in Neural Information Processing Systems-III", R.P. Lippmann, J. E. Moody and D. Touretzky, eds., Morgan Kaufmann Publishers, San Mateo, CA.

Deuser, L. and Middleton, D., 1979, On the classification of underwater acoustic signals: An environmentally adaptive approach, *J. Acoust. Soc. Am.*, 65:438.

Djuric, P. M., Kay, S. M., and Boudreaux-Bartels, G. F., 1992, Segmentation of nonstationary signals, *in*: *Proc. ICASSP*, 5:161.

Elman, J., 1990, Finding structure in time, *Cognit. Sci.*, 14:179.

Freeman, W. J., Yao, Y., and Burke, B., 1988, Central pattern generating and recognizing in olfactory bulb: a correlation learning rule, *Neural Networks*, 1:277.

Freitag, L. and Tyack, P., 1993, Passive acoustic localization of the Atlantic bottlenose dolphin using whistles and echolocation clicks, *J. Acoust. Soc. Am.*, 93:2197.

Ghosh, J., 1993, Representation and classification of temporal patterns, *in*: "Tutorial Notes, ANNIE '93", Publisher: Univ. of Missouri-Rolla, St. Louis, Nov. 1993.

Ghosh, J., Deuser, L., and Beck, S., 1990, Impact of feature vector selection on static classification of acoustic transient signals, *in*: "Government Neural Network Applications Workshop"

Ghosh, J., Deuser, L., and Beck, S., 1992, A neural network based hybrid system for detection, characterization and classification of short-duration oceanic signals, *IEEE J. Ocean Engineering*, 17:351.

Ghosh, J. and Gangishetti, N., 1993, Robust classification of variable length sonar sequences, *Proc. SPIE* 1965:96.

Ghosh, J. and Karamcheti, V., 1992, Sequence learning using recurrent networks: Analysis of internal representations, *SPIE Proc.* 1710:449.

Granger, R., Ambros-Ingerson, J., and Lynch, G., 1991, Derivation of encoding characteristics of layer II cerebral cortex, *J. Cognit. Neurosci.*, 61:78.

Grossberg, S., 1970, Some networks that can learn, remember, and reproduce any number of complicated space-time patterns, II, *Stud. App. Math.*, 49:135.

Haykin, S., 1994, "Neural Networks: A Comprehensive Foundation", Macmillan, New York, NY.

Hecht-Nielsen, R., 1987, Nearest matched filter classification of spatiotemporal patterns, *Applied Optics*, 26:1892.

Hecht-Nielsen, R., 1990, "Neurocomputing", Addison Wesley, Reading, MA.

Hermand, J.-P. and Nicolas, P., 1989, Adaptive classification of underwater transients, *Proc. ICASSP*, Vol. IV 2712-2715.

Hertz, J., Krogh, A., and Palmer, R. G., 1991, "Introduction to the Theory of Neural Computation", Addison-Wesley, Reading, MA.

Hopfield, J. and Tank, D., 1989, Neural architecture and biophysics for sequence recognition, *in*: "Neural Models of Plasticity", J. Byrne, and W. Berry., eds., Academic Press, San Diego, CA.

Jordan, M., 1989, Serial order: A parallel, distributed processing approach, *in*: "Advances in Connectionist Theory: Speech", J. Elman, and D. Rumelhart, eds., Lawrence Erlbaum Associates, Hillsdale, NJ.

Kohonen, T., 1988, The "neural phonetic typewriter", *IEEE Computer*, 21:11.

Kohonen, T., 1989, "Self-Organization and Associative Memory", Springer-Verlag, Berlin.

Kohonen, T., 1990, The self-organizing map, *Proc. IEEE*, 78:1464.

Kung, S., 1993, "Digital Neural Networks", Prentice Hall, Englewood Cliffs, NJ.

Kurogi, S., 1987, A model of neural network for spatiotemporal pattern recognition, *Biol. Cybern.*, 57:103.

Lang, K. J., Waibel, A. H., and Hinton, G. E., 1990, A time-delay neural network architecture for isolated word recognition, *Neural Networks*, 3:23.

Lefebvre, T., Nicolas, J., and Degoul, P., 1990, Numerical to symbolical conversion for acoustic signal classification using a two-stage neural architecture, *in*: "Proceedings of the International Neural Network Conference, Paris".

Lin, D., Dayhoff, J. E., and Ligomenides, P. A., 1992, Trajectory recognition with a time-delay neural network, *in*: "Proceedings of the International Joint Conference on Neural Networks, Baltimore", 3:197.

Lin, D.-T., Ligomenides, P. A., and Dayhoff, J. E., 1993, Learning spatiotemporal topology using an adaptive time-delay neural network, World Congress on Neural Networks, Oregon, 1:291.

Lippmann, R. P. , 1989, Review of neural networks for speech recognition, *Neural Computat.*, 1:1.

Maren, A., 1990, Neural networks for spatio-temporal recognition, *in*: "Handbook of Neural Computing Applications", A. Maren, C. Harston, and R. Pap, eds., Academic Press, San Diego, CA.

Mozer, M. C., 1993, Neural network architectures for temporal sequence processing, *in*: "Time Series Prediction", A.S. Weigend, and N. Gershenfeld, eds., Addison Wesley, Reading, MA.

Pao, Y., Hemminger, T., Adams, D., and Clary, S., 1991, An episodal neural-net computing approach to the detection and interpretation of underwater acoustic transients, *in*: "Conference on Neural Networks for Ocean Engineering", IEEE Press, New York, NY.

Payne, K., Tyack, P., and Payne, R., 1983, Progressive changes in the songs of humpback whales (*Megaptera novaegliae*), a detailed analysis of two seasons in Hawaii, *in*: "Behavior and Communication of Whales", Westview, Boulder, CO.

Pearlmutter, B. A., 1989, Learning state space trajectories in recurrent neural networks, *Neural Computat.*, 1:263.

Pineda, F. J., 1989, Recurrent backpropagation and the dynamical approach to adaptive neural computation, *Neural Computat.*, 1:161.

Principe, J. C., Kuo, J.-M., and Celebi, S., 1994, An analysis of the gamma memory in dynamic neural networks, *IEEE Trans. Neural Networks*, 5:331.

Rioul, O. and Vetterli, M., 1991, Wavelets and signal processing, *IEEE Signal Processing Magazine*, 14:38.

Sandberg, I., 1991, Structure theorems for nonlinear systems, *Multidimensional Systems and Signal Processing*, 2:267.

Sandberg, I., 1992a, Approximately finite memory and input-output maps, *IEEE Trans. Circuits Systems*, 39:549.

Sandberg, I., 1992b, Approximations for nonlinear functionals, *IEEE Trans. Circuits Systems*, 39:65.

Sandberg, I., 1994, General structures for classification, *IEEE Trans. Circuits Systems*, 41:372.

Sato, M., 1990, A real time learning algorithm for recurrent analog neural networks, *Biol. Cybern.*, 62:237.

Sayigh, L., Tyack, P., Wells, R., and Scott, M., 1990, Signature whistles of free-ranging bottlenose dolphins *Tursiops truncatus*: stability and mother-offspring comparisons, *Behav. Ecol. Sociobiol.*, 26:247.

Shamma, S., 1989, Spatial and temporal processing in central auditory networks, *in*: "Methods in Neuronal Modeling: From Synapses to Networks", C. Koch, and I. Segev, eds., MIT Press, Cambridge, MA.

Simpson, P., 1990, Neural networks for SONAR signal processing, *in*: "Handbook of Neural Computing Applications", A. Maren, C. Harston, and R. Pap, eds., Academic Press, San Diego, CA.

Stiles, B. and Ghosh, J., 1995, A habituation based mechanism for encoding temporal information in artificial neural networks, *Proc. SPIE* Vol. 2492, Orlando, April 1995, pp. 404-415.

Sun, G., Chen, H., Lee, Y., and Liu, Y., 1992, Time warping recurrent neural networks and trajectory classification, *in*: "Proceedings of the International Joint Conference on Neural Networks, Baltimore", 1:431.

Sutton, R. S., 1988, Learning to predict by the methods of temporal differences, *Machine Learning*, 3:9.

Tank, D. and Hopfield, J., 1987, Neural computation by time compression, *Proc. Natl. Acad. Sci. USA*, 84:1896.

Urick, R., 1975, "Principles of Underwater Sound", (2nd Ed.), McGraw-Hill, New York, NY .

Waibel, A., 1989, Modular construction of time-delay neural networks for speech recognition, *Neural Computat.*, 1:39.

Waibel, A. and Hampshire, J., 1989, Building blocks for speech, *Byte*, pp. 235-242.

Wan, E., 1990, Temporal backpropagation for FIR neural networks, International Joint Conference on Neural Networks, San Diego, 1:575.

Wang, D. and Arbib, M., 1990, Complex temporal sequence learning based on short-term memory, *Proc. IEEE*, 78:1536.

Wang, D., Buhmann, J., and von der Malsburg, C., 1990, Pattern segmentation in associative memory, *Neural Computat*, 2:94.

Weigend, A. S. and Gershenfeld, N., eds., 1993, "Time Series Prediction: Forecasting the Future and understanding the past", Addison Wesley, Reading, MA.

Werbos, P., 1988, Generalization of backpropagation with application to a recurrent gas market model, *Neural Networks*, 1:339.

Whitehead, H. and Weilgart, L., 1990, Click rates from sperm whales, *J. Acoust. Soc. Am.*, 87:1798.

Widrow, B. and Stearns, S., 1985, "Adaptive Signal Processing", Prentice-Hall, Englewood Cliffs, NJ.

INDEX